Advances in

MICROBIAL PHYSIOLOGY

Advances in

MICROBIAL PHYSIOLOGY

Edited by

A. H. ROSE

School of Biological Sciences
Bath University
England

and

D. W. TEMPEST

Laboratorium voor Microbiologie,
Universiteit van Amsterdam,
Amsterdam-C
The Netherlands

VOLUME 11

1974

ACADEMIC PRESS · LONDON and NEW YORK

A Subsidiary of Harcourt Brace Jovanovich, Publishers

ACADEMIC PRESS INC. (LONDON) LTD.
24/28 Oval Road
London NW1

United States Edition published by
ACADEMIC PRESS INC.
111 Fifth Avenue
New York, New York 10003

Library of Congress Catalog Card Number: 67-19850
ISBN: 0 12-027711-5

PRINTED IN GREAT BRITAIN BY
WILLIAM CLOWES AND SONS LIMITED
LONDON, COLCHESTER AND BECCLES

Contributors to Volume 11

A. R. Archibald, *Microbiological Chemistry Research Laboratory, The School of Chemistry, The University of Newcastle-upon-Tyne, England*

C. M. Brown, *Department of Microbiology, University of Newcastle-upon-Tyne, England (Present address: School of Biological Sciences, Dundee University, Dundee, Scotland)*

G. J. Dring, *Unilever Research Laboratory, Colworth House, Sharnbrook, Bedford, England.*

G. W. Gould, *Unilever Research Laboratory, Colworth House, Sharnbrook, Bedford, England.*

H. W. Jannasch, *Woods Hole Oceanographic Institution, Woods Hole, Mass., 02543, U.S.A.*

C. W. Jones, *Department of Biochemistry, University of Leicester, University Road, Leicester, England*

D. S. Macdonald-Brown, *Department of Microbiology, University of Newcastle-upon-Tyne, England*

R. I. Mateles, *Laboratory of Applied Microbiology, The Hebrew University, Jerusalem, Israel*

J. L. Meers, *Agricultural Division, Imperial Chemical Industries Ltd., Billingham, Teesside, England*

M. R. J. Salton, *Department of Microbiology, New York University School of Medicine, New York, N.Y. 10016 U.S.A.*

M. G. Yates, *A.R.C. Unit of Nitrogen Fixation, University of Sussex, Brighton, England*

Contents

Physiological Aspects of Microbial Inorganic Nitrogen Metabolism

C. M. BROWN, DEBORAH S. MACDONALD-BROWN AND J. L. MEERS

The Structure, Biosynthesis and Function of Teichoic Acid

A. R. ARCHIBALD

Respiration and Nitrogen Fixation in *Azotobacter*

M. G. YATES AND C. W. JONES

Mechanisms of Spore Heat Resistance

G. W. GOULD AND G. J. DRING

Experimental Bacterial Ecology Studied in Continuous Culture

H. W. JANNASCH AND R. I. MATELES

Membrane Associated Enzymes in Bacteria

MILTON R. J. SALTON

Physiological Aspects of Microbial Inorganic Nitrogen Metabolism

C. M. Brown, Deborah S. Macdonald-Brown and J. L. Meers*

*Department of Microbiology, University of Newcastle upon Tyne, NE1 7RU, and *Agricultural Division, Imperial Chemical Industries Ltd., Billingham, Teesside, England*

I. Introduction

In this article we have set out to provide a comparative review of the known mechanisms of inorganic nitrogen assimilation in free-living micro-organisms. We have excluded from consideration the assimilation of all other (organic) nitrogen sources but, since the breakdown of these compounds (e.g. amino acids, amides, urea) generally yields ammonia, these also may be assimilated by some of the mechanisms described below. To the best of our knowledge, the major pathways of inorganic nitrogen assimilation are those shown in Fig. 1, with ammonia occupying a central position as intermediate in the assimilation of both molecular nitrogen and nitrate. The direct assimilation of nitrate to form first nitropropionic acid is known to occur in some fungi (see Painter, 1970) but it is doubtful whether this, or analogous systems, are of widespread significance in micro-organisms.

Ammonium ions dissociate to form ammonia and hydrogen ions at alkaline pH values, the pK of this reaction being about 9·2. As the true substrate for many of the reactions described in this article is unknown, we have used the term "ammonia" throughout to denote the assimilated substrate, be it NH_3 or NH_4^+.

II. Assimilation of Molecular Nitrogen

Nitrogen fixation by free living organisms has been the subject of a number of reviews in recent years (Hardy and Burns, 1968; Chatt and Fogg, 1969; Postgate, 1970, 1971; Benemann and Valentine, 1972; Dalton and Mortenson, 1972). This section will therefore deal only with the overall physiology of the process.

Early workers used the measurement of nitrogen gain (determined by the Kjeldahl method) after growth in a medium free of fixed nitrogen as evidence for nitrogen fixation, but this was superseded by techniques employing enrichment with the stable isotope ^{15}N (Burris *et al.*, 1943). The ^{15}N method proved to be about 100-times more sensitive than the Kjeldahl procedure and, using this technique, the list of nitrogen-fixing organisms steadily increased. Efforts were also concentrated on the characterization of the "key intermediate" in nitrogen fixation (Wilson and Burris, 1953), defined as the inorganic product of the fixation reaction via which fixed nitrogen was assimilated into a carbon skeleton. Ammonia soon became the most likely candidate and the evidence for the involvement of this compound has been reviewed by Wilson and Burris (1953), Nicholas (1963a, b) and Wilson (1969). Such evidence includes the observation that ammonia may be used without lag by organisms fixing nitrogen. Furthermore, ammonia has been isolated as a product of ^{15}N fixation in cultures of *Clostridium pasteurianum* (Zelitch *et al.*, 1951) and *Azotobacter vinelandii* (Newton *et al.*, 1953). It is of interest to note that in *Cl. pasteurianum*, while about 50% of the ^{15}N fixed accumulated in the culture medium as ammonia (which always contained the highest labelling activity), the compound with the second highest activity was glutamine (amide nitrogen); this is in agreement with recent evidence concerning the mode of ammonia assimilation in this organism (Dainty and Peel, 1970; Dainty, 1972). Finally, cell-free extracts of a number of organisms showed conversion of ^{15}N to $^{15}NH_3$ and, indeed, nitrogen fixation has been assayed as ammonia production in several instances (e.g. Mortenson, 1962; Munson and Burris, 1969). In extracts of *Cl. pasteurianum* (Carnahan *et al.*, 1960a, b) incorporated ^{15}N was recovered quantitatively as ammonia and the normal yield of the product was of the order 30 μg ammonia nitrogen/ml. The possible pathways involved in the reduction of nitrogen to ammonia have been reviewed

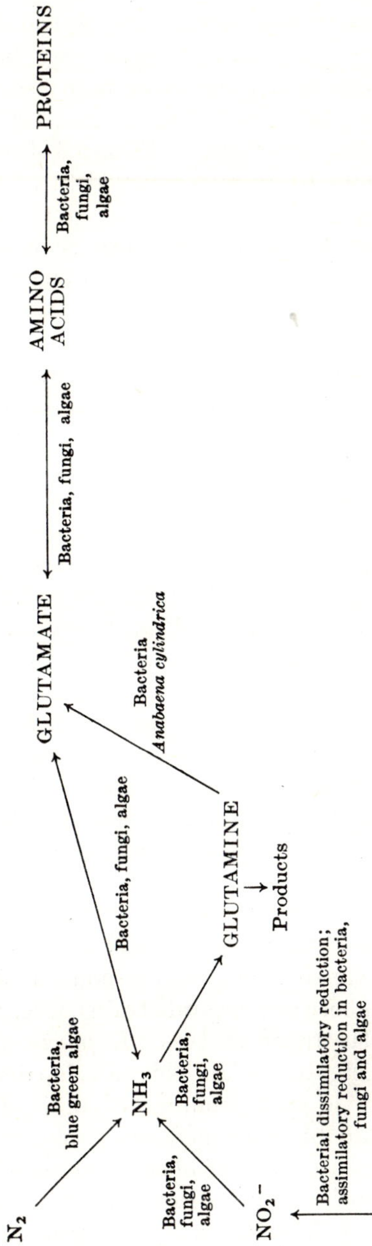

FIG. 1. Inorganic nitrogen assimilation in micro-organisms.

recently by Benemann and Valentine (1972) and will not be further elaborated here.

A useful property of the nitrogen-fixing enzyme complex (nitrogenase) is that substrates other than nitrogen also may be reduced. Such substrates include nitrous oxide, the azide and cyanide ions, methyl isocyanide and acetylene (see Hardy and Burns, 1968; Burris, 1969). The ATP-dependent reduction of any of these substrates is specifically associated with the ability to fix nitrogen and this property is absent from cultures in which synthesis of the enzyme system involved has been repressed by growth on fixed nitrogen. The reduction of acetylene (Schöllhorn and Burris, 1966; Dilworth, 1966) is of particular importance since the product (ethylene) may be detected in very small quantities using gas chromatography (Postgate, 1972) and the "acetylene reduction test" is now widely used for screening possible nitrogen-fixing organisms, for the assessment of nitrogen fixation in natural environments and for the assay of nitrogenase (the enzyme system involved in nitrogen fixation) in whole cells and cell extracts.

By the combined methods of ^{15}N incorporation and the acetylene reduction test, the present day list of *bona fide* free-living nitrogen fixing organisms is small, and restricted to a relatively few species of prokaryotic organisms (Stewart, 1969; Postgate, 1971). Reports of nitrogen-fixing yeasts and higher fungi have been discounted by the painstaking work of Millbank (1969, 1970). Nitrogen-fixing bacteria include obligate aerobes of the family Azotobacteriaceae and *Mycobacterium flavum*, facultative anaerobes such as *Klebsiella pneumoniae* and *Bacillus polymyxa* (which fix nitrogen only under anaerobic conditions) and obligate anaerobes such as *Cl. pasteurianum*, *Desulfovibrio desulfuricans* and *Desulfomaculum ruminis*. There are also authenticated reports of nitrogen fixation in photosynthetic bacteria such as *Rhodospirillum rubrum*, *Chromatium* and *Chloropseudomonas ethylicum*, while positive acetylene reduction has been reported in the acidophilic *Thiobacillus ferrooxidans* (Mackintosh, 1971). Dixon and Postgate (1971) succeeded in transferring the genes responsible for nitrogen fixation (nif) by conjugation between mutant strains of *K. pneumoniae*, and followed this with a similar transfer between *K. pneumoniae* and *Escherichia coli* (Dixon and Postgate, 1972). This development opens up the possibility of producing a large number of derived nitrogen-fixing organisms, and this could be of tremendous ecological and, possibly, economic significance. In a survey of blue-green algae, Stewart (1969) lists some 40 species known to fix nitrogen, all being filamentous and heterocystous members of the families Nostocaceae, Scytonemataceae, Stigonemataceae and Rivulariaceae. To this list must now be added the filamentous, non-heterocystous *Plectonema boryanum* (Stewart and Lex

1970) and two strains of the unicellular *Gleocapsa* (Wyatt and Silvey, 1969; Rippka *et al.*, 1971).

A. Bacterial Nitrogen Fixation

Nitrogen fixation is essentially an anaerobic process and, when resolved into its component proteins, the particulate nitrogenase of *Azotobacter* was found to be as sensitive to oxygen as the soluble enzyme preparations from anaerobes. *Azotobacter* must therefore possess mechanisms for allowing the oxygen-sensitive reactions of nitrogen fixation to occur during aerobic growth. However, although *Azotobacter* sp. are obligate aerobes, nitrogen-fixing cultures are inhibited by excessive aeration; and this sensitivity is not shown by cultures grown on ammonia (Dalton and Postgate, 1969a, b). In a similar way *Mycobacterium flavum* grows best at low oxygen tensions when fixing nitrogen (Biggins and Postgate, 1969), as does *Derxia gummosa* (Hill and Postgate, 1969).

As discussed by Postgate (1971) and Hill *et al.* (1972) two regulatory processes are thought to occur in *Azotobacter* sp. The first is the process of "respiratory protection" in which oxygen is probably excluded from the site of nitrogen fixation by the high rate of respiration characteristic of these organisms. The second process, which has been termed "conformational protection", occurs when, for some physiological reason (e.g. high aeration or carbon limitation), respiratory protection is inadequate. Under these conditions, a conformational change probably takes place in the enzyme complex so that the oxygen-sensitive sites become inaccessible to oxygen (but concomitantly lose their enzymic activity). Thus the particulate nitrogenase preparations from *A. vinelandii* may exist in an oxygen-insensitive form (Bulen *et al.*, 1965) unlike the soluble preparations from the anaerobe *Cl. pasteurianum* which are always oxygen sensitive. As the particulate nitrogenase enzyme system of *A. vinelandii* was oxygen sensitive when resolved into its component parts (Bulen and Le Compte, 1966), Dalton and Postgate (1969a, b) postulated that the oxygen-tolerant nitrogenase particle represented a model of the "conformationally protected" enzyme. Oppenheim and Marcus (1970a, b) have shown that *A. vinelandii* grown on molecular nitrogen possess an extensive internal network of membranes which is absent from cells grown on ammonia. The assumption is that this membrane system protects the particulate nitrogenase from damage by oxygen and, therefore, provides a site for the process of conformational protection. Drozd *et al.* (1972) have grown *A. chroococcum* in chemostat cultures in which nitrogenase synthesis was fully derepressed, partly repressed or fully repressed depending on the concentration of ammonia added to the medium. As in *A. vinelandii*, the convoluted internal membrane system present in nitrogen-fixing cells was absent from those repressed by ammonia although, perhaps surprisingly, the phospholipid content of both types of cell was similar. All members of partly repressed cultures possessed some internal membranes but in smaller amounts than in fully derepressed cells.

The first bacterial cell-free nitrogen-fixing system (Carnahan *et al.*, 1960a, b) was obtained with extracts of *Cl. pasteurianum* prepared either by Hughes press treatment or autolysis of dried cells. The soluble (not sedimented at 144,000 **g** after 2 hours) oxygen-sensitive system utilized pyruvate which was metabolized by phosphoroclastic cleavage. This process provided the enzyme system with the two prerequisites for nitrogen fixation, a source of reducing power and a source of ATP. Bulen *et al.* (1965) showed that sodium dithionite could act as electron donor in cell-free preparations from *A. vinelandii* and *Rhodospirillum rubrum*. Since this report, dithionite has been widely used in cell-free systems together with either ATP or an ATP-generating system (creatine phosphate and creatine kinase). The employment of dithionite, while being useful in determining many characteristics of nitrogenase, gave no information on the nature of the physiological electron donors. Mortenson *et al.* (1962), however, found that in extracts of *Cl. pasteurianum* an electron carrier of low potential was involved linking pyruvate utilization to nitrogen fixation. This carrier protein, which contained non-haem iron, was termed ferredoxin. Crude extracts of *K. pneumoniae* and *B. polymyxa* will also fix nitrogen in the presence of pyruvate while those of *A. vinelandii* will not, although all four organisms will utilize reduced nicotinamide nucleotides as electron donors. Ferredoxins have been implicated as electron carriers during nitrogen fixation in *B. polymyxa* and *A. vinelandii* as well as in *Cl. pasteurianum* (see Benemann and Valentine, 1972).

As pointed out by Postgate (1971), reduction of nitrogen to ammonia could in theory be exergonic, and yet nitrogen fixation has a strict requirement for ATP (the ATP/2e ratios obtained with purified extracts are 4·3 for *Azotobacter* and 3·0 for *Clostridum*; Dalton and Mortensen, 1972). The role of ATP is obscure although, with purified components of *Cl. pasteurianum* and *K. pneumoniae*, ATP binding to nitrogenase proteins could be demonstrated (Bui and Mortenson, 1968; Biggins and Kelly, 1970) and it was therefore assumed that this brought the ATP into a suitable complex for its utilization in the reduction process (perhaps for electron activation). This requirement for ATP may be demonstrated by the lower molar growth yields obtained in organisms fixing nitrogen relative to those utilizing ammonia. Hill *et al.* (1972) have compared the yields of *A. chroococcum*, *Kl. pneumoniae*, *Cl. pasteurianum* and *Desulfovibrio desulfuricans* in carbon and energy-limited chemostat cultures and, in every instance, the yield obtained when grown on molecular nitrogen was much less than when grown on ammonia.

Nitrogenase preparations from a number of bacteria have been purified and fractionated into their component proteins. Purification usually involved anaerobic ion-exchange chromatography and, using this method, two distinct protein moieties were resolved, one sensitive to oxygen and the other less so. Nitrogenase proteins from *A. vinelandii*, *K. pneumoniae* and *Cl. pasteurianum* were similar in their general properties and in the conditions they required for activity (for discussion see Postgate, 1971, Eady *et al.*, 1972). Protein 1 was the oxygen insensitive component which contained both molybdenum and non-haem iron in an average ratio of approximately 1:17. Protein 1 from all three organisms was made up of sub-units and, while the aggregate protein differed with respect to its molecular weight in different bacteria, there was a close similarity in specific activities. Protein 2 was the oxygen-sensitive component and, while containing non-haem iron, did not contain molybdenum. Protein 2 of *K. pneumoniae* and *Cl. pasteurianum* varied in size but both were made up of sub-units and contained iron and acid-labile sulphide in equivalent amounts. In *K. pneumoniae*, reduction of both nitrogen and acetylene was maximal when proteins 1 and 2 were present in a 1:1 molar ratio (Eady *et al.*, 1972). Proteins 1 and 2 from a number of bacteria showed some cross reactivity (Detroy *et al.*, 1969; Kelly, 1969). Thus proteins from *A. chroococcum* would substitute for those of *K. pneumoniae* and those of *K. pneumoniae* with those of *B. polymyxa* with about 80% maximum activity. The cross reactivity shown between proteins of *A. chroococcum* and *B. polymyxa*, and *A. vinelandii* and *B. polymyxa*, were much less, however, while those of *B. polymyxa* and *Cl. pasteurianum* were inactive. These cross reactivities may reflect evolutionary relationships between the organisms involved.

Nitrogen fixation in cell-free extracts of the photosynthetic bacterium *R. rubrum* was first reported by Schneider *et al.* (1960) and later Bulen *et al.* (1965) showed that enzyme activity was stimulated on addition of pyruvate. Burns and Bulen (1966) found that, in extracts of *R. rubrum* prepared by sonication or French press treatment, nitrogenase activity was present in a 144,000 g supernatant. Recently Schick (1971) determined nitrogen fixation in whole cells of *R. rubrum* manometrically, and concluded that a range of environmental conditions (including light intensity, pH value and temperature) influenced nitrogen uptake, and further reported that pyruvate stimulated uptake (10 mol pyruvate being consumed per mol nitrogen "fixed"). Cell-free nitrogen fixation has also been obtained with extracts of the purple sulphur bacterium *Chromatium* (Winter and Arnon, 1970, Yoch and Arnon, 1970) and in the green heterotrophic bacterium (*Chloropseudomonas ethylicum* (Evans and Smith, 1971). Winter and Arnon (1970) showed that, in extracts of *Chromatium* prepared by sonication, reduction of either nitrogen or acetylene required reducing power and ATP. Reducing power could be supplied by dithionite, reduced ferredoxin or by hydrogen in the presence of catalytic amounts of viologen dye. It was further reported (Yoch and Arnon, 1970) that the ATP requirement could be supplied by photosynthetic phosphorylation although the direct photoreduction of nitrogen was not demonstrated. Evans and Smith (1970) reported a soluble (150,000 **g**) nitrogenase in sonicated extracts of *Chloropseudomonas ethylicum*. In the intact organism, acetylene reduction was found to be light dependent; activity in the dark was only 10% of that in the light. In crude extracts, pyruvate produced a faster rate of reduction than did dithionite. Fractionation of the crude extract on DEAE cellulose removed ferredoxin and, in extracts so treated, pyruvate-dependent acetylene reduction occurred only in the presence of added ferredoxin. Ferredoxin photoreduced in the presence of photosynthetic particles from the same organism, or with illuminated spinach chloroplasts, served as electron donor for acetylene reduction in the presence of an ATP-generating system. Ferredoxins from *Chromatium* or *Cl. pasteurianum* were less effective than those from the parent organism.

The overall requirements for nitrogen fixation may be considered to be the presence of an adequate supply of substrates, the absence of inhibitors and a suitable environment. Substrates include molecular nitrogen, a supply of ATP provided by an active metabolism, and carbon skeletons to accept the product of nitrogenase action (ammonia). Inhibitors include ammonia, which represses synthesis of nitrogenase, and ADP which, in *Cl. pasteurianum* at least, inhibits nitrogenase activity. Environmental factors include the oxygen tension, for, even with aerobic bacteria and blue-green algae (see below), nitrogen fixation is most efficient at low dissolved oxygen tension. In a natural aquatic or soil environment, fixation by non-photosynthetic bacteria is probably limited by the availability of sources of carbon and energy, and for this reason, nitrogen fixation by photosynthetic bacteria and blue-green algae is usually considered to be of greater significance.

There is general agreement that some fixed nitrogen sources (including nitrate and ammonia) repress the synthesis of nitrogenase in bacteria and that nitrogen itself, at least in more than trace amounts, is not required for nitrogenase synthesis (Hill *et al.*, 1972; Dalton and Mortenson, 1972; Benemann and Valentine, 1972; Drozd *et al.*, 1972). Thus nitrogenase was absent from ammonia-grown cultures of *Azotobacter* (Bulen *et al.*, 1964) and did not appear in cultures of *Azotobacter* or *Cl. pasteurianum* until ammonia was exhausted (Strandberg and Wilson, 1967; Daesch and Mortenson, 1968). In both *K. pneumoniae* and *A.*

vinelandii diauxic growth occurred when cultures were grown first on ammonia and then on molecular nitrogen and it was assumed that nitrogenase synthesis occurred during the diauxic lag (Yoch and Pengra, 1966; Strandberg and Wilson, 1967). Addition of certain amino acids (especially aspartate) stimulated enzyme formation in the absence of ammonia in *K. pneumoniae* possibly by providing preformed organic nitrogen for enzyme synthesis. As in *K. pneumoniae*, nitrogenase synthesis in *A. chroococcum* was not repressed by the presence of aspartate, glutamate or glutamine nor were these compounds metabolized (Drozd *et al.*, 1972). Even during active nitrogen fixation, synthesis of nitrogenase was partially repressed by the intracellular pools of ammonia accumulated under these conditions. In ammonia-limited chemostat cultures therefore (with presumably much lower pool levels of ammonia) both *Cl. pasteurianum* and *Azotobacter chroococcum*, growing in the presence of an inert gas phase (but lacking nitrogen), contained higher nitrogenase activities than did nitrogen-fixing populations (Dalton and Postgate, 1969). Munson and Burris (1969) obtained similar results with fixed-nitrogen-limited chemostat cultures of the photosynthetic bacterium *R. rubrum*. In further experiments with *A. chroococcum* (Hill *et al.*, 1972), the nitrogenase activity of sulphate-limited chemostat cultures growing on molecular nitrogen decreased as the content of ammonia in the input medium was increased. Free ammonia could be detected in the medium only when nitrogenase synthesis was totally repressed. It was possible to obtain stable populations at different states of repression which suggested that nitrogen fixation and utilization of exogenous ammonia occurred at the same time. The degree of repression by a particular concentration of ammonia was a function of the culture population density and, in populations of low cell density, only low concentrations of ammonia were required to repress nitrogenase synthesis. This is of significance in natural environments in which only small populations of cells are normally detected. Furthermore, the addition of ammonia to nitrogen-fixing chemostat cultures of *A. chroococcum* was shown to curtail the culture nitrogenase activity at an exponential rate; the culture enzyme activity therefore decreased faster than predicted for the wash-out of stable enzyme (Drozd *et al.*, 1972). This indicated that, as in *A. vinelandii* (Hardy *et al.*, 1968) ammonia had a double effect, bringing about repression of nitrogenase synthesis and, concomitantly, a small decrease in activity of existing nitrogenase. In *Cl. pasteurianum* and *K. pneumoniae*, the effect of ammonia may be solely one of repression since preformed nitrogenase remained active in the presence of ammonia (Daesch and Mortenson, 1972; Mahl and Wilson, 1968). In *A. vinelandii* the synthesis of both of the constituent proteins of nitrogenase was repressed co-ordinately in the presence of ammonia and derepressed in its absence (Shah *et al.*, 1972).

At the present time, the only physiological compound known to affect nitrogenase activity *in vitro* is ADP. With extracts of *Cl. pasteurianum*, and in the presence of 18 m*M* ATP, 3 m*M* ADP reduced nitrogenase activity by 42% (and 5 m*M* ADP by 53%) of the activity determined in the absence of ADP. With an ATP/ADP ratio of 0·5, complete inhibition of nitrogenase activity was observed (Moustafa and Mortenson, 1967).

B. Nitrogen Fixation by Blue-Green Algae

Blue-green algae are ubiquitous both in aquatic environments and soils and are found in greatest abundance in tropical regions. Their ecological significance in relation to nitrogen fixation is especially established in aquatic habitats (Dugdale and Dugdale, 1962; Stewart *et al.* 1967; Stewart, 1969; Horne and Fogg, 1970). Many species of filamentous blue-green algae contain characteristic types of cells with thick refractive walls, termed heterocysts. Early experiments by Fogg (1942, 1949) established that fixed nitrogen repressed both heterocyst formation and the ability of organisms to fix nitrogen. These observations were subsequently repeated by a number of authors (e.g. Ogawa and Carr, 1969; Kulasooriya *et al.*, 1972). When non-heterocystous filaments of *Anabaena cylindrica* (grown in the presence of ammonia) were transferred to a medium free of fixed nitrogen, and incubated in light, it was found that nitrogenase activity and heterocyst development increased in parallel. Blue-green algae, alone amongst nitrogen-fixing organisms, produce oxygen during photosynthesis and heterocystous algae possess the ability to fix nitrogen aerobically. High oxygen tensions however inhibited acetylene reduction by cultures of *Anabaena flos-aquae* while this activity was markedly increased at oxygen tensions below 30 mm Hg (Stewart and Pearson, 1970). As discussed below, algal nitrogenases are oxygen-sensitive and, as with aerobic bacteria, some mechanisms must exist for protecting the enzyme from inhibition by oxygen. Fay *et al.* (1968) reviewed the data available at that time and postulated that heterocysts functioned as the sites of nitrogen fixation. An attractive feature of their proposal was that heterocysts could supply an environment suitable for nitrogen fixation by virtue of their high levels of respiratory activity and their virtual lack of both photosynthetic $^{14}CO_2$ fixation and oxygen evolution (Fay and Walsby, 1966; Lang and Fay, 1971). Therefore within the thick wall of the heterocyst there is a reducing environment which can be readily demonstrated by histochemical methods; for example, Stewart *et al.* (1969) reported reduction of silver salts in a photographic emulsion, and of triphenyl tetrazolium chloride, by heterocysts but not by vegetative cells of *A. cylindrica*. In similar experiments, $^{14}CO_2$ fixation was recorded in vegetative cells but not in heterocysts. Stewart *et al.* (1969) prepared fractions of filaments of *A. cylindrica* by sonication under anaerobic conditions, and were able to show acetylene reduction with a particulate fraction rich in heterocysts to which ATP and dithionite had been added. Little or no activity was found with soluble protein or photosynthetic lamellae fractions of the vegetative cells. In contrast to these findings, however, the non-heterocystous *Plectonema boryanum* reduced acetylene, incorporated $^{15}N_2$ and grew readily in a medium free from fixed nitrogen when incubated in a gas phase lacking oxygen; cultures grown on ammonia failed to reduce acetylene (Stewart and Lex, 1970). Thus, in this latter organism, nitrogenase activity was associated with vegetative cells. Smith and Evans (1970, 1971) have also reported nitrogenase activity in vegetative cells of *A. cylindrica* exposed to micro-aerophilic conditions; with a French press treatment of 15,000 p.s.i., most of the vegetative cells, but no heterocysts, were broken, while at 30,000 p.s.i. both types of cell were disintegrated. The nitrogenase specific activity of both of these extracts was similar, and the authors concluded that there was no evidence to suggest that heterocysts were the primary sites of nitrogenase activity. Fay and Lang (1971), however, found that, when viewed in the electron microscope, heterocysts prepared by French press treatment were invariably damaged and may therefore have leaked out some of their enzymic constituents. In a comparative study with cell-free extracts from *A. cylindrica* and *P. boryanum*, prepared by sonication, Haystead *et al.* (1970) found that nitrogenase from both organisms was inhibited by oxygen. Therefore, while the nitrogenase of both organisms was oxygen sensitive, the heterocystous *A. cylindrica* fixed nitrogen aerobically while the non-heterocystous *P. boryanum* would do so only under micro-aerophilic conditions. This situation was complicated by reports of light-dependent, nitrate repressible, acetylene reduction in separate isolates of the unicellular *Gleococapsa* (Wyatt and Silvey, 1969; Rippka *et al.*, 1971). Electron micrographs of *Gleococapsa* 6501 (Rippka *et al.*, 1971) showed no apparent differences in the structure of cells grown under nitrogen-fixing conditions and those grown on fixed nitrogen. According to Stanier (cited by Postgate, 1971) *Gleococapsa* was very photosensitive and would therefore tolerate only low oxygen tensions. Thus, while nitrogenase is certainly present in vegetative cells of blue-green algae (at least under microaerophilic conditions), the presence of the enzyme in heterocysts appears to be necessary for aerobic nitrogen fixation.

Studies on nitrogen fixation in cell-free extracts of blue-green algae are not so advanced as those in bacteria. The first reports of cell-free preparations by Schneider *et al.* (1960) showed a soluble (not sedimented by centrifugation at 45,000 **g** for 45 min) nitrogenase in *Mastigocladus laminosus*. This was followed by reports by Cox *et al.* (1964) that, in *A.*

cylindrica, extracts prepared by sonication, or by treatment in a French press, contained a nitrogenase associated with a particulate fraction sedimenting between 5,000 and 35,000 **g**. In their study on nitrogenase in heterocysts, Stewart *et al.* (1969) used anaerobic conditions during disruption of filaments by sonication and showed that both ATP and reducing power (dithionite) were required for acetylene reduction. Haystead *et al.* (1970) prepared cell-free extracts from *A. cylindrica* and *P. boryanum* and reported nitrogenase activity in the 40,000 **g** supernatant fractions. Adenosine triphosphate and dithionite were again required for activity, and the extracts were cold sensitive. Smith and Evans (1970, 1971) however found that, in *A. cylindrica* extracts prepared by sonication or by French press treatment, nitrogenase activity was particulate (sedimenting within 3h at 158,000 **g**). These authors confirmed the requirement for ATP, and the sensitivity to oxygen, and also found that dithionite supported higher rates of acetylene reduction than did reduced ferredoxin. Haystead and Stewart (1972) have recently partially purified the nitrogenase from *A. cylindrica* and, in studies involving inhibitors, showed the enzyme to be a metalloprotein containing iron and requiring thiol groups for activity. They also confirmed the findings of Bothe (1970) concerning the involvement of ferredoxin in the transfer of electrons from hydrogen to the enzyme. Centrifugation of crude extracts at 144,000 **g** for 3 h sedimented most of the enzyme activity although only 30% could be resuspended. With their partially purified (25-fold) extract, however, no activity was sedimented and they maintained that, in crude extracts, nitrogenase may be part of, or adsorbed to, a larger complex of protein molecules, thus explaining sedimentation in these systems.

Nitrogen fixation in extracts of blue-green algae requires for activity both ATP and a reducing system. Most workers have used dithionite as reductant but recently attempts have been made to determine the nature of the physiological electron donors. Bothe (1970) found that, in *A. cylindrica*, both ferredoxin from spinach and phytoflavin from *Anacystis nidulans* stimulated light-dependent nitrogenase activity. Smith and Evans (1971) and Smith *et al.* (1971) confirmed that ferredoxin was involved in photosynthetic electron transport in *Anabaena* and concluded that direct photoreduction of ferredoxin by photosynthetic electron transport provided the bulk of the reductant for nitrogenase activity. Haystead and Stewart (1972) were able to provide reductant for nitrogenase using a hydrogenase preparation from *Cl. kluyveri* and showed that ferredoxin from *A. cylindrica* or *Cl. pasteurianum* stimulated hydrogen-dependent acetylene reduction in cell-free extracts of *A. cylindrica*. Many blue-green algae will fix nitrogen, at a low rate, in the dark. Fay and Cox (1966) and Cox and Fay (1967) found that, under these conditions, pyruvate supported nitrogen fixation. These results were confirmed by Smith *et al.* (1971) with extracts of *A. cylindrica.* Bothe (1970) and Smith *et al.* (1971) have also reported that NADP-linked reductions can serve as electron donors functioning via a ferredoxin-NADP reductase. The rates of acetylene reduction recorded however were low (~5% light-saturated rate) and this system may be of limited physiological significance.

As in bacterial systems, the function of ATP in algal nitrogen fixation is unknown. Cox and Fay (1969) have presented evidence that cyclic phosphorylation provides ATP for nitrogenase activity.

As mentioned above there is a great deal of evidence indicating that nitrogenase synthesis, and the development of heterocysts in blue-green algae, are both repressed by growth on fixed nitrogen sources such as ammonia and nitrate. As in bacteria, the presence of nitrogen is not required for nitrogenase synthesis since this occurred in an inert gas phase and in the absence of a fixed nitrogen source in cultures of both *A. cylindrica* (Smith and Evans, 1970; Neilson *et al.*, 1971) and *A. flos-aquae* (Bone, 1971). Therefore, synthesis of nitrogenase is controlled by derepression rather than induction. Stewart *et al.* (1968) showed that, while nitrate repressed nitrogenase synthesis in *Nostoc muscorum*, it did not inhibit the activity of preformed enzyme.

III. Nitrate Reduction

Nitrate reduction may be accomplished by two distinct physiological mechanisms (see Painter, 1970). The first is the process of nitrate assimilation, widespread in micro-organisms, in which nitrate is reduced via nitrite and probably hydroxylamine to ammonia. Ammonia is then assimilated by mechanisms discussed in the next section. The second process is one in which nitrate serves as an alternative electron acceptor to oxygen (i.e. anaerobic respiration) and is consequently reduced. This process is common only in anaerobic bacteria, and in facultatively anaerobic bacteria at low oxygen tensions. In *Aerobacter aerogenes*, the molar growth yield of anaerobically grown cells almost doubled when nitrate was added as electron acceptor; about 0.5 mol nitrate was reduced to ammonia per mol of glucose utilized (Hadjipetrou and Stouthamer, 1965). Ammonia production during this process of nitrate respiration is not common and either nitrite or molecular nitrogen are the usual products.

Assimilatory nitrate reduction is catalysed by at least two enzymes. Nitrate reductase reduces nitrate to nitrite while nitrite reductase reduces nitrite to ammonia. The most widely accepted route of nitrate assimilation is the inorganic pathway proposed by Fewson and Nicholas

(1961) in which nitrite, nitric oxide and hydroxylamine are established intermediates. There is speculation, however, as to the intermediate compound(s) existing between nitric oxide and hydroxylamine, although it is possible that these compounds are enzyme-bound and therefore will be difficult to identify (Painter, 1970).

A. Nitrate Reduction in Bacteria

There have been many reports of respiratory nitrate reduction in bacteria, but few of the assimilatory process. In the former process, nitrite may be the usual product, although this compound might be further reduced to ammonia by other organisms and thence assimilated. Consequently we have included a brief account of bacterial nitrate reduction in this article.

In most instances, nitrate reductase has been assayed in cell-free extracts using methyl or benzyl viologen (reduced by dithionite) as electron donor. Two respiratory nitrate reductases may be assayed by this means, one which will utilize chlorate as substrate and a second which is inhibited by chlorate (Pichinoty *et al.*, 1971). The respiratory enzyme is membrane-bound in bacteria such as *Aerobacter aerogenes* (Van't Riet *et al.*, 1968), *Escherichia coli* (Showe and De Moss, 1968; Cole and Wimpenny, 1968) and *Haemophilus parainfluenzae* (Sinclair and White, 1970).

In coliform bacteria, respiratory nitrate reductase is thought to be induced in the presence of nitrate, while enzyme synthesis is repressed and activity inhibited in the presence of oxygen (Pichinoty, 1965, 1969a, b). Pichinoty and Ornano (1961) reported that, in *A. aerogenes*, nitrate reductase was synthesized in the presence of nitrate under anaerobic conditions and that this synthesis was repressed by air, even in the presence of nitrate. Van't Riet *et al.* (1968) further showed that the activity of respiratory nitrate reductase from *A. aerogenes* was low or absent during anaerobiosis in the absence of nitrate and confirmed that oxygen both repressed synthesis of this enzyme and inhibited its activity. *Aerobacter aerogenes* carries out both nitrate respiration and nitrate assimilation. The particulate respiratory activity of crude extracts was not sensitive to sonic oscillations nor to ammonia, while the soluble assimilatory enzyme was very sensitive to sonic oscillations but was insensitive to air (but not to pure oxygen). The synthesis of assimilatory nitrate reductase was repressed by ammonia under both aerobic and anaerobic conditions. The anaerobic respiratory enzyme did not function in the assimilatory process under aerobic conditions, since a growth lag occurred when cultures grown anaerobically in the presence of both nitrate and ammonia were transferred to aerobic conditions in the presence of nitrate as nitrogen source. The anaerobic enzyme, when partially purified (and solubilized), was more sensitive to sonic oscillations than the activity of crude particulate preparations. Van't Riet *et al.* (1968) showed that, while the respiratory and assimilatory activities might be brought about by separate enzymes, similar kinetic parameters and similar pH value optima could be demonstrated. These authors therefore proposed the presence of one enzyme which was part of two different structural complexes with different electron transfer systems, subject to different regulation and with perhaps separate cellular locations. Recently, cytochrome *b* has been implicated in the respiratory complex, but plays no part in nitrate assimilation (Van't Riet *et al.*, 1972). In *Escherichia coli* the membrane-bound respiratory nitrate reductase was also found to be induced by nitrate under anaerobic conditions (Cole and Wimpenny, 1968; Azoulay *et al.*, 1969). Showe and De Moss (1968) further reported that enzyme activity was low (or absent) during anaerobiosis in the absence of nitrate but increased some 20-fold on nitrate addition. Enzyme synthesis was repressed by aeration. A regulatory system proposed by Showe and De Moss (1968) contained two repressors, one sensitive to nitrate and the other to the intracellular redox potential. The respiratory nitrate reductase in *E. coli* required $NADH_2$ as electron donor (Nason, 1962; Cole and Wimpenny, 1968) but Cole and Wimpenny showed that this could be replaced with a range of compounds including formate, lactate and pyruvate.

Pichinoty (1970) reported that, in a number of facultative anaerobic bacteria including strains of *Aeromonas* and *Hafnia*, respiratory nitrate reductase was induced by nitrate and repressed by air. The *Pseudomonas putida* enzyme was constitutive, as was that of *Micrococcus denitrificans*, which was unaffected by oxygen or ammonia. In all the bacteria studied, enzyme activity was inhibited by air. In *Proteus mirabilis* (De Groot and Stouthamer, 1970) nitrate reductase activity was low during anaerobiosis in the absence of nitrate and was decreased in presence of air. A two-repressor system, similar to that in *E. coli*, has been proposed.

In *B. stearothermophilus* (Downey *et al.*, 1969), respiratory nitrate reductase was induced by nitrate and inactivated by oxygen. In *B. licheniformis*, however, this enzyme was induced at low oxygen tensions in the presence or absence of nitrate (Schulp and Stouthamer, 1970). Therefore, in this organism, the intracellular redox potential alone appeared to control enzyme synthesis, in contrast to the two repressor systems of *E. coli* and *P. mirabilis*.

In a marine psychrophylic strain of *Pseudomonas* (D. S. Macdonald-Brown and C. M. Brown, unpublished) the assimilatory nitrate reductase was soluble and, like that of *A. aerogenes*, sensitive to sonic oscillations.

This enzyme required $NADH_2$ and FAD for activity and was repressed in chemostat cultures containing an excess of ammonia or glutamate. It was synthesized, however, in the absence of nitrate, in ammonia-limited cultures and in nitrogen-limited cultures with glutamate as nitrogen source.

Cole (1968) listed three distinct nitrite reductase activities in *E. coli*; one described by Lazzarini and Atkinson (1961) was soluble, NADP-linked, and reduced nitrite to ammonia; one was particulate and one NAD-linked. Cole (1971) and Ward and Cole (1971) presented evidence to show that the NADP-linked activity is associated with sulphite reductase and probably only contributes about 5% total nitrite reduction *in vivo*. Little is known of the particulate activity and the soluble, NAD-linked respiratory nitrite reductase is assumed to be of greater physiological significance. Synthesis of this latter enzyme was repressed by air (Cole, 1968), was stimulated when the nitrate level of the environment was low and increased with increasing medium concentration of nitrite.

B. NITRATE REDUCTION IN FUNGI

Many filamentous fungi and a small number of yeasts (Campbell, 1971) utilize nitrate as a nitrogen source and the assimilatory reduction of nitrate to ammonia is well characterized in these organisms. Fungal nitrate reductases are in general soluble molybdo-flavoproteins requiring reduced nicotinamide nucleotide coenzymes as electron donors. Nicholas *et al.* (1954) showed that cell-free extracts of molybdenum-deficient *Neurospora crassa* and *Aspergillus niger* contained much less nitrate reductase activity than those from control cultures, while Garrett and Nason (1969) have established the presence of molybdenum in purified nitrate reductase from *N. crasssa*. The nicotinamide nucleotide coenzyme electron donor in *N. crassa* (Nicholas *et al.*, 1960; Garrett and Nason, 1969) and *A. nidulans* (Cove, 1966) is $NADPH_2$, while that of the yeast *Hansenula anomala*, which may be mitochondrial in origin, was either $NADPH_2$ or $NADH_2$, although the latter was more active *in vitro* (Silver, 1957; Pichinoty and Mettnier, 1967). In *Candida utilis*, the soluble nitrate reductase required $NADH_2$ as electron donor and activity was slightly stimulated by added FMN and molybdenum (V. J. Wiles, and C. M. Brown, unpublished data). The *Neurospora* nitrate reductase was purified by Nason and Evans (1953) and found to contain FAD and to require thiol groups for activity. This was confirmed by Garrett and Nason (1969), who also established the presence in this enzyme of a *b* cytochrome together with molybdenum. Garrett and Nason (1969) also reported the presence of a second heavy metal ion which was thought by Pichinoty (1969) to be involved in the binding of substrate to enzyme.

Nitrate reductase activity in *N. crassa* was closely associated with a NADP-cytochrome *c* reductase and the sequence of electron transport in this organism is thought to be:

$$\text{NADPH} \rightarrow \text{FAD} \rightarrow \text{Cyt}\ b\ 557 \rightarrow \text{Mo} \rightarrow \text{NO}_3^-$$

$$\searrow \text{Cyt}\ c$$

(Kinsky, 1961; Garrett and Nason, 1969). Nason (1962) has proposed that two types of NADP-cytochrome *c* reductase exist, one being a constitutive enzyme with no nitrate reductase activity and the other an inducible enzyme closely associated with nitrate reductase. Repression of nitrate reductase synthesis in *N. crassa* by ammonia is well established (Nason and Evans, 1953; Nicholas, *et al.*, 1954; Kinsky, 1961). Subramanian and Sorger (1972) have further shown that both NADP-linked nitrate reductase and the related NADP-cytochrome reductase and reduced benzyl viologen nitrate reductase activities were all induced following transfer from ammonia to nitrate medium. After induction, the addition of ammonia or the removal of nitrate resulted in rapid inactivation of all three enzymes. This inactivation was much slower in the presence of cycloheximide indicating the possible involvement of some inactivating protein that was synthesized *de novo*. Ammonia did not repress uptake of nitrate in this organism. The nitrate-inducible NADP-cytochrome *c* reductase and viologen nitrate reductase activities of non nitrate-utilizing mutants (*nit*^{1} and *nit*^{3}) were not inactivated by removal of nitrate, or addition of ammonia, suggesting that the integrity of the nitrate reductase complex may be required for the *in vivo* inactivation of nitrate reductase and associated activities.

The repression of nitrate reduction by ammonia has also been demonstrated in a number of other fungi. Morton and MacMillan (1954) reported that *Scopulariopsis brevicaulis* assimilated ammonia more rapidly than nitrate; at low concentrations, ammonia completely inhibited nitrate assimilation in the presence of both substrates. Similar results were obtained in *Myrothecium verrucaria*, *Penicillium chrysogenum*, *Aspergillus repens*, and *Mucor rammanianus*. Morton (1956) further showed that, in *Scopulariopsis brevicaulis*, the mycelial nitrate reductase activity fell to a low value within one hour of ammonia addition and remained low until all the ammonia was assimilated. Nitrate reductase was formed in the absence of nitrate, providing that ammonia was also absent; similar results were obtained with *Penicillium griseofulvum*. In the basidiomycete *Ustilago maydis*, nitrate reductase was synthesized when nitrate was the sole nitrogen source; this synthesis was repressed, and enzyme activity rapidly lost, on addition of ammonia.

Ammonia did not inhibit, and amino acids only partially inhibited, the *in vitro* nitrate reductase activity and the rapid loss of activity in the presence of ammonia suggested that the enzyme probably was broken down. In mycelia of *Aspergillus nidulans* (Cove and Pateman, 1963) the nitrate reductase activity, with nitrate as nitrogen source, was 20-times that with glutamate or urea. Cove (1966) further demonstrated that enzyme activity declined rapidly in the absence of nitrate and that, in the presence of nitrate, ammonia repressed enzyme synthesis. As in other fungi there were no *in vitro* effects of ammonia on enzyme activity. Downey (1971) purified nitrate reductase from *A. nidulans* and showed that the enzyme was a flavoprotein which catalysed the $NADPH_2$-dependent reduction of both nitrate and cytochrome *c*. In contrast to the purified *Neurospora* enzyme, the nitrate reductase of *Aspergillus* did not contain a cytochrome. In *Candida utilis* and *Hansenula anomala*, ammonia was assimilated preferentially to nitrate in cultures containing both nitrogen sources. Neither ammonia nor glutamate showed any *in vitro* effects on enzyme activity in extracts of *Candida utilis* but enzyme synthesis was repressed in ammonia- and glutamate-grown cultures (P. Turner, V. J. Wiles, and C. M. Brown, unpublished data).

In *Aspergillus nidulans*, nitrate reductase, nitrite reductase and hydroxylamine reductase activities were all repressed by ammonia (Pateman *et al.*, 1967). The growth and enzyme characteristics of a total of 123 mutants, involving nine different genes, indicated that only two possible structural genes were involved in the reduction of nitrate to ammonia. The first specified nitrate reductase and the second the reduction of nitrite to ammonia. Thus both nitrite and hydroxylamine reductase activities may be carried on the same protein. The nitrite reductase of *N. crassa* was similar to the nitrate reductase of that organism in that it was induced in the presence of either nitrate or nitrite (Nason and Evans, 1953; Garrett, 1972) and was repressed by the presence of ammonia and a mixture of amino acids (Cook and Sorger, 1969; Garrett, 1972). In the absence of nitrate and nitrite, nitrite reductase activity was lost. Garrett (1972) has presented genetic evidence for the co-ordinate control of both nitrate and nitrite reductase in *Neurospora* and *Aspergillus*.

C. Nitrate Reduction in Algae

Ammonia is probably the most readily utilized source of inorganic nitrogen by algae although many organisms will also utilise nitrate (see Naylor, 1970). The route of nitrate assimilation is less well characterized in algae than in bacteria and fungi, although ammonia is the product of nitrate reduction; for example, Bongers (1956) showed that carbon-starved cultures of *Scenedesmus* reduced nitrate quantitatively to ammonia in the absence of carbon dioxide. It was assumed that the lack of carbon prevented ammonia assimilation and therefore accounted for its accumulation. Intermediates are rarely observed unless the normal course of the reaction is interrupted; for example, Kessler (1959) was able to demonstrate the involvement of nitrite as an intermediate in *Ankistrodesmus braunii* only after lowering the pH value of the medium well below the optimum for growth. The generally accepted scheme involves a two-enzyme system, as in fungi, consisting of nitrate and nitrite reductases. Nitrate reductase of *Chlorella vulgaris* is a complex protein, which has been extensively purified (Solomonson and Vennesland, 1972), requires $NADH_2$ as electron donor and, like the *Neurospora* enzyme, contains a *b*-type cytochrome (cytochrome *b* 557 *Chlorella*). The participation of molybdenum in electron transport during nitrate reduction has been demonstrated in *Anabaena cylindrica* (Wolfe, 1954a, b) and in green algae such as *Chlorella vulgaris* (Solomonson and Vennesland, 1972). In the latter organism, tungsten was a physiological antagonist to molybdenum, being incorporated into the nitrate reductase and rendering it inactive. Reduction of nitrite to ammonia is catalysed by nitrite reductase which, in some organisms, requires reduced ferredoxin as electron donor (Zumft, 1972). That light stimulates nitrate reduction in algae has been known for some time although the mechanisms involved are not fully understood. Grant (1967, 1968) reported that, in the marine phytoflagellate *Dunaliella tertiolecta*, light stimulated both nitrate and nitrite assimilation some 20-fold in the presence of carbon dioxide. Glucose, glycerol, acetate, pyruvate and 2-oxoglutarate were ineffective as substitutes for carbon dioxide. Grant proposed that, in this organism nitrate reductase was located in the chloroplasts and linked to photosynthesis. Grant and Turner (1969) surveyed a number of organisms and reported that, in three species of *Chlorella* and in *Tetraselmis suecica*, *Dunaliella tertiolecta* and *Phaeodactytum tricornutum*, light stimulated both nitrate and nitrite assimilation in the presence of carbon dioxide. These authors also concluded that, in the light, the rate-limiting step in nitrogen assimilation was ammonia assimilation rather than the reduction of nitrate or nitrite to ammonia. Cultures of *Chlamydomonas rheinhardii* were unable to assimilate nitrate in the dark unless acetate was provided, and even under these conditions a period of adaptation was required following removal of the light source (Thacker and Syrett, 1972a). As in *Dunaliella*, light-dependent nitrate reduction in *Chlamydomonas* also required a source of carbon although, in the latter organism, both carbon dioxide and acetate could fill this role. If cultures of *Chlamydomonas rheinhardii* were allowed to accumulate an internal reserve of carbon (by growth in nitrogen-deficient medium) then dark nitrate assimilation occurred in the presence of an exogenous carbon source.

Light stimulation has also been observed in the blue-green *Anabaena cylindrica* (Hattori, 1962). In this organism, assimilation of nitrate, nitrite and hydroxylamine was stimulated by light, and photosynthetically-reduced ferredoxin participated in the reduction of nitrite to ammonia. Eppley and Coatsworth (1968) similarly suggested that, in *Ditylum brightwellii*, both nitrate and nitrite were photoreduced. Thacker and Syrett (1970a), however, proposed that the role of light in stimulating nitrogen assimilation was to provide ATP by photophosphorylation.

Ludwig (1938) reported that ammonia inhibited algal nitrate assimilation; similarly Proctor (1957) reported that *Haematococcus pluvialis* assimilated ammonia at a faster rate than nitrate and that, in *Chlamydomonas*, ammonia was assimilated in preference to nitrate. In *Anabaena cylindrica*, the nitrate and nitrite reducing systems were induced by nitrate and nitrite but not by atmospheric nitrogen. The enzyme responsible for hydroxylamine reduction was induced by nitrate, nitrite and nitrogen. Nitrate, nitrite and hydroxylamine reductase systems were repressed by ammonia and glutamate and the induction of nitrite reductase was inhibited by chloramphenicol, by anaerobiosis and in the dark (Hattori, 1962). Thacker and Syrett (1972a) reported that, in *Chlamydomonas rheinhardii*, assimilation of nitrate was inhibited by the presence of nitrite and ammonia, and that of nitrite by ammonia assimilation. They suggested that, as nitrite reduction occurred at a faster rate than nitrate reduction, the inhibitory effect of nitrite on nitrate assimilation was due to nitrite successfully competing for the available electron donors. As with cultures of *Chlorella* (Syrett and Morris, 1963) products of ammonia assimilation rather than ammonia *per se* were thought to inhibit nitrate assimilation since inhibition did not occur if ammonia assimilation was prevented by an inadequate supply of carbon. Thacker and Syrett (1972b) further showed that ammonia-grown *Chlamydomonas* did not contain nitrate reductase, but that this enzyme was present on incubation with nitrate and lost on ammonia addition. Nitrate reductase activity declined rapidly if photosynthesis was prevented by the absence of carbon dioxide or light, or by the presence of DCMU which is an inhibitor of photosynthesis. The decline of nitrate reductase activity in the dark was prevented by addition of acetate to acetate-adapted cells or if prior nitrogen starvation had lead to the accumulation of carbon reserves. Thus, a supply of organic carbon (perhaps for ammonia assimilation) was required for both nitrate assimilation by whole cells and the appearance of nitrate reductase in cell extracts. These results did not show whether the changes in nitrate reductase activity of extracts were due to repression/derepression of enzyme synthesis or to activation as in *Chlorella* (see p. 22).

Eppley *et al.* (1969) found that, in a number of organisms derived from marine phytoplanton including *Dunaliella*, *Ditylum*, *Cocoolithus*, *Cyclotella* and *Goryanlax*, nitrate reductase was synthesized during ammonia assimilation provided that the ammonia concentration was sufficiently low (0·5 to 1·0 μM). Similar results were obtained in a marine species of *Pseudomonas* as discussed on p. 16. Synthesis of nitrate reductase was repressed, however, by growth on higher concentrations of ammonia and synthesized during growth on nitrate. Thus, when presented with both nitrate and ammonia, ammonia was utilized preferentially. In cultures of *Ditylum brightwellii* both nitrate and nitrite were taken up simultaneously. Synthesis of the $NADH_2$-linked nitrate reductase was induced by nitrate and repressed by ammonia. Synthesis of nitrite reductase was induced by nitrate and nitrite and repressed by ammonia. Enzyme activity was inhibited by nitrate but ammonia, glutamate and aspartate did not show any effect *in vitro*. The cellular levels of both nitrate and nitrite reductases decreased in the absence of their substrates and ammonia decreased the rate of nitrate uptake by whole cells. Marine phytoplankton, in common with marine bacteria, must be able to scavenge nutrients present at low concentrations in the sea. Thus, in *Ditylum brightwellii* the K_m of nitrate reductase for nitrate was found to be as low as 0·11 mM and that of nitrite reductase, for nitrite, 0·19 mM.

The bulk of the published work on control of algal nitrate assimilation has been concerned with species of *Chlorella*. Pratt and Fong (1940) reported that, in *Chlorella vulgaris*, ammonia was utilized in preference to nitrate, a finding confirmed by many subsequent workers. Cramer and Myers (1949) reported that, in cultures of *Chlorella pyrenoidosa*, nitrate assimilation was markedly decreased in the presence of ammonia, or by carbon starvation. Syrett and Morris (1963) showed that nitrate assimilation by cultures of *Chlorella vulgaris* was completely inhibited by addition of small quantities of ammonia and that this inhibition was relieved only when the ammonia had itself been assimilated. Ammonia only partially inhibited nitrite assimilation and it was therefore concluded that ammonia in some way affected the reduction of nitrate to nitrite. Ammonia had little effect in carbon-starved cells when ammonia assimilation was restricted and it seemed likely that some product of ammonia, rather than ammonia itself, was responsible for the observed inhibition. Morris and Syrett (1963) then reported that, in cell-free extracts, an ammonium sulphate concentration of $3 \times 10^{-2} M$ had no effect on enzyme activity although this was some 30-times the inhibitory concentration in intact cells. Cultures grown on ammonia contained little nitrate reductase activity but this activity increased rapidly on transfer to a nitrate-containing medium and was only partially prevented by chloramphenicol or *p*-fluorophenyl alanine. Nitrate stimulated the development of nitrate reductase activity but its presence was not completely essential since ammonia grown cells, starved of nitrogen (and cultures grown on urea or glycine, and to a lesser extent alanine or arginine, as nitrogen source), contained some enzyme activity although this was less than that produced in the presence of nitrate. In a further report Morris and Syrett (1965) showed that ammonia-grown cells acquired nitrate reductase and the ability to assimilate nitrate after a short period of nitrogen starvation, thus confirming their previous results. They further demonstrated that both nitrate and ammonia grown cells lost their nitrate reductase activity on prolonged nitrogen starvation. These authors also pointed out that the nitrate reductase activities they were able to measure were too low to account for the culture nitrate assimilation rates. Losada *et al.* (1970) showed that the loss of nitrate reductase activity in the presence of ammonia was due, at least in part, to enzyme inactivation. Vennesland and Jetschmann (1971) reported that, in freshly prepared extracts, nitrate reductase was present as a pro-enzyme with diminished catalytic activity. Activation (some 100-fold) was accelerated by addition of nitrate, or by phosphate buffer of low pH value, and also by partial purification. Solomonson and Vennesland (1972) purified the enzyme and showed that the enzyme from *Chlorella fusca* differed from that of *Chlorella vulgaris* in requiring FAD for activity and in not existing as a pro-enzyme. Jetschmann *et al.* (1972) found that activation of nitrate reductase pro-enzyme required an oxidizing agent and that, while oxygen itself caused slow and incomplete conversion to active enzyme, ferricyanide caused complete activation within a few minutes even at 0°C. Monerno *et al.* (1972) have confirmed oxidative activation and proposed that the interconversion of active and inactive forms of the enzyme was determined by the redox state of the cell.

IV. Ammonia Assimilation

Ammonia holds a central position in the growth of micro-organisms on inorganic sources of nitrogen. Ammonia is probably used by most micro-organisms capable of growth on inorganic nitrogen and, as discussed earlier, is itself the product of the reduction of both molecular nitrogen and nitrate. Moreover ammonia is used preferentially in the presence of nitrogen or nitrate, and its presence in many instances prevents the reduction of these compounds. It has been claimed that ammonia assimilation may proceed via the synthesis of alanine, glutamate, valine, leucine or carbonyl phosphate, and it is the purpose of this section to discuss the relative significance of these potential routes of ammonia assimilation in micro-organisms.

A. Pathways of Ammonia Assimilation in Bacteria

The various routes by which ammonia could possibly be assimilated in different bacteria include pathways involving aspartase, the amino acid dehydrogenases and the glutamine synthetase/glutamate synthase pathway.

Vender and Rickenberg (1964) obtained a mutant of *E. coli* lacking glutamate dehydrogenase but which still grew well in a minimal salts glycerol medium with ammonia as nitrogen source. It was suggested that ammonia assimilation in these mutant organisms proceeded via aspartate ammonia lyase (aspartase). Recently, however, an alternative pathway of ammonia assimilation involving glutamine synthetase and glutamate synthase has been found (Tempest *et al.*, 1970a) and detected in *E. coli* (Berberich, 1972); thus, a biosynthetic role for aspartase is in doubt. Aspartase has, however, been implicated in the degradation of glutamate especially when that amino acid is used as carbon and nitrogen source. Thus aspartase activity was increased in glutamate-grown cultures of *E. coli* (Leiss *et al.*, 1966); mutants lacking aspartate amino transferase and aspartase would no longer utilize glutamate as sole source of carbon and a mutant with a thermosensitive aspartase could not use glutamate at elevated temperatures (Marcus and Halpern, 1969).

Amino acid dehydrogenases are enzymes which catalyse the reductive amination of 2-oxo acids by ammonia to yield the corresponding 2-amino acids. Alanine dehydrogenase has been reported to occur in *Bacillus* species (Fairhurst *et al.*, 1956; Shen *et al.*, 1959; Goldman, 1959; Freese and Oosterwyk, 1965; Germano and Anderson, 1968) and in some actinomycetes including *Mycobacterium tuberculosis* (Goldman, 1959) and *Streptomyces erythreus* (Roszkowski *et al.*, 1969). Shen *et al.* (1959) reported that glutamate dehydrogenase-deficient mutants of some *Bacillus* species could assimilate ammonia provided that they contained alanine dehydrogenase. Assuming this to be the only other enzyme system in these organisms capable of an assimilatory role, they concluded that these organisms synthesized alanine by direct amination of pyruvate and then produced glutamate by transamination. Freese *et al.* (1964), however, noted that mutants of *B. subtilis* lacking both glutamate dehydrogenase and alanine dehydrogenase would still grow readily with ammonia as nitrogen source. Meers *et al.* (1970a, b) and Elmerich and Aubert (1971) extended these observations and detected the glutamine synthetase/glutamate synthase route of ammonia assimilation in *B. subtilis* and *B. megaterium*. Elmerich and Aubert (1971) were further able to show that mutant organisms lacking both glutamate dehydrogenase and either glutamine synthetase or glutamate synthase were unable to utilize ammonia as nitrogen source even though they contained an active alanine dehydrogenase. Alanine dehydrogenase when present in bacterial extracts, inevitably showed a high K_m for ammonia (Wiame *et al.*, 1962; Yoshida and Freese, 1965); indeed, in *B. licheniformis*, Meers and Kjaergaard Pedersen (1972) found a value of 300 mM. The latter organism grew and assimilated ammonia at concentrations <1 mM at which concentrations aminating alanine dehydrogenase activity was not recorded in cell extracts. When bacteria are grown in a chemostat under conditions of ammonia-limitation, it is to be expected that those enzymes directly involved in ammonia assimilation would be derepressed. However, under such growth conditions, the specific activity of alanine dehydrogenase in *B. licheniformis* was comparatively small (see Table 1). In contrast, alanine dehydrogenase

TABLE 1. Influence of Energy Source and Growth-Limiting Nutrient on the Level of Alanine Dehydrogenase in *Bacillus licheniformis*

Growth-Limiting Substrate	Energy Source	Nitrogen Source	Specific Activity of Alanine Dehydrogenase
Carbon	Glucose	NH_3	134
Carbon	Alanine	Alanine	4700
Nitrogen	Glucose	NH_3	37
Nitrogen	Glucose/ Alanine	Alanine	152

The bacteria were grown in chemostats under carbon- or nitrogen-limited conditions with various substrates as carbon and nitrogen sources. The dilution rate was $0 \cdot 2\ h^{-1}$. Enzyme activities are expressed as n moles of $NADH_2$ oxidized/mg. protein/min. Data from Meers and Kjaergaard Pedersen (1972).

was synthesized in abundance when the organisms were grown in the presence of alanine (Freese, 1964). Meers and Kjaergaard Pedersen (1972) found that the highest activity for this enzyme was present in cultures provided with alanine as the sole energy source for a carbon-limited culture. The availability of alanine as a substrate was not the sole factor prescribing high alanine dehydrogenase levels, since nitrogen-limited organisms metabolizing alanine as the growth limiting nitrogen source had a comparatively low specific activity for alanine dehydrogenase (Table 1). These observations suggest that synthesis of alanine dehydrogenase was repressed by the presence of catabolites and induced by alanine. Berberich *et al.* (1968) have provided evidence to show that D-alanine is in fact the inducer molecule; D-alanine regulating its own biosynthesis via alanine racemase by the induction of L-alanine dehydrogenase. The catabolic activity of this enzyme is further subject to severe end-product inhibition by pyruvate (Goldman, 1959; Meers and Kjaergaard Pedersen, 1972). Freese and Oosterwyk (1963) concluded that, in addition to its main role in ammonia assimilation, a physiological function of alanine dehydrogenase was to catalyse the catabolism of alanine. Now that an alternative route of ammonia assimilation involving glutamate synthase can be proposed, it is concluded that the essential physiological function of alanine dehydrogenase is in the degradation of L-alanine to produce pyruvate, which can then be readily used as a carbon and energy source.

Sanwal and Zink (1961) isolated amino-acid dehydrogenases from *B. cereus* and *B. subtilis* which were distinct from the alanine and glutamate dehydrogenases that had been isolated from these species, and which oxidatively deaminated leucine, isoleucine and valine to their respective keto acids. Norleucine and norvaline were oxidized more slowly. Sanwal and Zink (1961) considered that this enzyme might play a biosynthetic role during synthesis of branched-chain amino acids. Porella (1971), however, studied the physiological role of this enzyme in *B. subtilis* and found that it was induced after the addition of branched-chain amino acids to growing cultures, and not repressed as would have been predicted had this enzyme been directly involved in amino acid biosynthesis. Furthermore, the organisms contained a repressible transaminase which, in common with the enzymes from other bacterial species, had a biosynthetic function. Raunio (1966) observed that, when isoleucine, leucine and valine were added to cultures of *E. coli*, the corresponding keto acids accumulated. These observations are consistent with a physiological role for the non-specific bacterial isoleucine dehydrogenase in anteiso fatty-acid synthesis (Porella, 1971). Thus keto-methylvalerate is a precurser of both isoleucine and anteiso fatty acids, and induction of isoleucine dehydrogenase is the means whereby the organisms produce this compound from isoleucine during feedback inhibition of synthesis of isoleucine *de novo*.

The presence of glutamate dehydrogenase in bacteria is well established. For example, Adler *et al.* (1938) described the presence of an NADP-linked enzyme in *Bacterium* (*Escherichia*) *coli*. Many bacteria such as *E. coli*, *B. subtilis*, *Aerobacter aerogenes* (Meers *et al.*, 1970b) and *B. licheniformis* (Meers and Kjaergaard Pedersen, 1972) contain only one type of glutamate dehydrogenase with a specific requirement for NADP. The main physiological role of this enzyme has been assumed to be biosynthetic but, due to the high K_m for ammonia of these enzymes, it appears unlikely that they function efficiently, in ammonia assimilation, except when the environmental ammonia concentration is high.

For example, in chemostat cultures of *A. aerogenes* growing with glucose or phosphate as limiting substrate (and therefore in the presence of an excess of ammonia), appreciable concentrations of NADP-linked glutamate dehydrogenase were synthesized (see Table 2). In ammonia-limited cultures, however, when the intracellular ammonia level was less than 0·5 m*M* (less than one-tenth the K_m for ammonia), the glutamate dehydrogenase content fell to about 3% its original level and could

TABLE 2. Influence of the Growth-Limiting Substrate on the Concentration of Free Glutamate, in *Aerobacter aerogenes*, and on the Cellular Activities of Glutamate Dehydrogenase and Glutamate Synthase

Growth-Limiting Substrate	Pool Glutamate Content (m*M*)	Specific activity of Glutamate Dehydrogenase	Specific activity of Glutamate Synthase
Glucose	4·3	560	<1
Nitrogen (NH_3)	5·8	19	66
Nitrogen (glutamate)	5·0	<1	<1
Nitrogen (NH_3 + 2%, w/v, NaCl)	30·2	36	32
Phosphate	1·1	600	<1
Phosphate (+50 m*M* glutamate)	>20	10	<1

The details of the growth conditions were essentially as described in Table 1 except that the dilution rate was 0·3 h^{-1}. Data from Meers *et al.* (1970b).

not therefore adequately fulfil a biosynthetic role. Moreover as *A. aerogenes* did not appear to synthesize aspartase or other amino-acid dehydrogenases under these conditions, it was assumed that some other system was responsible for ammonia assimilation when the concentration of that substrate was low. As discussed below, the system involved was glutamine synthetase/glutamate synthase (Tempest *et al.*, 1970a; Meers *et al.*, 1970a). In *A. aerogenes*, then, NADP-linked glutamate dehydrogenase played a biosynthetic role at high ammonia concentration and its synthesis was severely repressed when organic nitrogen (glutamate) was present in the medium. Glutamate *per se* was not responsible for the repression of synthesis of glutamate dehydrogenase since, as shown in Table 2, there was no relationship between pool glutamate and glutamate dehydrogenase activity. Thus, when sodium chloride (2% w/v) was added to an ammonia-limited culture, the pool glutamate level was greatly increased, but the level of glutamate dehydrogenase doubled. Furthermore ammonia-limited and N(glutamate)-limited cultures had similar pool glutamate and ammonia contents, but produced quite different contents of glutamate dehydrogenase. In the presence of glutamate, especially under carbon-limited conditions, some mechanism may exist in order to deaminate the amino acid and render 2-oxoglutarate available as a carbon source. Meers and Kjaergaard Pedersen (1972) found that, in *B. licheniformis*, the NADP-linked glutamate dehydrogenase served in biosynthesis at high ammonia concentrations (as in *A. aerogenes*) but must also be of catabolic significance since an increase in enzyme activity occurred when cultures were grown in the presence of glutamate. Some bacteria, incapable of growth on an inorganic source of nitrogen also contain glutamate dehydrogenase activities. For example, a marine psychrophylic *Micrococcus sp.* (C. M Brown and S. O. Stanley, unpublished data) contained an NAD-linked glutamate dehydrogenase and a strain of *Streptococcus mutans* (D. C. Ellwood and C. M. Brown, unpublished data) an NADP-linked glutamate dehydrogenase when grown on a mixture of amino acids. The latter organism contained ten-times more glutamate dehydrogenase when grown under nitrogen- than under carbon-limited conditions.

The glutamate dehydrogenase from *Mycoplasma laidlawii* has dual coenzyme specificity, unlike bacterial enzymes and similar glutamate dehydrogenases from mammalian sources (Frieden, 1965). This enzyme, which was resolved as a single band on polyacrylamide gel electrophoresis, showed measurable activity with alanine and aspartate, as well as glutamate, and had a molecular weight of about 250,000 daltons (subunit size about 48,000 daltons). Unlike mammalian enzymes (but similar to most bacterial enzymes) the *Mycoplasma* enzyme was unaffected by nicotinamide nucleotides at concentrations below 100 μM. The K_m for ammonia of the NADP-linked activity (5·5 mM) was much less than that of the NAD-linked activity (30 mM) while the glutamate K_m values for the reverse reaction were 20 mM and 32·5 mM for the NAD- and NADP-linked activities, respectively. Thus, *in vivo*, subject to coenzyme availability, it may be possible that NADP-linked enzyme activity plays a biosynthetic role and both coenzyme linked activities catabolic roles.

Some bacteria do show two distinct glutamate dehydrogenase activities, one linked to NAD and one to NADP. Le John and McCrea (1968) reported that cultures of the facultative lithotroph *Thiobacillus novellus*, growing autotrophically in a thiosulphate mineral salts medium, produced two such activities. These were shown to be distinct enzymes by purification on DEAE cellulose, and differed in their pH value optima and heat stability. Under autotrophic conditions early logarithmic phase cultures had a NADP/NAD activity ratio of five to one but this fell to about two to one in late logarithmic phase. Glutamate-grown cultures contained the highest content of NAD-linked enzyme which, according to the authors, was controlled by the intracellular glutamate concentration. Lower levels of the NAD-linked enzyme were found in cells grown heterotrophically on carbon sources such as arginine, alanine, glucose, glycerol and carboxylic acids. Arginine, histidine and aspartate caused repression of the NADP-linked enzyme. The physiological roles of these enzymes is obscure and it is unfortunate that more defined environments were not employed. The NAD-linked enzyme is so far unique in bacteria in showing allosteric activation with both AMP and ADP (Le John and McCrea, 1968). Kramer (1970) has also reported the presence of chromatographically distinct glutamate dehydrogenases in *Hydrogenomonas* H 16, one specific for NAD and one for NADP, which differed in their thermolability. The lowest content of NAD-linked enzyme was found in cultures grown with glutamate as nitrogen source and in the presence of high concentrations of ammonia, and the highest contents at low concentrations of ammonia or in a nitrogen free medium. There was evidence in this organism that the synthesis of NAD-linked glutamate dehydrogenase and glutamine synthetase were subject to co-ordinate control. The highest content of NADP-linked glutamate dehydrogenase was found in cultures grown in the presence of an excess of ammonia. Kramer suggested that this NADP-linked enzyme had a predominantly biosynthetic function. Brown *et al.* (1972), in a study of ammonia assimilation in a number of psychrophylic marine pseudomonads, found that every organism studied (twelve in all) contained an NAD-linked glutamate dehydrogenase the content of which could be increased markedly by growth in the presence of amino acids (hydrolysed casein) as source of nitrogen, as opposed to growth on nitrate. In a more detailed chemostat study with a *Vibrio* strain (SW_2) it was shown that this organism synthesized only a NAD-linked glutamate dehydrogenase and that the activity of this enzyme was highest when organic nitrogen (glutamate or hydrolysed casein) served as nitrogen source. As discussed below the assimilation of ammonia in *Vibrio* strain SW_2, grown with nitrate or with limiting concentrations of ammonia, proceeded via glutamine synthetase/glutamate synthase. In the presence of an excess of ammonia, however, the enzymes of this system were repressed and it is assumed that ammonia assimilation proceeded via glutamate dehydrogenase. Thus the NAD-linked glutamate dehydrogenase of this organism served a dual physiological function. Brown *et al.* (1973) extended this chemostat study to include five strains of *Pseudomonas* (three marine psychrophiles, *Ps. fluorescens* and *Ps. aeruginosa*) and confirmed the presence of NAD-linked glutamate dehydrogenase in all organisms. All five organisms, however, also synthesized a NADP-linked glutamate dehydrogenase activity but only when ammonia was present in the culture fluid in excess of requirement. This suggested a biosynthetic role for glutamate dehydrogenase under these conditions, the alternative biosynthetic route (glutamine synthetase/glutamine synthase) being repressed. The role of the NAD-linked activity was assumed to be catabolic. While not subjected to purification procedures, these glutamate dehydrogenase activities appeared to be due to different enzymes, and in one marine organism (and *Ps. aeruginosa*) showed slightly dissimilar kinetic parameters and pH activity profiles; also they had markedly different temperature characteristics. Distinct glutamate dehydrogenase activities of this type have also been recorded in a freshwater psychrophylic strain of *Pseudomonas* and in *Flavobacterium* sp. (B. Johnson, B. Gibson and C. M. Brown, unpublished).

Glutamic acid is produced commercially by growing biotin-requiring auxotrophs in a culture medium containing a growth-limiting amount of biotin. Under such growth conditions, the cells become permeable to glutamate which accumulates in the culture medium. Since glutamate does not accumulate when the cells are grown in the presence of excess biotin it seems reasonable to suggest that glutamate formation in the glutamate producing strains is repressed by glutamate accumulation in the intracellular pool (see Demain, 1971, 1972). Shiio and Ozaki (1970) found that the NADP-linked glutamate dehydrogenase from a glutamate-producing strain (*Brevibacterium flavum*) was inhibited by glutamate in the aminating direction and by ammonium ion and 2-oxoglutarate in the reverse direction. Meers *et al* (1970b) also found that glutamate inhibited the glutamate dehydrogenase enzymes from several species. Kitano *et al.* (1972) investigated the accumulation of glutamate by acetate-grown bacteria and found that glutamate in the medium inhibited further glutamate accumulation. The reasons for this are as yet unclear.

Glutamine is involved in the synthesis of a number of important nitrogen containing metabolites (amino sugars, nicotinamide, nucleotides, histidine, tryptophan, carbamoyl phosphate and, thereby, nicotinamide nucleotides). Recently it was demonstrated that the amide nitrogen of glutamine could be transferred to 2-oxoglutarate, a reaction that plays an important part in glutamate synthesis in bacteria. It is not surprising therefore that glutamine synthetase has been studied in depth (notably by Holzer and Stadtman and their colleagues) and found to be subject to complex control mechanisms. This enzyme has been the subject of several recent reviews (Holzer, 1969; Shapiro and Stadtman, 1970).

The enzyme glutamine synthetase catalyses the irreversible reaction by which glutamine is formed from glutamate and ammonia in the presence of a divalent cation (magnesium or manganese) and ATP.

The activity of glutamine synthetase is regulated in three different ways: by control of enzyme synthesis, by cumulative feedback inhibition, and by a complicated system of chemical modifications to the enzyme structure which have subtle effects on enzyme activity. Several authors have found variations in the activity of glutamine synthetase in extracts of organisms grown on different nitrogen-containing substrates. Thus, when organisms were grown on amino acids or in the presence of excess ammonia, low levels of enzyme activity were found in *E. coli* (Mecke and Holzer, 1966), *B. subtilis* (Robello and Strauss, 1969), *Lactobacillus arabinosus* (Ravel *et al.*, 1965) and a number of *Pseudomonas* spp. (Brown *et al.*, 1972, 1973). Meers and Tempest (1971) used a chemostat to define which factors led to high glutamine synthetase activity in cultures of *A. aerogenes*. From data such as that presented in Table 3 it was concluded that, at least with several Gram-negative bacteria, the pool ammonia concentration had a profound effect on glutamine synthetase activity *in vivo*.

TABLE 3. Influence of Environment on the Concentrations of Free Ammonia, Glutamate and Glutamine in *Aerobacter aerogenes*, and on the Cellular Content of Glutamine Synthetase

Growth Condition	Pool Concentration (m*M*) of			Glutamine Synthetase* activity
	NH_3	Glutamate	Glutamine	
Glucose-limited (NH_3)	10	3·4	0·1	0·1
Glucose-limited (NH_3) + 2% NaCl	10	37·4	1·3	0·1
NH_3-limited	1·0	5·8	0·1	1·6
NH_3-limited + 2%, w/v, NaCl	1·2	30·2	0·4	1·7
Nitrogen (glutamate)-limited	0·7	5·0	0·2	2·8
Phosphorus-limited (N-source glutamate)	10	20·0	4·0	0·2

The bacteria were grown in chemostats at a dilution rate of 0·3 h^{-1}. Data from Meers and Tempest (1970).

* Glutamine synthetase activity is expressed in arbitrary units per mg protein.

The above-mentioned results may be criticized if they are interpreted as showing quantitative changes in the rate of synthesis of the enzyme glutamine synthetase, without taking into account the possibility that chemical modifications to the enzyme could lead to similar changes in enzyme activity. Wu and Yuan (1968), however, have confirmed that changes in enzyme synthesis occur in response to changes in growth conditions and, although the real significance of repression of enzyme synthesis in the regulation of glutamine production remains unclear, it does seem certain that an active form of glutamine synthetase is produced in greatest quantities under conditions of nitrogen limitation. Such an observation is consistent with the view that the two enzymes (glutamine synthetase and glutamate synthase) together provide a route for the incorporation of ammonia into bacteria when the concentration of ammonia is low. The main way in which glutamine synthesis is controlled in *E. coli* is probably by the enzyme catalyzed chemical modification of the glutamine synthase molecule. It has for some years been known that glutamine synthetase could become "inactivated", but it is only in recent years that the enzyme from *E. coli* has been shown to exist in two distinct forms. The biosynthetically active form has been termed "glutamine synthetase *a*" by Holzer and his colleagues, whereas Stadtman and his co-workers used the term "glutamine synthetase I". This form of the enzyme was active in the presence of magnesium (but not in the presence of manganese ions) and its activity was resistant to feedback inhibition. Glutamine synthetase *b* (or II) had the reverse metal-ion specificity and was subject to feedback inhibition. It had the lower specific activity and corresponded to the "inactivated" form of the enzyme (see Shapiro and Stadtman, 1970, for details). The conversion of glutamine synthetase *a* to glutamine synthetase *b* was facilitated by an enzyme (ATP: glutamine synthetase adenyltransferase), which brought about esterification of a phenolic hydroxyl group of tyrosine with adenylic acid (Shapiro and Stadtman, 1968). This adenylation reaction was stimulated by glutamine (and several end products of glutamine metabolism) and inhibited by 2-oxoglutarate and high concentrations of ATP (Holzer *et al.*, 1969). Glutamine synthetase *b* was de-adenylated by a complex reaction involving at least two protein fractions (Shapiro, 1969). This enzyme-activating reaction was inhibited by glutamine and AMP, and stimulated by ATP and 2-oxoglutarate (Shapiro, 1969). The regulation of glutamine synthesis by enzyme-catalysed enzyme modification occurred in *E. coli* and other Gram-negative organisms. However, in *B. subtilis*, the control of glutamine synthesis is more direct. The enzyme from this species could not be shown to occur in adenylated forms, but its activity was directly inhibited by glutamine and end products of glutamine metabolism such as AMP, histidine and tryptophan (Deuel and Stadtman 1970). The *E. coli* enzyme was also subject to feedback inhibition by end products of glutamine metabolism (Woolfolk and Stadtman, 1964). The kinetics of inhibition were, however, unusual in that each substance produced only partial inhibition, the individual inhibitory effects being cumulative. Hubbard and Stadtman (1967) have shown that cumulative feedback inhibition occurred in bacteria and algae and suggested that this form of regulation is possibly a general control mechanism for this enzyme in micro-organisms. However, when considered in conjunction with the adenylation of glutamine synthetase, many points with regard to feedback inhibition remain obscure. The complex regulation system outlined above ensures that the rate of glutamine synthesis reflects substrate and product availability, the energy state of the cell, and the availability, of ammonia. Thus, the system was found to be substrate-activated by 2-oxoglutarate (the precursor of glutamate) and product-inhibited by glutamine and its derivatives. When energy in the form of ATP is available, the cell should be capable of synthetic activity, and it is reasonable to adduce that ATP should stimulate (and AMP inhibit) glutamine production, and this they do. The reason why glutamine synthesis should be promoted by low concentrations of ammonia is now clarified by the discovery of the enzyme glutamate synthetase. This latter enzyme functions in conjunction with glutamine synthetase when the availability of ammonia is limited (Tempest *et al.*, 1970a; Meers and Tempest, 1971).

When bacteria were grown in a chemostat under conditions of nitrogen limitation, the extracellular ammonia concentration was found to be negligible and the intracellular free ammonia concentration was below 0·5 m*M* in *A. aerogenes* (Tempest *et al.*, 1970a, b). However, amino acid dehydrogenases generally have K_m values for ammonia greater than 4 m*M*, and it seems unlikely that these enzymes could be responsible for nitrogen incorporation into bacteria grown under ammonia-limited conditions, unless their synthesis was derepressed when ammonia was in short supply. Meers *et al.* (1970b) found that, under nitrogen-limited conditions, the rate of glutamate dehydrogenase synthesis in *A. aerogenes*, *Erwinia carotovora*, *P. fluorescens*, *B. subtilis* and *B. megaterium* was low (see Tempest *et al.*, 1973). Furthermore, in *A. aerogenes* no other known pathways of ammonia assimilation could be discovered that would account for the known rate of ammonia incorporation into this organism. The chemical modifications of the enzyme glutamine synthetase are such that glutamine synthesis is favoured when growth of bacteria is limited by the supply of ammonia (see p. 31). Meers and Tempest (1971) showed that the activity of this enzyme was far greater when the growth of cultures of *A. aerogenes* was limited by ammonia than by carbon, phosphorus or magnesium (Table 3). Therefore, it seemed reasonable to consider glutamine as a possible intermediate in the ammonia incorporation route used by ammonia-limited cells. This suggestion seemed all the more tenable since a small pulse of ammonia, when added to an ammonia-limited culture of *A. aerogenes*, caused a 25-fold increase in the pool glutamine level within two minutes (Figure 2). However, the synthesis of glutamine could not in itself account for the increased net synthesis of other amino acids (such as glutamate and alanine) since it required an amino acid (glutamate) to make the amino acid (glutamine). Clearly, to provide a functional pathway of ammonia assimilation, organisms would need to possess an enzyme system capable of transferring the amide nitrogen of glutamine to the 2-position of a 2-oxo acid (e.g. pyruvic or 2-oxoglutaric acid).

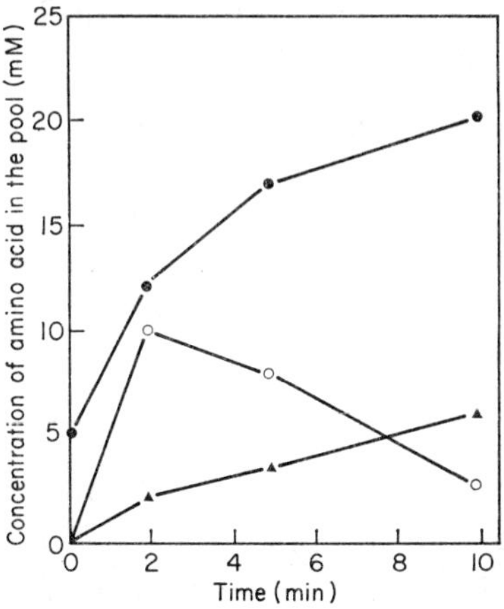

FIG. 2. Transient changes in amino acid "pool" concentrations of (●) glutamate, (○) glutamine, and (▲) alanine, in ammonia-limited *Aerobacter aerogenes* organisms, following addition of a pulse of ammonia to steady-state chemostat culture (35°C, pH 6·8, $D = 0{\cdot}3\ h^{-1}$). Data from Tempest *et al.* (1970b, 1973).

Such a reaction, which is analogous to that catalysed by glutamate or alanine dehydrogenase, would have to include a coupled oxido-reduction step. This reasoning led Tempest *et al.* (1970a) to incubate cell-free extracts of ammonia-limited *A. aerogenes* organisms with various combinations of substrates and observe whether or not a net synthesis of amino acids was obtained. It was found that incubation of bacterial extracts with glutamine, $NADPH_2$ and 2-oxoglutarate led to a considerable synthesis of glutamate. The possibility that glutaminase activity was responsible for this observation was excluded by further experimentation. It was subsequently shown that the reaction could be conveniently followed spectrophotometrically by measuring the change in $E_{340\,nm}$ due to coenzyme oxidation. It was concluded that the net synthesis of glutamate by ammonia-limited *A. aerogenes* organisms was effected by the two-stage process shown in Fig. 3. In this process, the synthesis of glutamine is followed by the reductive transfer of the amide group to the 2-position of 2-oxoglutarate. Each turn of this cycle leads to the net synthesis of one glutamate molecule. The relationship between this pathway and that mediated by glutamate dehydrogenase is also shown in Fig. 3. The net results of the two pathways are the same except that the route involving glutamine requires expenditure of energy in the form of ATP. Presumably this energy expenditure is the "price that organisms pay" in order to assimilate low concentrations of ammonia (Tempest *et al.*, 1970a). It is interesting to note that this pathway was almost absent in glucose-limited organisms where ammonia was present in quantities sufficient for glutamate dehydrogenase to function but

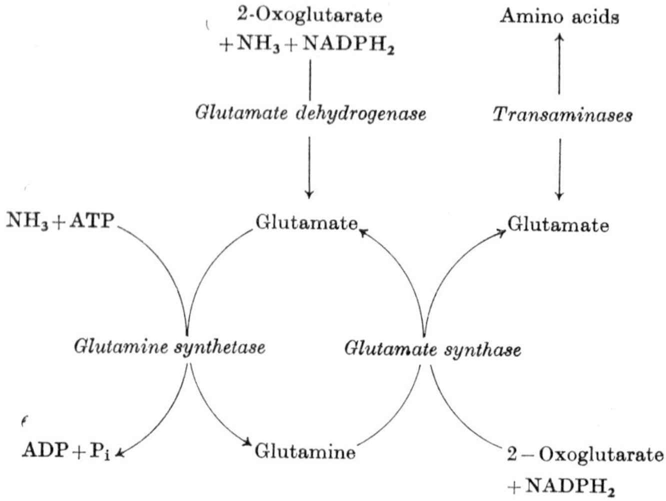

FIG. 3. Pathways of ammonia assimilation in prokaryotic organisms.

where energy supply was restricted. Under these glucose-limited conditions, glutamate dehydrogenase was formed in greatly increased quantities (Table 2). These early results, obtained with *A. aerogenes*, suggested that alternative pathways for ammonia incorporation were used, depending on the growth conditions. Such a view is close to the almost prophetic suggestions made by Umbarger (1969) who was of the opinion that, under conditions of nitrogen-limitation, the ATP-driven conversion of ammonia to an amide group could act as a "pump" which could scavenge the last traces of ammonia from the environment. In this connection it is relevant to note that the K_m of bacterial glutamine synthetases for ammonia is usually low (Meers and Tempest, 1971; Brown *et al.*, 1973), an observation consistant with Umbarger's "scavenging" role for this enzyme under conditions where glutamate dehydrogenase (by virtue of its high K_m for ammonia) would be inadequate.

This alternative pathway of glutamate synthesis described above was first reported in *A. aerogenes* by Tempest *et al.* (1970a) and these initial results were amplified in a succeeding report (Meers *et al.*, 1970a). Since then this pathway has been shown to operate in many other bacterial species (Meers *et al.*, 1970b; Elmerich and Aubert, 1971; Nagatani *et al.*, 1971; Meers and Kjaergaard Pedersen, 1972; Dainty, 1972; Brown *et al.*, 1972, 1973; Elmerich, 1972; Brown and Stanley, 1972; Berberich, 1972; Brooks and Meers, 1973).

The original name given by Tempest *et al.* (1970a) to the enzyme catalysing glutamate synthesis from glutamine was glutamine (amide): 2-oxoglutarate amino transferase (oxido-reductase NADP). This remains the appropriate systematic name and it is unfortunate that other names such as glutamine:2 oxoglutarate amidotransferase (NADP oxido-reductase) and glutamate synthetase have since been used by other authors. A trivial name would be convenient, and it is suggested that glutamate synthase (Prusiner *et al.*, 1972) is more appropriate than glutamate synthetase since the reaction it catalyses does not require ATP.

Glutamate synthase appears to catalyse a unidirectional reaction although at first sight it seems to be analogous to that catalysed by glutamate dehydrogenase. The recent purification of glutamate synthase from *E. coli* and the determination of some of its properties (Miller and Stadtman, 1973) goes some way to resolving this anomaly. The *E. coli* glutamate synthase is made up of eight subunits four of each of two types of dissimilar subunits with molecular weights of 135,000 and 53,000 daltons. Each complex also contains 32 iron atoms, 32 labile sulphide atoms and eight non-covalently linked flavine molecules. The purified enzyme may consist of four identical, catalytically active subunits, each consisting of one unit of 53,000 daltons and one of 135,000 daltons, eight iron atoms, eight labile sulphide atoms and two flavine molecules. From spectroscopic measurements of the purified enzyme it is suggested that the nature of the iron-protein bonding might be similar to that found in ferredoxin. The reaction mechanism is thought to occur in two stages, the first involves the reduction of the enzyme (flavine) with $NADPH_2$ or sodium dithionite as electron donor and the second the reductive transfer of the glutamine amide nitrogen to 2-oxoglutarate forming two molecules of glutamate. It will be of considerable interest to compare the characteristics of the *E. coli* enzyme with those of other bacteria and blue-green algae when these data become available.

Glutamate synthase has been shown to have a high degree of specificity for glutamine, 2-oxoglutarate and either $NADH_2$ or $NADPH_2$. Meers *et al.* (1970b) and Meers and Kjaergaard Pedersen (1972) found that the enzymes from a variety of organisms could not accept an alternative electron donor to $NADPH_2$, nor a nitrogen donor other than glutamine, nor a keto-acid other than 2-oxoglutarate. Nagatani *et al.* (1971), working with *Klebsiella pneumoniae*, and Brown *et al.* (1972) obtained similar results, but found that some species synthesized an enzyme that was specific for $NADH_2$. C. M. Brown and J. L. Meers (unpublished data) have found that a *Pseudomonas* sp. growing on one-carbon compounds produced a $NADH_2$-linked enzyme, as does *Mycobacterium smegmatis*. No organism has yet been isolated that produces both $NADH_2$- and $NADPH_2$-linked enzymes, but Dainty (1972) has reported that the enzymes from *Clostridium pasteurianum* has dual coenzyme specificity. The reaction catalysed by glutamate synthase appears to be irreversible, but is inhibited by some metal ions and glutamate (Meers *et al.*, 1970b) and by the glutamine analogue, 6-diazo-5-oxo L-norleucine (Nagatani *et al.*, 1971). More detailed kinetic data regarding glutamate synthase activity are given in the papers by Meers *et al.* (1970b) and Brown *et al.* (1972). The work of Nagatani *et al.* (1971) is of particular interest since it clearly demonstrated that a number of nitrogen-fixing species (including the photosynthetic organism *Chromatium*) can utilize the glutamine synthetase/glutamate synthase amination route. These results have more recently been confirmed by Dainty (1972) and Drozd *et al.* (1972). Brown *et al.* (1972, 1973) have demonstrated that marine pseudomonads, grown on nitrate as their sole nitrogen source, synthesize glutamate via glutamine, whether grown as carbon- or nitrogen-limited environments. These marine species synthesized glutamate dehydrogenases when grown on either excess ammonia or amino acids as their nitrogen source. The importance of the glutamine route in the physiology of marine bacteria was confirmed by Brown *et al.* (1972) when it was observed that there was a growth lag when organisms were transferred from a high-ammonia to a nitrate-containing medium, but no lag when the organisms were transferred from a low-ammonia to a nitrate-containing medium. The lag in the former case was interpreted as being due to the time required to synthesize the two enzymes, glutamine synthetase and glutamate synthase. The conversions of molecular nitrogen or nitrate to ammonia require the expenditure of energy by the cell, and, furthermore, ammonia is a repressor of both conversions. Therefore, for organisms to continue either to fix nitrogen or reduce nitrate, a mechanism must exist that will efficiently remove free ammonia from the intracellular pool as soon as it is formed. Presumably the glutamine pathway for ammonia assimilation performs this function, because the low K_m of the enzyme glutamine synthetase enables the organisms to assimilate ammonia before it accumulates at levels where it would inhibit growth. The importance of high rates of growth (and enzymes with low K_m values for primary metabolites), with regard to competitive success in natural environments, has been discussed elsewhere (Meers, 1972).

In all of the species so far examined by the authors, ammonia-limited organisms contained high glutamate synthase activities. In some species, such as *Erwinia carotovora* (Meers *et al.*, 1970b) and *B. megaterium* (Elmerich and Aubert, 1971, 1972), which lack the enzyme glutamate dehydrogenase, glutamate synthase was synthesized constitutively. Savageau *et al.* (1972) found that, in *E. coli*, glutamate dehydrogenase and glutamate synthase activities varied in parallel and concluded that these activities were associated with a single complex. These authors presented a hypothesis which proposed that both of the ammonia-assimilation routes shown in Fig. 3 occurred in *E. coli*. It was suggested that high levels of ammonia in the environment would, by competition with glutamine, inhibit the activity of glutamate synthase. The converse would be true in the presence of low concentrations of ammonia. Such a view now seems improbable as a consequence of the recent findings of Berberich (1972) and Prusiner *et al.* (1972) with this organism (see p. 38).

Although ammonia-limited *A. aerogenes* contained high levels of glutamate synthase activity, N(glutamate)-limited organisms contained little activity of this enzyme despite the fact that the pool ammonia and glutamate levels in the two types of organisms were similar (Table 2). It therefore seemed that neither of these compounds acted directly as repressors. In support of this conclusion it has been observed that, when the pool glutamate concentration of *A. aerogenes* or *B. licheniformis* was increased by adding either sodium chloride (Meers *et al.*, 1970b) or glutamate (Meers and Kjaergaard Pederson, 1972), glutamate synthase activity was not repressed to any significant extent. However, when alanine was pulsed into a carbon-limited culture of *B. licheniformis*, glutamate synthase activity decreased at a rate far greater than could be explained by simple "wash out". Meers and Kjaergaard Pedersen (1972) suggested that, since the pool glutamate level was invariably high in *B. licheniformis*, quantitatively large, but proportionately small, changes in glutamate concentration did not influence enzyme synthesis; on the other hand quantitatively similar, but proportionately larger, increases in the initially low-alanine pool could have significant effects. Such a suggestion was tentative and the true nature of the mechanisms regulating the synthesis of this enzyme is still not clear. In *A. aerogenes*, *Erwinia cartovora*, and a number of pseudomonads (Meers *et al.*, 1970b; Brown and Stanley, 1972; Brown *et al.*, 1972, 1973), the glutamine synthetase and glutamate synthase activities of cell extracts were controlled independently of one another and of glutamate dehydrogenase; this suggested that no co-ordinate control was exerted over these enzymes. Berberich (1972) has also shown that, in *E. coli* K12, a glutamate-dependent phenotype was the result of two independent mutations (involving glutamate synthase and glutamate dehydrogenase) and that the genes involved were not closely linked. Berberich, however, concluded that these two enzymes could be indirectly linked in some organization complex concerned with the overall control of nitrogen metabolism. It is apparent, therefore, that ammonia, when at low concentrations in the environment, is normally assimilated via glutamine synthetase and glutamate synthase. At higher concentrations, ammonia is assimilated via glutamate dehydrogenase. The cellular contents of some or of all three enzymes is controlled via repression and derepression, and the activity of glutamine synthetase by the complex series of enzyme modifications outlined above.

The fact that the interconversion of glutamine and glutamate lies at a common point where the pathways of nitrogen and carbohydrate metabolism intersect led Prusiner *et al.* (1972) to study the effect of cyclic AMP (c-AMP) on the cellular contents of the enzymes involved. Thus a further control circuit is involved in which c-AMP, when added to a culture of *E. coli*, increased the content of glutamate dehydrogenase and glutamine synthetase and decreased the content of glutaminase A (see Prusiner and Stadtman, 1971) and glutamate synthase. The level of glutaminase B was unaffected. These alterations in enzyme contents required c-AMP receptor protein since c-AMP had no effect in a mutant organism lacking this receptor; the presence of chloramphenicol also abolished the effects of c-AMP suggesting that protein synthesis was required.

Carbamoyl phosphate is formed either from ammonia or glutamine, and its synthesis is therefore related to the other reactions discussed in this section. This compound is an essential intermediate in the synthesis of arginine and pyrimidines. Cultures of *E. coli* contain the enzyme carbamoyl phosphate synthetase which catalyses the following reaction (Anderson and Meister, 1965):

$$\text{L-Glutamine} + 2\,\text{ATP} + HCO_3^- + H_2O \rightarrow \text{Carbamoyl phosphate} + 2\,\text{ADP} + \text{Pi} + \text{L-Glutamate}$$

In *E. coli*, synthesis of this enzyme is subject to feedback repression by arginine and uracil, activation by ornithine, and to feedback inhibition of its activity by UMP (Anderson and Meister, 1966; Pierard, 1966). The microbial enzyme has a lower affinity for ammonia than for glutamine, but some synthesis of carbamoyl phosphate from ammonia can be obtained (Anderson *et al.*, 1970). This observation contrasts the *E. coli* enzyme with that obtained from liver, because liver carbamoyl phosphate synthetase is active with ammonia, but not with glutamine. It seems likely that *in vivo* glutamine is the substrate for carbamoyl phosphate synthetase in micro-organisms, due to the high K_m value of this enzyme for ammonia (93 mM), and it is doubtful if this enzyme contributes to ammonia assimilation.

B. Ammonia Assimilation by Fungi

The only clearly demonstrable pathway of ammonia assimilation in moulds and yeasts is the synthesis of glutamic acid via glutamate dehydrogenase. Fincham (1951) reported that amination-deficient mutants of *Neurospora crassa* did not contain a biosynthetic glutamate dehydrogenase, and concluded that in this organism the α-amino groups of all of the amino acids which supported growth of the mutants were derived to a large extent by glutamate synthesis from ammonia. Nicholas and Mabey (1960) showed that glutamate dehydrogenase activity in extracts of *N. crassa* could be linked to either $NADH_2$ or $NADPH_2$, and Sanwal and Lata (1961) extended this observation demonstrating the presence of two distinct enzymes in this organism with different pH optima. Sanwal and Lata (1961, 1962) further proposed that the NAD-linked enzyme fulfilled a catabolic role while the NADP-linked enzyme was biosynthetic and that, to this end, growth in the presence of glutamate derepressed the synthesis of the NAD-linked enzyme and repressed the synthesis of the NADP-linked enzyme. The NADP-linked alanine dehydrogenase activity of *Neurospora* extracts was shown by Burk and Pateman (1962) to reside in the same protein as NADP-linked glutamate dehydrogenase activity, since both activities were absent from amination-deficient mutants. The K_m value for ammonia of the alanine dehydrogenase activity was 29 mM, and it seems unlikely that this enzyme contributes to any extent in ammonia assimilation. Barratt and Strickland (1963) purified the NADP-linked *Neurospora* enzyme and found that it would reductively aminate and oxidatively deaminate several 2-oxo and α-amino acids but with an activity of only about 5% that shown with 2-oxoglutarate or glutamate. The K_m value for 2-oxoglutarate was 0.2 mM while that for other keto acids was higher (e.g. >3 mM for pyruvate). Barratt (1963) found that the level of the NADP-linked glutamate dehydrogenase in mycelia increased after nitrogen starvation, and proposed that excess ammonia repressed in part the synthesis of this enzyme. The NADP-linked *Neurospora* enzyme, in common with most other microbial glutamate dehydrogenases was not inhibited by purine nucleotides (Frieden, 1965; Stachow and Sanwal, 1964) but was subject to activation by substrates in an apparently co-operative manner (Tuveson *et al.*, 1967). Thus this enzyme was inactive at pH 7·2 yet fully active at pH 8·05; at the lower pH value, 2-oxoglutarate and $NADPH_2$ activated the enzyme synergistically. In the absence of $NADPH_2$, pre-incubation with substrates (2-oxoglutarate or glutamate) or with a number of carboxylic acids (including citrate, isocitrate and succinate) lead to activation. The NADP-linked glutamate dehydrogenase of *Neurospora* was inhibited in the presence of D-glutamic acid which therefore inhibited the growth of this organism (Arkin and Grossowicz, 1970). Sanwal (1961), working with cultures of *Fusarium*, found two distinct glutamate dehydrogenases, reminiscent of *N. crassa*. He found that, in a synthetic medium containing ammonium nitrate as nitrogen source, the activity of the NADP-linked enzyme was highest during 24 to 48 hours of growth, after which time the NAD-linked enzyme predominated. These enzymes were both soluble proteins with requirements for thiol groups for activity and, as in *Neurospora*, had distinct pH optima. While Sanwal did not comment on proposed roles for these enzymes it is noticeable that the NADP-linked enzyme had the lower K_m value for ammonia and the NAD-linked enzyme the lower K_m value for glutamate thus suggesting biosynthetic and catabolic roles, respectively. Pateman and Cove (1967) working with *A. nidulans* reported that organisms grown on nitrate contained only low internal concentrations of ammonia and synthesized maximal amounts of NADP-linked glutamate dehydrogenase. Pateman (1969), in a further study with *A. nidulans, N. crassa* and *E. coli*, showed that in all three organisms glutamate repressed the synthesis of NADP-linked glutamate dehydrogenase and glutamine that of glutamine synthetase. There was no evidence in the fungi for separate forms of glutamine synthetase, and cultures of these organisms grown either on glutamate, high concentrations of ammonia, or urea contained only low amounts of biosynthetic glutamate dehydrogenase but highest amounts of glutamine synthetase.

The unicellular "water-mould" *Blastocladiella emersonii* (Le John and Jackson, 1968) contains an NAD-specific glutamate dehydrogenase which shows strong allosteric purine nucleotide effects, AMP and ADP being positive and ATP negative effectors. Sanner (1971) extended these observations and reported that a 20-fold activation could be achieved in the presence of 1 m*M* AMP. This activation was associated with an increase in the K_m values for NAD, $NADH_2$ and 2-oxoglutarate, but a decrease in the K_m value for ammonia (to 25 m*M*). The K_m value for ammonia in the absence of AMP was not measured since the reaction rate increased linearly with concentration up to 0·4 *M* ammonium sulphate.

In *Saccharomyces cerevisiae* (Jones *et al.*, 1969), as in *Candida utilis* (Sims and Folkes, 1964), experiments using ^{15}N have demonstrated that only glutamate and glutamine derived their α-amino nitrogen directly from ammonia and were synthesized at a rate sufficient to provide all the α-amino nitrogen required for growth; all other amino acids obtained their α-amino nitrogen through transamination reactions. Adler *et al.* (1938) reported the presence of an NADP-specific glutamate dehydrogenase in yeast, and Holzer and Schneider (1952) extended this observation to show that an NAD-specific enzyme was also synthesized under appropriate conditions. Holzer and co-workers (Hierholzer and Holzer, 1963; Westphal and Holzer, 1964) further showed that in *Sacch. cerevisiae* the NAD-linked enzyme was degradative, being repressed by growth on ammonia. Polakis and Bartley (1965) and Thomulka and Moat (1972) have also reported increased levels of the NAD-linked enzyme in organisms grown on glutamate, while the content of NADP-linked enzyme was highest in organisms grown in a defined medium containing ammonia. Lamminmaki and Pierce (1969) have further shown that the synthesis of this biosynthetic enzyme was repressed by growth in the presence of an amino acid mixture (brewer's malt wort). In both *Sacch. cerevisiae* (Brown and Johnson, 1970) and *Candida utilis* (P. Turner and C. M. Brown, unpublished), as in *N. crassa*, the cellular content of the NADP-linked enzyme was higher under conditions of ammonia limitation than ammonia excess. Growth of *Sacch. cerevisiae* on glutamate resulted in the content of this enzyme being decreased to about 30% of that produced by growth on ammonia. In *C. utilis*, however, growth on glutamate had little effect on the content of the NADP- or the NAD-linked enzymes relative to growth on ammonia.

Both alanine dehydrogenase and aspartase have been implicated in ammonia assimilation in yeasts, but Lamminmaki and Pierce (1969) have shown that in *Sacch. cerevisiae* alanine was not produced by direct amination of pyruvate. Thomulka and Moat (1972) found only low activities of NADP-linked alanine dehydrogenase in *Sacch. cerevisiae* and concluded that, as in *N. crassa*, this activity resided in the same protein as glutamate dehydrogenase (NADP) since the two activities showed identical migration on polyacrylamide-gel electrophoresis. No aspartase activity was detected in these experiments. Cultures of *Sacch. cerevisiae* (Kohlaw *et al.*, 1965) and *C. utilis* (Ferguson and Sims, 1971), when grown on glutamate or another amino acid as nitrogen source, showed the derepression not only of NAD-linked glutamate dehydrogenase but also of glutamine synthetase. Conversely, synthesis of both enzymes was repressed by growth on ammonia, results similar to those obtained with *N. crassa* and *A. nidulans* (Pateman, 1969). In *C. utilis*, *Sacch. cerevisiae* and *Torulopsis candida* the addition of ammonia or glutamine to cultures adapted to growth on glutamate resulted in the extensive inactivation of both NAD-linked glutamic dehydrogenase and glutamine synthetase. The possible significance of these results in the control of yeast nitrogen assimilation has been discussed by Ferguson and Sims (1971).

What is apparent from these results is that the control of glutamine synthetase synthesis and activity in fungi differs markedly from that in bacteria. The significance of this probably lies in the fact that in bacteria, but not in these eukaryotic organisms, glutamine synthetase serves as the first enzyme of the glutamine synthetase/glutamate synthase pathway of ammonia assimilation and as such is particularly active when the ammonia concentration in the culture is low. Clearly no such function is apparent in fungi. Indeed Brown *et al.* (1970) failed to demonstrate the presence of glutamate synthase in *Sacch. cerevisiae* or *C. utilis* or any yeast studied to date (see Tempest *et al.*, 1973). Thus fungal ammonia assimilation depends upon glutamate dehydrogenase, and it is significant that in a number of these organisms the content of the biosynthetic NADP-linked enzyme is highest under ammonia-limited chemostat (or nitrogen-starved batch) culture conditions. Brown and Stanley (1972) have discussed possible mechanisms in yeasts which enables ammonia assimilation to occur efficiently via glutamate dehydrogenase in these organisms.

C. Ammonia Assimilation by Algae

It appears that the route of ammonia assimilation in eukaryotic algae might well differ from that in the prokaryotic blue-green organisms. In *Chlorella vulgaris*, Morris and Syrett (1965) reported that the activity of NADP-linked glutamate dehydrogenase was higher in ammonia-grown cells than in nitrate-grown cells and increased during nitrogen starvation. They proposed that a nitrogen excess repressed in part the formation of glutamate dehydrogenase in this organism. The enzyme assay used with extracts of *Chlorella* employed ammonia at 2 mM, a much lower concentration than required for the bacterial or fungal enzymes. The K_m value for ammonia in this organism is about 0·5 mM which obviates the requirement for the glutamine synthetase/glutamate synthase pathway; indeed the latter enzyme was not detected in extracts of ammonia-grown cells (C. M. Brown, unpublished observation).

Kretovitch *et al.* (1970) reported the presence of two glutamate dehydrogenase activities in *Chlorella*, one requiring NAD and the other NADP for activity. Synthesis of the NADP-linked enzyme was induced in the presence of ammonia but enzyme activity was low in extracts of nitrate-grown organisms. Talley *et al.* (1972) detected similar glutamate dehydrogenase isoenzymes in a thermophilic strain of *Chlorella pyrenoidosa*. In this organism, only the NAD-specific enzyme was detected in nitrate-grown cells while synthesis of the NADP-specific enzyme was induced by ammonia.

In the marine plankton diatom *Ditylum brightwellii* (Eppley and Rogers, 1970) an NADP-linked glutamate dehydrogenase was present in cultures grown on nitrate, nitrite or ammonia. The enzyme content remained high during (and increased following) nitrogen exhaustion from the medium, underlining a similarity shown in this respect to the biosynthetic glutamate dehydrogenases of other eukaryotic algae and fungi. The K_m value for ammonia for the *Ditylum* enzyme may be deduced from the data presented by Eppley and Rogers (1970) to be about 10 mM. The internal ammonia concentration of cells grown on nitrate, nitrite or ammonia, however, was 5–10 mM indicating that this organism possessed an ability to accumulate this substrate perhaps to aid glutamate dehydrogenase activity. Recent work with marine phytoplankton, however, showed that there was little correlation between the cellular glutamate dehydrogenase content and ammonia assimilation. It will be of interest, therefore, to await the results of a survey of such organisms for an alternative pathway of ammonia assimilation such as glutamine synthetase/glutamate synthase.

Glutamate dehydrogenase, alanine dehydrogenase and the glutamine synthetase/glutamate synthase system have all been implicated in ammonia assimilation in blue-green algae. Extracts of *Anabaena variabilis* contain NADP-linked glutamate dehydrogenase activity (Pearce *et al.*, 1969), although only low levels of the enzyme were detected. Dharmawardne *et al.* (1972) have detected glutamine synthetase activity in *Anabaena cylindrica*; enzyme activity was lower in nitrate- or ammonia-grown cells than in those grown on molecular nitrogen (or nitrogen starved). The activity of this enzyme as a function of Mg^{2+} and Mn^{2+} concentration suggested that it may exist in different forms as in *Escherichia coli*. Extracts of nitrogen-fixing cells were also shown to contain NADP-linked glutamate synthase activity. Neilson and Doudoroff (1973), however, have surveyed the possible enzymes of ammonia assimilation in a number of blue-green algae, and concluded that either alanine dehydrogenase or glutamate dehydrogenase but not glutamate synthase were involved. This assimilatory role of alanine dehydrogenase contrasts with the catabolic function of this enzyme in *Bacillus* sp.

V. Conclusions and Future Prospects

From the account given in this review, it is clear that the conversion of ammonia into glutamate is a key reaction in assimilation of inorganic nitrogen, whether the nitrogen source be molecular nitrogen, nitrate or ammonia itself. In this connection the recent discovery of the enzyme glutamate synthase is most significant. The related enzyme glutamine synthetase has long been known to be subject to elaborate control mechanisms, and to be produced in greater quantities than would seem to be required, were its functions solely those known prior to the realization that this enzyme was involved (with glutamate synthase) in ammonia assimilation into α-amino compounds. The position is now much clearer, and it seems that regulation of glutamine synthetase is the means by which prokaryotic organisms control incorporation of inorganic nitrogen into glutamate, and hence most other amino acids. Miller and Stadtman (1973) have recently purified the *Escherichia coli* glutamate synthase, and it will be of considerable interest to compare the characteristics of this enzyme with enzymes isolated from other bacteria and blue-green algae. The glutamine synthetase/glutamate synthase pathway of ammonia incorporation into amino acids is apparently restricted to prokaryotes, but the reason for this is obscure. Eukaryotes seem to incorporate ammonia solely by the glutamate dehydrogenase reaction and adapt to conditions of ammonia deficiency by producing increased quantities of the biosynthetic form of this enzyme. Prokaryotes on the other hand adapt to such conditions by regulating glutamine synthetase in such a way that ammonia assimilation is facilitated.

Growth on molecular nitrogen or nitrate is in some ways analogous to nitrogen-limited conditions in chemostat cultures, and molecular nitrogen (or nitrate) grown cells thus incorporate ammonia via the cyclic route mentioned above. But growth on molecular nitrogen or nitrate is more costly, in energetic terms, than growth on ammonia as a nitrogen source. The energy requirements for nitrogen fixation have been mentioned earlier, and reduction of nitrate to ammonia also requires energy (the ΔF values for the reduction of nitrate to nitrite, and of nitrite to ammonia, are 22·3 and 59·3 K cal/mole, respectively; Syrett, 1954). On this basis the growth yield obtained with nitrate as nitrogen source would be expected to be less than that obtained on ammonia. B. Watts and J. L. Meers (unpublished) obtained such a result in carbon-limited cultures of methanol-grown pseudomonads when the growth yields were 43 g/g and 26 g/g methanol utilized, respectively, for ammonia and nitrate. Thus it is not surprising that ammonia is generally a preferred nitrogen source for micro-organisms, and that control mechanisms exist by which nitrogenase or nitrate reductase (assimilatory) are repressed by the presence of ammonia. However, the actual repressing molecules involved are unknown. In many algal systems it is apparent that some product of ammonia assimilation, rather than ammonia itself, is involved. It is possible to speculate as to whether in organisms capable of growth on all three major inorganic nitrogen sources some common metabolites are involved in the control of both molecular nitrogen and nitrate reduction. This point will not be resolved until further firm data accumulates concerning intracellular pool levels of those low molecular-weight compounds which influence enzyme synthesis and activity. Stadtman and Holzer and their colleagues have gone a long way towards explaining the regulation, at a molecular level, of glutamine synthetase. What is now required is further study of the other enzymes involved in nitrogen assimilation. Only then will a clear understanding of the physiological significance of the various control mechanisms be forthcoming.

It may then be possible to apply the results of laboratory studies to microbial growth in natural environments. No doubt the elaborate control mechanisms mentioned above were acquired because selective pressures in natural ecosystems enabled mutants with more efficient control systems to compete effectively with parent strains. For example, organisms able to utilize only ammonia in an environment containing limited quantities of energy and, say, both nitrate and ammonia as nitrogen sources would replace species constitutive for nitrate reductase; also, organisms able to use glutamate dehydrogenase (rather than the energy-requiring glutamine synthetase/glutamate synthase route to accumulate ammonia in a carbon-limited environment) would be at a selective advantage (see Meers, 1972, for a detailed discussion of this topic). Many other such examples could be cited, and it seems tenable to suggest that organisms able to adapt their metabolism to changing conditions would be best able to compete in environments that are subject to variations. For example, the River Tees is, in some parts, alternately aerobic and anaerobic, depending on the state of the tide, and also contains an appreciable concentration of inorganic nitrogen. An understanding of the microbial transformations that occur in situations such as this is of considerable practical interest, but this understanding remains at best fragmentary. A similarly neglected area concerns transformations in the sea where, presumably, a large part of the global nitrogen cycle occurs. Stewart (1971) has suggested that the concentrations of fixed nitrogen in the sea are unlikely to inhibit nitrogen fixation. To what extent this is so in coastal or inland waters and in cultivated soils where, through pollution and chemical fertilization, the concentrations of nitrogenous compounds are likely to be higher, is unknown. Neither is it known to what extent it matters if man, through his activities, causes major changes in the nitrogen economy of this planet.

References

Adler, E., Günther, G. and Everett, J. E. (1938). *Hoppe-Seyler's Zeitschrift für Physiologische Chemie*, **255**, 29.
Anderson, P. M. and Meister, A. (1965). *Biochemistry* **4**, 2803.
Anderson, P. M. and Meister, A. (1966). *Biochemistry* **5**, 3164.

Anderson, P. M., Wellner, V. P., Rosenthal, G. A. and Meister, A. (1970). *In* "Methods in Enzymology", Vol. **12** (H. Taber and C. W. Taber, eds.) Academic Press, New York.

Arkin, H. and Grossowicz, N. (1970). *Journal of General Microbiology* **61**, 255.

Azoulay, E., Puig, J. and Rosada de Souza, M. (1969). *Annales de l'Institut Pasteur* **117**, 474.

Barratt, R. W. (1963). *Journal of General Microbiology* **33**, 33.

Barratt, R. W. and Strickland, W. (1963). *Archives of Biochemistry and Biophysics* **102**, 66.

Benemann, J. R. and Valentine, R. C. (1972). *Advances in Microbial Physiology* **8**, 59.

Berberich, R. *Biochemical and Biophysical Research Communications* **47**, 1498.

Berberich, R., Kaback, M. and Freese, E. (1968). *Journal of Biological Chemistry* **243**, 1006.

Biggins, D. R. and Postgate, J. R. (1969). *Journal of General Microbiology* **56**, 181.

Biggins, D. R. and Kelly, M. (1970). *Biochimica et Biophysica Acta* **205**, 288.

Bone, D. H. (1971). *Archiv für Mikrobiologie* **80**, 234

Bongers, L. H. J. (1956). *Mededeling Landbouwhogeschool Wageningen* **56**, 1.

Bothe, H. (1970). *Berliner deutsche botanische Gesellschaft* **23**, 421.

Brooks, J. W. and Meers, J. L. (1973). *Journal of General Microbiology* (in press).

Brown, C. M. and Johnson, B. (1970). *Journal of General Microbiology* **64**, 279.

Brown, C. M., Macdonald-Brown, D. S. and Stanley, S. O. (1972). *Journal of the Marine Biological Association of the United Kingdom* **52**, 793.

Brown, C. M., Macdonald-Brown, D. S. and Stanley, S. O. (1973). *Antonie van Leeuwenhoek* **39**, 89.

Brown, C. M., Meers, J. L. and Tempest, D. W. (1970). *Journal of General Microbiology* **61**, vii.

Brown, C. M. and Stanley, S. O. (1972). *Journal of Applied Chemistry and Biotechnology* **22**, 363.

Bui, P. T. and Mortenson, L. E. (1968). *Proceedings of the National Academy of Sciences of the United States of America* **61**, 1021.

Bulen, W. A., Burns, R. C. and Le Compte, J. R. (1964). *Biochemical and Biophysical Research Communications* **17**, 265.

Bulen, W. A., Burns, R. C. and Le Compte, J. R. (1965). *Proceedings of the National Academy of Sciences of the United States of America* **53**, 532.

Bulen, W. A. and Le Compte, J. R. (1966). *Proceedings of the National Academy of Sciences of the United States of America* **56**, 979.

Burk, R. and Pateman, J. (1962). *Nature, London* **196**, 450.

Burns, R. C. and Bulen, W. A. (1966). *Archives of Biochemistry and Biophysics* **113**, 461.

Burris, R. H., Eppling, F. J., Wahlin, H. B. and Wilson, P. W. (1943). *Journal of Biological Chemistry* **148**, 349.

Burris, R. H. (1969). *Proceedings of the Royal Society* B **172**, 339.

Campbell, I. (1971). *Journal of General Microbiology* **67**, 223.

Carnahan, J. E., Mortenson, L. E., Mower, H. F. and Castle, J. E. (1960a). *Biochimica et Biophysica Acta* **38**, 188.

Carnahan, J. E., Mortenson, L. E., Mower, H. F. and Castle, J. E. (1960b). *Biochimica et Biophysica Acta* **44**, 520.

Chatt, J. and Fogg, G. E. (1969). *Proceedings of the Royal Society* B **172**, 317.

Cole, J. A. (1968). *Biochimica et Biophysica Acta* **162**, 356.

Cole, J. A. (1971). *Journal of General Microbiology* **65**, vii.

Cole, J. A. and Wimpenny, J. W. T. (1968). *Biochimica et Biophysica Acta* **162**, 39.
Cook, K. A. and Sorger, G. J. (1969). *Biochimica et Biophysica Acta* **177**, 412.
Cove, D. J. (1966). *Biochimica et Biophysica Acta* **113**, 51
Cove, D. J. and Pateman, J. A. (1963). *Nature, London* **198**, 262.
Cox, R. M., Fay, P. and Fogg, G. E. (1964). *Biochimica et Biophysica Acta* **88**, 208.
Cox, R. M. and Fay, P. (1967). *Archiv für Mikrobiologie* **58**, 537.
Cox, R. M. and Fay, P. (1969). *Proceedings of the Royal Society* B **172**, 357.
Cramer, M. and Myers, J. (1949). *Journal of General Physiology* **32**, 93.
Daesch, G. and Mortenson, L. E. (1968). *Journal of Bacteriology* **96**, 346.
Daesch, G. and Mortenson, L. E. (1972). *Journal of Bacteriology* **110**, 103.
Dainty, R. H. (1972). *Biochemical Journal* **126**, 1055.
Dainty, R. H. and Peel, J. L. (1970). *Biochemical Journal* **117**, 573.
Dalton, H. and Mortenson, L. E. (1972). *Bacteriological Reviews* **36**, 231.
Dalton, H. and Postgate, J. R. (1969a). *Journal of General Microbiology* **54**, 463.
Dalton, H. and Postgate, J. R. (1969b). *Journal of General Microbiology* **56**, 307.
De Groot, J. N. and Stouthamer, A. H. (1970). *Biochimica et Biophysica Acta* **208**, 114.
Demain, A. L. (1971). *Symposium of the Society for General Microbiology* **21**, 77.
Demain, A. L. (1972). *Journal of Applied Chemistry and Biotechnology* **22**, 345.
Detroy, R. W., Witz, D. F., Parejko, R. A. and Wilson, P. W. (1969). *Proceedings of the National Academy of Sciences of the United States of America* **51**, 537.
Deuel, T. F. and Stadtman, E. R. (1970). *Journal of Biological Chemistry* **245**, 5206.
Dharmawardene, M. W. N., Stewart, W. D. P. and Stanley, S. O. (1972). *Planta* **108**, 133.
Dilworth, M. J. (1966). *Biochimica et Biophysica Acta* **127**, 285.
Dixon, R. A. and Postgate, J. R. (1971). *Nature, London* **234**, 47.
Dixon, R. A. and Postgate, J. R. (1972). *Nature, London* **237**, 102.
Downey, R. J. (1971). *Journal of Bacteriology* **105**, 759.
Downey, R. J., Kiszkiss, D. J. and Nuner, J. H. (1969). *Journal of Bacteriology* **98**, 1056.
Drozd, J. W., Tubb, R. S. and Postgate, J. R. (1972). *Journal of General Microbiology* **73**, 221.
Dugdale, V. A. and Dugdale, R. C. (1962). *Limnology and Oceanography* **7**, 170.
Eady, R. R., Smith, B. E., Cook, K. A. and Postgate, J. R. (1972). *Biochemical Journal* **128**, 655.
Elmerich, C. (1972). *European Journal of Biochemistry* **27**, 216.
Elmerich, C. and Aubert, J. P. (1971). *Biochemical and Biophysical Research Communications* **42**, 371.
Elmerich, C. and Aubert, J. P. (1972). *Biochemical and Biophysical Research Communications* **46**, 892.
Eppley, R. W. and Rogers, J. N. (1970). *Journal of Phycology* **6**, 344.
Eppley, R. W. and Coatsworth, J. L. (1968). *Journal of Phycology* **4**, 151.
Eppley, R. W. Coatsworth, J. L. and Solorzano, L. (1969). *Limnology and Oceanography* **14**, 194.
Evans, M. C. W. and Smith, R. V. (1971). *Journal of General Microbiology* **65**, 95.
Fairhurst, A. S., King, H. K. and Sewell, C. E. (1956). *Journal of General Microbiology* **15**, 105.
Fay, P. and Cox, R. M. (1966). *Biochimica et Biophysica Acta* **126**, 402.
Fay, P. and Lang, N. J. (1971). *Proceedings of the Royal Society* B **178**, 185.
Fay, P., Stewart, W. D. P., Walsby, A. E. and Fogg, G. E. (1968). *Nature, London* **220**, 810.

Fay, P. and Walsby, A. E. (1966). *Nature, London* **209**, 94.
Ferguson, A. R. and Sims, A. P. (1971). *Journal of General Microbiology* **69**, 423.
Fewson, C. A. and Nicholas, D. J. D. (1961). *Nature, London* **190**, 2.
Fincham, J. R. S. (1951). *Journal of General Microbiology* **5**, 793.
Fogg, G. E. (1942). *Journal of Experimental Biology* **19**, 78.
Fogg, G. E. (1949). *Annals of Botany* **13**, 241.
Freese, E. and Oosterwyk, J. (1965). *Biochemistry* **2**, 1212.
Freese, E. (1964). *Biochimica et Biophysica Acta* **81**, 442.
Freese, E., Park, S. W. and Cashel, M. (1964). *Proceedings of the National Academy of Sciences of the United States of America* **51**, 1164.
Frieden, C. (1965). *Journal of Biological Chemistry* **238**, 3286.
Garrett, R. H. (1972). *Biochimica et Biophysica Acta* **264**, 481.
Garrett, R. H. and Nason, A. (1969). *Journal of Biological Chemistry* **244**, 2870.
Germano, J. and Anderson, K. E. (1968). *Journal of Bacteriology* **96**, 55.
Goldman, D. S. (1959). *Biochimica et Biophysica Acta* **34**, 527.
Grant, B. R. (1967). *Journal of General Microbiology* **48**, 379.
Grant, B. R. (1968). *Journal of General Microbiology* **54**, 327.
Grant, B. R. and Turner, J. M. (1969). *Comparative Biochemistry and Physiology* **29**, 995.
Hadjipetrov, L. P. and Stouthamer, A. H. (1965). *Journal of General Microbiology* **38**, 29.
Hardy, R. W. F. and Burns, R. C. (1968). *Annual Review of Biochemistry* **37**, 331.
Hattori, A. (1962). *Plant Cell Physiology* (*Tokyo*) **3**, 355.
Haystead, A., Robinson, R. and Stewart, W. D. P. (1970). *Archiv für Mikrobiologie* **74**, 235.
Haystead, A. and Stewart, W. D. P. (1972). *Archiv für Mikrobiologie* **82**, 325.
Hierholzer, G. and Holzer, H. (1963). *Biochemische Zeitschrift* **339**, 175.
Hill, S., Drozd, J. W. and Postgate, J. R. (1972). *Journal of Applied Chemistry and Biotechnology* **22**, 541.
Hill, S. and Postgate, J. R. (1969). *Journal of General Microbiology* **58**, 277.
Holzer, H. and Schneider, S. (1952). *Biochemische Zeitschrift* **329**, 361.
Holzer, H. (1969). *Advances in Enzymology* **32**, 97.
Holzer, H., Meck, D., Leiss, K., Wulff, K., Heilmeyer, C., Ebner, E., Gancedo, C., Schutt, H., Battig, F. A., Heinrich, P. and Wolf, D. (1969). *Federation of European Biochemical Society Symposium* **19**, 171.
Horne, A. J. and Fogg, G. E. (1970). *Proceedings of the Royal Society* B **175**, 351.
Hubbard, J. S. and Stadtman, E. R. (1967). *Journal of Bacteriology* **94**, 1007.
Jetschmann, K., Solomonson, L. P. and Vennesland, B. (1972). *Biochimica et Biophysica Acta* **275**, 276.
Jones, M., Pragnell, M. J. and Pierce, J. R. (1969). *Journal of the Institute of Brewing* **75**, 520.
Kelly, M. (1969). *Biochimica et Biophysica Acta* **171**, 9.
Kessler, E. (1959). *Symposium of the Society for Experimental Biology* **13**, 87.
Kinsky, S. L. (1961). *Journal of Bacteriology* **82**, 898.
Kitano, K., Sugiyama, T. and Kanzaki, T. (1972). *Journal of Fermentation Technology* **50**, 182.
Kohlow, A., Dragert, W. and Holzer, H. (1965). *Biochemische Zeitschrift* **341**, 224.
Kramer, J. (1970). *Archiv für Mikrobiologie* **71**, 226.
Kretovich, W. L., Evstigneeva, Z. G. and Tamora, N. G. (1970). *Canadian Journal of Botany* **48**, 1179.

Kulasooriya, S. A., Lang, N. J. and Fay, P. (1972). *Proceedings of the Royal Society* B **181**, 199.
Lamminmaki, O. A. and Pierce, J. S. (1969). *Journal of the Institute of Brewing* **75**, 515.
Lang, N. J. and Fay, P. (1971). *Proceedings of the Royal Society* B **178**, 193.
Lazzarini, R. A. and Atkinson, D. E. (1961). *Journal of Biological Chemistry* **236**, 3330.
Le John, H. B. and Jackson, A. (1968). *Journal of Biological Chemistry* **243**, 3447.
Le John, H. B. and McCrea, B. E. (1968). *Journal of Bacteriology* **95**, 87.
Leiss, K., Mecke, D. and Holzer, H. (1966). *Biochemische Zeitschrift* **346**, 244.
Losada, M., Paneque, A., Aparico, P. J., Vega, J. M., Cardenas, J. and Herrera, J. (1970). *Biochemical and Biophysical Research Communications* **38**, 1009.
Ludwig, C. A. (1938). *American Journal of Botany* **25**, 448.
Mackintosh, M. E. (1971). *Journal of General Microbiology* **66**, i.
Mahl, M. C. and Wilson, P. W. (1968). *Canadian Journal of Microbiology* **14**, 33.
Marcus, M. and Halpern, Y. S. (1969). *Biochimica et Biophysica Acta* **177**, 314.
Mecke, D. and Holzer, H. (1966). *Biochimica et Biophysica Acta* **122**, 341.
Meers, J. L. (1972). *Critical Reviews in Microbiology* (in press).
Meers, J. L. and Kjaergaard Pedersen, L. (1972). *Journal of General Microbiology* **70**, 277.
Meers, J. L. and Tempest, D. W. (1971). *Biochemical Journal* **119**, 603.
Meers, J. L., Tempest, D. W. and Brown, C. M. (1970a). *Journal of General Microbiology* **60**, x.
Meers, J. L., Tempest, D. W. and Brown, C. M. (1970b). *Journal of General Microbiology* **64**, 187.
Millbank, J. W. (1969). *Archiv für Mikrobiologie* **68**, 32.
Millbank, J. W. (1970). *Archiv für Mikrobiologie* **72**, 375.
Miller, R. E. and Stadtman, E. R. (1973). *Journal of Biological Chemistry* **247**, 7407.
Monerno, C. E., Aparico, P. J., Palacien, E. and Losada, M. (1972). *Federation of European Biochemical Societies Letters* **26**, 11.
Morris, I. and Syrett, P. J. (1963). *Archiv für Mikrobiologie* **47**, 32.
Morris, I. and Syrett, P. J. (1965). *Journal of General Microbiology* **38**, 21.
Mortenson, L. E. (1962). *Analytical Biochemistry* **2**, 216.
Mortenson, L. E., Valentine, R. C. and Carnahan, J. E. (1962). *Biochemical and Biophysical Research Communications* **7**, 448.
Morton, A. G. (1956). *Journal of Experimental Botany* **7**, 97.
Morton, A. G. and MacMillan, A. (1954). *Journal of Experimental Botany* **5**, 232.
Moustafa, E. and Mortenson, L. E. (1967). *Nature, London* **216**, 1241.
Munson, T. O. and Burris, R. H. (1969). *Journal of Bacteriology* **97**, 1093.
Nagatani, H., Shimizu, M. and Valentine, R. C. (1971). *Archiv für Mikrobiologie* **79**, 164.
Nason, A. (1962). *Bacteriological Reviews* **26**, 16.
Nason, A. and Evans, H. J. (1953). *Journal of Biological Chemistry* **202**, 655.
Naylor, A. W. (1970). *Annales of the New York Academy of Science* **175**, 511.
Neilson, A. H., Rippka, R. and Kunisawa, R. (1971). *Archiv für Mikrobiologie* **76**, 139.
Neilson, A. H. and Doudoroff, M. (1973). *Archiv für Mikrobiologie* **89**, 15.
Newton, J. W., Wilson, P. W. and Burris, R. H. (1953). *Journal of Biological Chemistry* **204**, 445.
Nicholas, D. J. D. (1963a). *Symposium of the Society for General Microbiology* **13**, 92.
Nicholas, D. J. D. (1963b). *Biological Reviews* **38**, 530.

Nicholas, D. J. D. and Mabey, G. L. (1960). *Journal of General Microbiology* **22**, 184.
Nicholas, D. J. D., Medina, A. and Jones, O. T. G. (1960). *Biochimica et Biophysica Acta* **37**, 468.
Nicholas, D. J. D., Nason, A. and McElroy, W. D. (1954). *Journal of Biological Chemistry* **207**, 341.
Ogawa, R. E. and Carr, J. F. (1969). *Limnology and Oceanography* **14**, 342.
Oppenheim, J. and Marcus, L. (1970a). *Journal of Bacteriology* **101**, 286.
Oppenheim, J. and Marcus, L. (1970b). *Bacteriological Proceedings* 148.
Painter, H. A. (1970). *Water Research* **4**, 393.
Pateman, J. A. (1969). *Biochemical Journal* **115**, 769.
Pateman, J. A. and Cove, D. J. (1967). *Nature, London* **215**, 1234.
Pateman, J. A., Rever, B. M. and Cove, D. J. (1967). *Biochemical Journal* **104**, 103.
Pearce, J., Leach, C. K. and Carr, N. G. (1969). *Journal of General Microbiology* **55**, 371.
Pichinoty, F. (1965). *Centre Nationale de Recherche Scientifique Symposea* **124**, 507.
Pichinoty, F. (1969a). *Archiv für Mikrobiologie* **68**, 51.
Pichinoty, F. (1969b). *Archiv für Mikrobiologie* **68**, 75.
Pichinoty, F. (1970). *Archiv für Mikrobiologie* **71**, 116.
Pichinoty, F. and Metenier, M. (1967). *Annales de l'Institut Pasteur Paris*, **112**, 701.
Pichinoty, F., Muchhielli, A. and Petatin, C. (1971). *Archiv für Mikrobiologie* **75**, 353.
Pichinoty, F. and Ornano, L. (1961). *Nature, London* **191**, 879.
Pierard, A. (1966). *Science* **154**, 1572.
Polakis, E. and Bartley, W. (1965). *Biochemical Journal* **97**, 284.
Porella, K. (1971). *Archiv für Mikrobiologie* **77**, 339.
Postgate, J. R. (1970). *Nature, London* **226**, 25.
Postgate, J. R. (1971). *In* "The Chemistry and Biochemistry of Nitrogen Fixation". Plenum Press, London.
Postgate, J. R. (1972). *Methods in Microbiology* **6B**, 343.
Pratt, R. and Fong, J. (1940). *American Journal of Botany* **27**, 735.
Proctor, V. W. (1957). *American Journal of Botany*, **44**, 141.
Prusiner, S., Miller, R. E. and Valentine, R. C. (1972). *Proceedings of the National Academy of Sciences of the United States of America* **69**, 2922.
Prusiner, S. and Stadtman, E. R. (1971). *Biochemical and Biophysical Research Communications* **45**, 1474.
Raunio, R. (1966). *Acta Chemica Scandinavia* **20**, 11.
Ravel, J. M., Humphreys, J. S. and Shrine, W. (1965). *Archives of Biochemistry and Biophysics* **111**, 720.
Rebello, J. L. and Strauss, N. (1969). *Journal of Bacteriology* **98**, 683.
Rippka, R., Neilson, A., Cohen-Bazire, G. and Kunisawa, R. (1971). *Archiv für Mikrobiologie* **76**, 341.
Roszkowski, J., Ratalski, A. and Raczynska-Bojanowska, K. (1969). *Acta Microbiologica Polonica Series B* **18**, 59.
Sanner, T. (1971). *Biochimica et Biophysica Acta* **250**, 297.
Sanwal, B. D. (1961). *Archives of Biochemistry and Biophysics* **93**, 377.
Sanwal, B. D. and Lata, M. (1961). *Canadian Journal of Microbiology* **7**, 319.
Sanwal, B. D. and Lata, M. (1962). *Archives of Biochemistry and Biophysics* **97**, 582.
Sanwal, B. D. and Zink, M. W. (1961). *Journal of Biological Chemistry* **244**, 4437.

Savageau, M. A., Kotre, M. A. and Sakamoto, N. (1972). *Biochemical and Biophysical Research Communications* **48**, 41.
Schick, H. J. (1971). *Archives für Mikrobiologie* **75**, 89.
Schneider, K. C., Bradbeer, C., Singh, R. N., Wang, L. C., Wilson, P. W. and Burris, R. H. (1960). *Proceedings of the National Academy of Sciences of the United States of America* **46**, 726.
Schöllhorn, R. and Burris, R. H. (1966). *Federation Proceedings, Federation of American Societies for Experimental Biology* **25**, 710.
Schulp, J. A. and Stouthamer, A. H. (1970). *Journal of General Microbiology* **64**, 195.
Shah, V. K., Davis, L. C. and Brill, W. J. (1972). *Biochimica et Biophysica Acta* **261**, 63.
Shapiro, B. M. (1969). *Biochemistry* **8**, 659.
Shapiro, B. M. and Stadtman, E. R. (1968). *Journal of Biological Chemistry* **243**, 3769.
Shapiro, B. M. and Stadtman, E. R. (1970). *Annual Review of Microbiology* **24**, 501.
Shen, S. C., Hong, M. M. and Braunstein, A. E. (1959). *Biochimica et Biophysica Acta* **36**, 290.
Shiio, I. and Ozaki, H. (1970). *Journal of Biochemistry* **68**, 633.
Showe, M. K. and De Moss, J. A. (1968). *Journal of Bacteriology* **85**, 1305.
Silver, W. S. (1957). *Journal of Bacteriology* **73**, 241.
Sims, A. and Folkes, B. (1964). *Proceedings of the Royal Society*, B. **159**, 479.
Sinclair, P. R. and White, D. C. (1970). *Journal of Bacteriology* **101**, 365.
Smith, R. V. and Evans, M. C. W. (1970). *Nature, London* **225**, 226.
Smith, R. V. and Evans, M. C. W. (1971). *Journal of Bacteriology* **105**, 913.
Smith, R. V., Noy, R. J. and Evans, M. C. W. (1971). *Biochimica et Biophysica Acta* **253**, 104.
Solomonson, L. P. and Vennesland, B. (1972). *Biochimica et Biophysica Acta*, **267**, 544.
Stachow, C. S. and Sanwal, B. D. (1964). *Biochemical and Biophysical Research Communications* **17**, 368.
Stewart, W. D. P. (1969). *Proceedings of the Royal Society*, B, **172**, 367.
Stewart, W. D. P., Fitzgerald, G. P. and Burris, R. H. (1967). *Proceedings of the National Academy of Sciences, United States of America* **580**, 2071.
Stewart, W. D. P., Fitzgerald, G. P. and Burris, R. H. (1968). *Archiv für Mikrobiologie* **62**, 336.
Stewart, W. D. P., Haystead, A. and Pearson, H. W. (1969). *Nature, London* **224**, 226.
Stewart, W. D. P. and Lex, M. (1970). *Archiv für Mikrobiologie* **73**, 250.
Stewart, W. D. P. and Pearson, H. W. (1970). *Proceedings of the Royal Society* B, **175**, 293.
Stewart, W. D. P. (1971). *In* "Fertility of the Sea" (J. D. Costolow, ed.) Gordon and Breach, London.
Strandberg, G. W. and Wilson, P. W. (1967). *Canadian Journal of Microbiology* **14**, 25.
Subramanian, K. N. and Sorger, G. J. (1972). *Journal of Bacteriology* **110**, 547.
Syrett, P. J. (1954). *Symposium of the Society for General Microbiology* **4**, 126.
Syrett, P. J. and Morris, I. (1963). *Biochimica et Biophysica Acta* **67**, 566.
Talley, D. J., White, L. H. and Schmidt, R. R. (1972). *The Journal of Biological Chemistry* **247**, 7972.

Tempest, D. W., Meers, J. L. and Brown, C. M. (1970a). *Biochemical Journal* **117**, 405.

Tempest, D. W., Meers, J. L. and Brown, C. M. (1970b). *Journal of General Microbiology* **64**, 171.

Tempest, D. W., Meers, J. L. and Brown, C. M. (1973). *In* "Enzymes of Glutamine Metabolism" (S. Prusiner and E. R. Stadtman, eds). Academic Press, New York.

Thacker, A. and Syrett, P. J. (1972a). *New Phytologist* **71**, 423.

Thacker, A. and Syrett, P. J. (1972b). *New Phytologist* **71**, 435.

Thomulka, K. W. and Moat, A. G. (1972). *Journal of Bacteriology* **109**, 25.

Tuveson, R. W., West, D. J. and Barratt, R. W. (1967). *Journal of General Microbiology* **48**, 235.

Umbarger, H. E. (1969). *Annual Review of Biochemistry* **38**, 323.

Van't Riet, J., Stouthamer, A. H. and Planta, R. L. (1968). *Journal of Bacteriology* **96**, 1455.

Van't Riet, J., Knook, D. L. and Planta, R. L. (1972). *Federation of European Biochemical Societies Letters* **23**, 44.

Vender, J. and Rickenberg, H. V. (1964). *Biochemica et Biophysica Acta* **90**, 218.

Vennesland, B. and Jetschmann, C. (1971). *Biochimica et Biophysica Acta* **227**, 554.

Ward, F. B. and Cole, J. A. (1971). *Journal of General Microbiology* **68**, xiii.

Westphal, H. and Holzer, H. (1964). *Biochimica et Biophysica Acta* **89**, 42.

Wiame, J. M., Pierard, A. and Ramos, F. (1962). *In* "Methods in Enzymology", Vol. 5. Academic Press, London & New York, p. 637.

Wilson, P. W. and Burris, R. H. (1953). *Annual Review of Microbiology* **7**, 415.

Wilson, P. W. (1969). *Proceedings of the Royal Society* B **172**, 319.

Winter, H. C. and Arnon, D. J. (1970). *Biochimica et Biophysica Acta* **197**, 170.

Woolfolk, C. A. and Stadtman, E. R. (1964). *Biochemical and Biophysical Research Communications* **17**, 313.

Wolfe, M. (1954a). *Annals of Botany* **18**, 299.

Wolfe, M. (1954b) *Annals of Botany* **18**, 309.

Wu, C. and Yuan, L. (1968). *Journal of General Microbiology* **51**, 57.

Wyatt, J. T. and Silvey, J. K. G. (1969). *Science* **165**, 908.

Yoch, D. C. and Arnon, D. J. (1970). *Biochemica et Biophysica Acta* **197**, 180.

Yoch, D. C. and Pengra, M. (1966). *Journal of Bacteriology* **92**, 618.

Yoshida, A. and Freese, E. (1965). *Biochimica et Biophysica Acta* **96**, 248.

Zelitch, J., Rosenblum, E. D., Burris, R. H. and Wilson, P. W. (1951). *Journal of Biological Chemistry* **119**, 295.

Zumft, W. G. (1972). *Biochimica et Biophysica Acta* **276**, 363.

The Structure, Biosynthesis and Function of Teichoic Acid

A. R. ARCHIBALD

Microbiological Chemistry Research Laboratory, The School of Chemistry, The University of Newcastle upon Tyne, Newcastle upon Tyne, NE4 7RU, England

I. Introduction

Teichoic acids are phosphorylated polymers which occur in probably all Gram-positive bacteria. They are located exclusively in the outer layers—membrane, wall and capsule—of the cell and can account for more than 10% of its dry weight. Several different types of teichoic acid have been encountered and their structure, biosynthesis and cellular location are described in this review.

The widespread distribution, quantitative importance and, indeed, chemical nature of teichoic acids suggest that they are likely to be of importance in the cell. Recent studies have indicated that teichoic acids are functionally involved in cation binding and in the provision of the correct cationic environment at the cytoplasmic membrane. Teichoic acids may also be involved in the regulation of autolytic activity. These topics are discussed in the final sections of the review.

II. Structure

Several different structural types of teichoic acid have been described. They all contain glycerol or ribitol phosphate and, usually, a sugar or acetamidosugar and D-alanine (Baddiley, 1972; Archibald *et al.*, 1968a). These components are present in repeating structures which are held together by phosphodiester linkages to form linear polymers which usually contain between 10 and 50 units. All Gram-positive bacteria so far examined contain simple 1,3-poly(glycerol phosphate) teichoic acids (Fig. 1) as "intracellular" components which are apparently associated

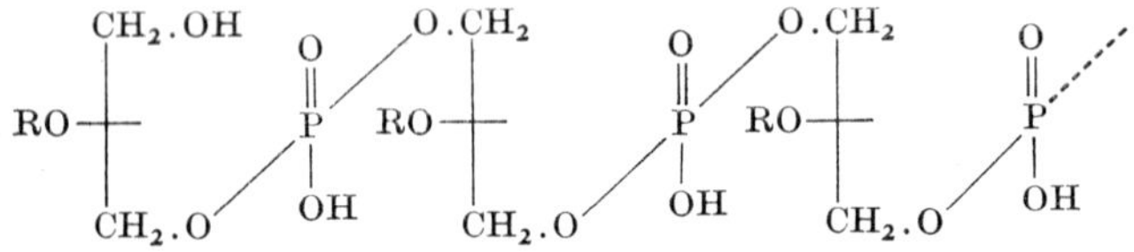

FIG. 1. Structure of 1,3-poly(glycerol phosphate) teichoic acids. R = H, D-alanyl or glycosyl.

with the cell membrane. These teichoic acids may be isolated after extraction with trichloroacetic acid (Critchley *et al.*, 1962) or phenol (Burger and Glaser, 1964) from preparations of disrupted bacteria. Many Gram-positive bacteria also contain teichoic acids as major components of the cell wall. Such teichoic acids show greater structural diversity than those present in association with membrane, and conditions employed for their extraction depend in part on their structure. Most structural studies have been carried out on teichoic acids which had been extracted from washed walls by treatment with cold dilute trichloroacetic acid solution (Armstrong *et al.*, 1958). In some cases extraction into dilute sodium hydroxide results in less depolymerization of the teichoic acid chain, although the alanyl ester residues are hydrolysed (Archibald *et al.*, 1969b; Hughes and Tanner, 1968). Extraction has also been effected by dilute aqueous dimethylhydrazine (Anderson *et al.*, 1969) and by enzymic dissolution of walls (Ghuysen and Strominger, 1963). An account of various procedures for the isolation and preliminary characterization of teichoic acids has been given recently (Archibald, 1971), and a detailed treatment of the chemistry of teichoic acids has also been published

(Archibald and Baddiley, 1966). The different structural types of teichoic acid can be classified in different ways, but perhaps the most fundamental structural difference is between those teichoic acids in which the sugar substituents form an integral part of the polymer chain and those in which the sugars are attached to a poly(alditol phosphate) "backbone". Glycerol and ribitol teichoic acids of both types are known and are described below.

A. Teichoic Acids which Contain a Poly(alditol phosphate) Chain

1. Ribitol Teichoic Acids

1-5 Poly(ribitol phosphate) teichoic acids (Fig. 2) occur as major components of walls of various bacilli, lactobacilli and staphylococci. Walls of all wild-type strains of *Staphylococcus aureus* so far examined normally contain ribitol teichoic acids in which, with one reported exception (Karakawa and Kane, 1971), α or β N-acetylglucosaminyl substituents are attached to the 4(D) positions of the ribitol units

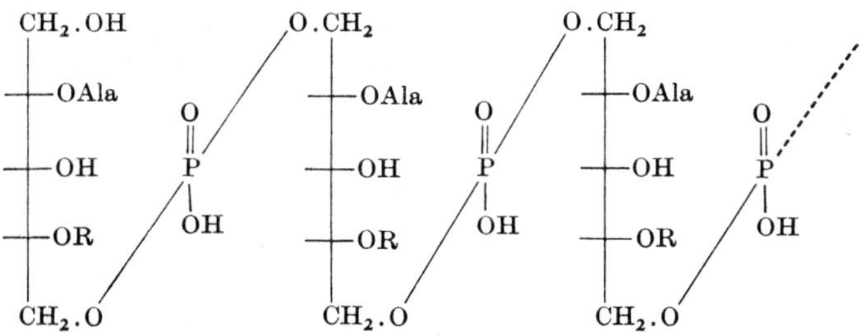

Fig. 2. Structure of 1,5-poly(ribitol phosphate) teichoic acids. R = glycosyl.

(Baddiley *et al.*, 1961; 1962a, b; Davison and Baddiley, 1963; Davison *et al.*, 1964; Sanderson *et al.*, 1961, 1962). In some strains all of the N-acetylglucosaminyl substituents in the teichoic acid have the same anomeric configuration (α or β) whereas, in others, both anomers are present (Nathenson *et al.*, 1966). In at least some of these latter strains, the different anomers are present on separate homogeneously substituted chains (Torii *et al.*, 1964). The D-alanyl ester residues are attached directly to ribitol, and studies on a mutant lacking glucosamine in its wall teichoic acid have shown that they are attached to the 2(D) positions (Mirelman *et al.*, 1970). These ester linkages are flanked by both phosphate and hydroxyl groups and so (Shabarova *et al.*, 1962) are very labile towards alkali and organic bases. Detailed accounts of the structural and serological properties of staphylococcal teichoic acids have been published recently (Archibald, 1972; Oeding and Grov, 1972).

Similar teichoic acids occur in certain strains of *Bacillus subtilis* and *Bacillus licheniformis*. These contain β-D-glucopyranosyl substituents at the 4(D) positions and D-alanyl substituents at the 3 or 2(D) positions of ribitol. Glucosyl substituents were attached to every ribitol residue in the teichoic acid isolated by Armstrong *et al.* (1960, 1961) from a laboratory strain of *B. subtilis*, but the teichoic acids in walls of *B. subtilis* W23 (Chin *et al.*, 1966) and *B. licheniformis* NCTC 6346 (Hughes, 1965) are only partially glucosylated. Chin *et al.* (1966) showed that the teichoic acid isolated from *B. subtilis* W23 was a mixture of fully glucosylated poly(ribitol phosphate) and glucose-free poly(ribitol phosphate). The amount of glucose present appears to be influenced by the conditions under which the bacteria are grown, and Ellwood and Tempest (1972b) report that *B. subtilis* W23 contains only a fully glucosylated teichoic acid when grown in continuous culture.

The proportion of glucose present in the ribitol teichoic acid isolated from *Lactobacillus plantarum* 17-5 also varied in different preparations (Archibald *et al.*, 1961b). This teichoic acid contains a mixture of unsubstituted ribitol and mono- and diglucosyl ribitol residues; these may be present in separate chains (Knox and Wicken, 1972).

Ribitol teichoic acids have also been found in walls of certain actinomycetes. *Actinomyces violaceus* contains a 1,5-poly(ribitol phosphate) teichoic acid in which a proportion of the ribitol residues bear β-glucosyl substituents. This teichoic acid does not contain D-alanyl ester residues but O-acetyl groups are present (Naumova and Zaretskaya, 1964a; Naumova and Belozersky, 1966; Naumova *et al.*, 1969). The ribitol teichoic acid from *Actinomyces streptomycini* contains both glucosyl and N-acetylglucosaminyl substituents. This teichoic acid is particularly interesting in that, although it does not contain D-alanine, succinate is present as a mono-ester (Naumova *et al.*, 1962, 1963). Succinic acid half-esters are also present in the ribitol teichoic acid of *Streptomyces griseus* NCIB 9001. A unique feature of this polymer is the presence of glycerol phosphate linked to the hydroxyl group at position 3 of some of the ribitol residues (Bews, 1967).

2. *Glycerol Teichoic Acids*

All Gram-positive bacteria so far examined contain 1,3-poly(glycerol phosphate) teichoic acids (Fig. 1, p. 54) as "intracellular" components which are distinct from the teichoic acids in the wall (Baddiley, 1972; Sharpe *et al.*, 1973). As discussed later, these "intracellular" teichoic acids are associated with the cytoplasmic membrane and, at least in some cases, are covalently attached to lipid. 1,3-Poly(glycerol phosphate) teichoic acids are also present in walls of several species. The glycerol phosphate chain usually carries glycosyl and D-alanyl substituents attached to the 2-hydroxyl groups of certain of the glycerol residues. The most common sugar substituents are glucose, N-acetylglucosamine and N-acetylgalactosamine (Archibald *et al.*, 1968a; Davison and Baddiley, 1963; Oeding *et al.*, 1967; Partridge *et al.*, 1973; Sharpe *et al.*, 1964; Wicken and Knox, 1971) but the proportion of glycerol residues bearing sugar substituents varies widely. Thus certain membrane teichoic acids are devoid of sugars (Kelemen and Baddiley, 1961; McCarty, 1959) whereas, in others, every glycerol residue carries a sugar substituent (Wicken and Baddiley, 1963). Fully glycosylated glycerol teichoic acids also occur in walls of certain bacteria (Burger, 1966; Young, 1966) and both wall and membrane teichoic acids are known in which only a proportion of the glycerol residues bears glycosidic substituents (Archibald *et al.*, 1968c; Critchley *et al.*, 1962; Ellwood *et al.*, 1963; Raj-Bhandary and Baddiley, 1963; Shaw and Baddiley, 1964; Knox and Wicken, 1970). In many of these teichoic acids, glycosylated glycerol and unsubstituted glycerol residues are present in the same chain (Archibald *et al.*, 1969a) but in other cases the partially glycosylated teichoic acid is a mixture of fully glycosylated chains and chains which are devoid of sugar substituents (Burger, 1966). In most membrane teichoic acids only a few of the glycerol residues are glycosylated.

The D-alanyl ester residues in the fully glycosylated membrane teichoic acid of *Streptococcus faecalis* NCIB 8191 are attached to the glycosidic substituents (Wicken and Baddiley, 1963). In incompletely glycosylated teichoic acids, the alanyl esters are attached to the secondary hydroxyl groups of a proportion of those glycerol residues which do not bear sugar substituents; such alanyl ester residues exhibit normal stability in acid but, as a consequence of the neighbouring phosphate groups, are highly susceptible to base-catalysed hydrolysis (Kelemen and Baddiley, 1961). The amount of alanine present in isolated teichoic acid is therefore greatly influenced by the conditions used for its extraction and purification.

Walls of *Bacillus stearothermophilus* B65 (Wicken, 1966) and *Actinomyces antibioticus* (Naumova and Zaretskaya, 1964b; Zaretskaya *et al.*, 1971) have been reported to contain poly(glycerol phosphate) teichoic acids which differ from those described above in that the adjacent glycerol residues are connected by phosphodiester linkages between hydroxyl groups at positions 2 and 3 (L) rather than positions 1 and 3 (Fig. 3). The 1-position of each glycerol residue in the teichoic acid from *Actinomyces antibioticus* is substituted by a disaccharide composed of galactose and N-acetylgalactosamine.

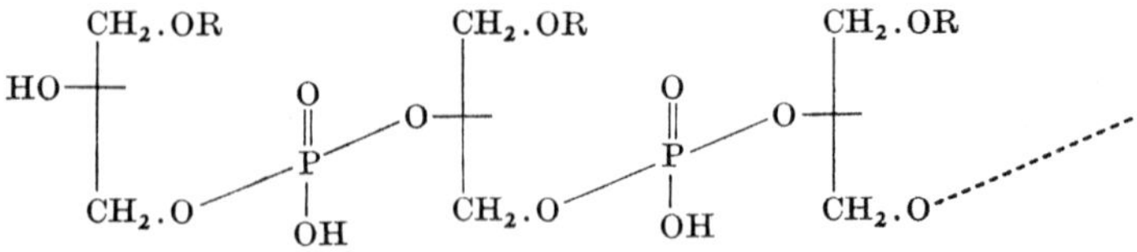

FIG. 3. Structure of 1,2-poly(glycerol phosphate) teichoic acids. R = H or glycosyl.

Alkali hydrolysis of the wall teichoic acid of *B. stearothermophilus* B65 gives the monophosphates of 1-O-α-D-glucopyranosyl glycerol and small amounts of the free glucoside, glycerol and glycerol mono- and di-phosphates. These products could equally well be derived from a mixture of 1,3-poly(glycerol phosphate) and poly(glucosylglycerol phosphate)

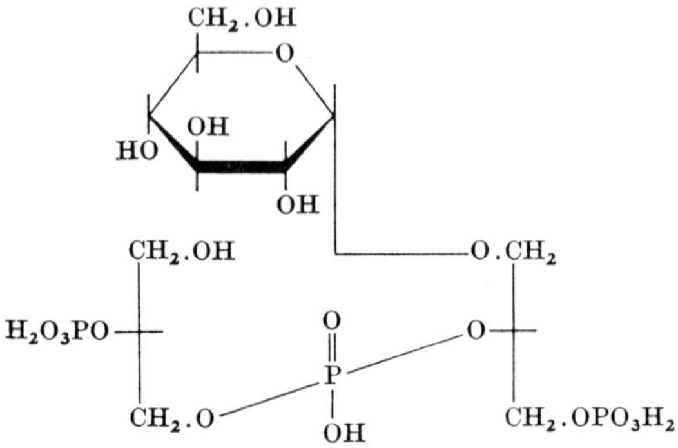

FIG. 4. Proposed structure of a compound formed on alkali hydrolysis of the wall teichoic acid of *Bacillus stearothermophilus*.

but the proposed structure (Fig. 3, R = α-D-glucopyranosyl) was supported by the isolation, from alkali-hydrolysed teichoic acid, of a small amount of material which was identified as the compound shown in Fig. 4; this could be derived only from a partially glucosylated 2,3-poly-(glycerol phosphate).

B. Teichoic Acids in which Sugar Residues form an Integral Part of the Polymer Chain

1. *Ribitol Teichoic Acids*

The specific capsular materials of several strains of *Diplococcus pneumoniae* are ribitol teichoic acids in which complex ribitol glycosides are held together by phosphodiester linkages between ribitol and one of the sugars of the adjacent unit. The structures of several of these capsular teichoic acids have been established. They all contain rather complex oligosaccharides, many of which, like those in the Type 13 material (Fig. 5), are composed of both pyranosyl and furanosyl components together with N-acetylaminosugars (Chittenden *et al.*, 1968; Dixon *et al.*, 1968; Kennedy *et al.*, 1969; Rao *et al.*, 1966, 1969; Rebers and Heidelberger, 1961; Watson *et al.*, 1972).

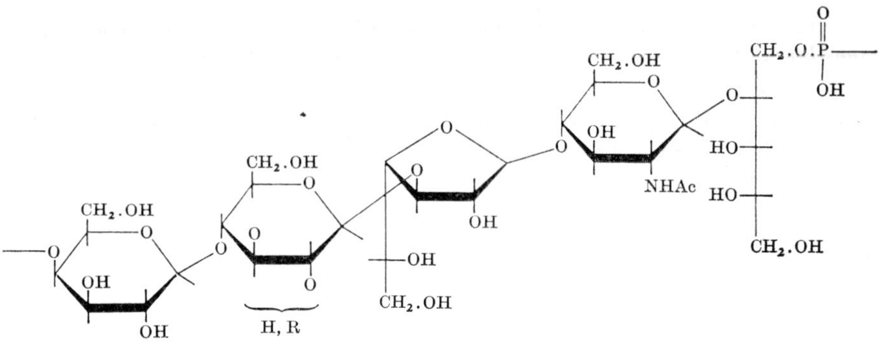

FIG. 5. Structure of the Type 13 pneumococcal capsular teichoic acid.

The pneumococcal somatic antigen known as C substance has also been shown to be a ribitol teichoic acid. This is a component of the walls of the organism and is present in all strains. The structure of this teichoic acid has not been fully elucidated but it has been established that repeating units containing ribitol phosphate, choline phosphate, glucose, N-acetylgalactosamine and an N-acetyldiaminotrideoxyhexose are held together by phosphodiester linkages between ribitol and sugar so that the oligosaccharide forms part of the polymer chain (Brundish and Baddiley, 1968). Walls of certain streptococci also contain complex ribitol teichoic acids which may be of the same structural type as those of pneumococci (Bleiweis *et al.*, 1967; Miller, 1969; Montague, 1964).

2. *Glycerol Teichoic Acids*

Poly(glucosyl glycerol phosphate) and poly(galactosyl glycerol phosphate) (Fig. 6) teichoic acids occur together in walls of *Bacillus licheniformis* ATCC 9945. Alanine was not present in the teichoic acid isolated by Burger and Glaser (1966) but the conditions used for isolation would almost certainly have hydrolysed all labile ester linkages. A similar mixture of related teichoic acids is present in walls of *Lactobacillus plantarum* NIRD C106 (Archibald and Coapes, 1971a). Three different repeating structures are present (Fig. 7), and studies with concanavalin

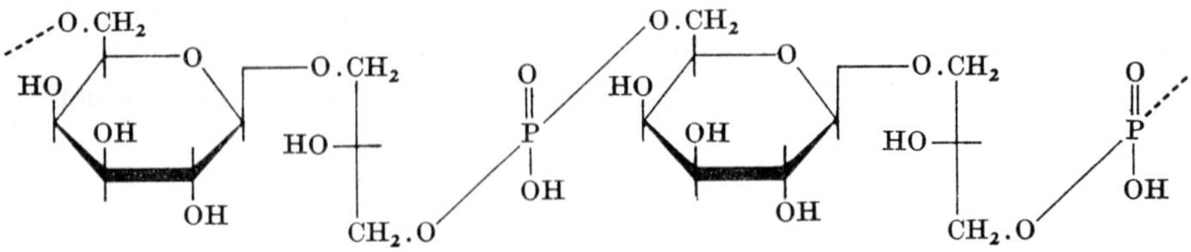

FIG. 6. Structure of the poly(galactosyl glycerol phosphate) teichoic acid from *Bacillus licheniformis.*

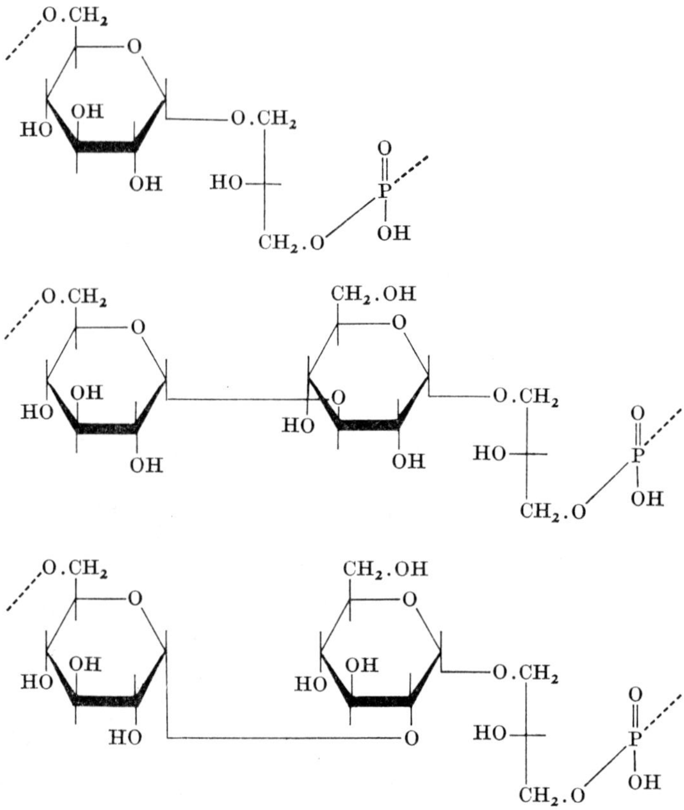

FIG. 7. Structures of the three repeating structures in the wall teichoic acids of *Lactobacillus plantarum* C106. Alanyl esters are not shown.

A have shown that these are present in separate chains containing the different glycerol glycosides. The phosphodiester linkages are located as shown in Fig. 7. D-Alanine is present in this teichoic acid, and its lability in alkali suggests that it is attached to glycerol rather than to the sugar residues.

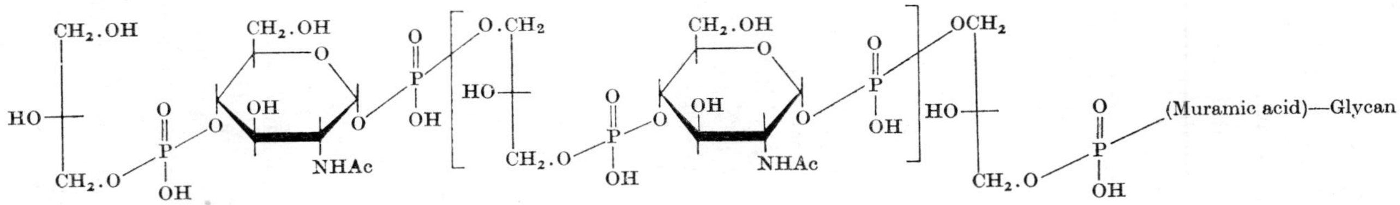

FIG. 8. Structure of the wall teichoic acid of *Staphylococcus lactis* I3, showing the mode of the linkage to the glycan; alanyl esters are omitted.

Walls of *Staphylococcus lactis* I3 (Archibald *et al.*, 1968, 1971) and *Micrococcus* sp 24 (Archibald and Heptinstall, 1971) contain glycerol teichoic acids in which N-acetylglucosaminyl residues form part of the polymer chain (Fig. 8). These teichoic acids have the same structure, though a small amount of a poly(ribitol phosphate) teichoic acid is also present in walls of *Micrococcus* sp 24. The N-acetylglucosaminyl residues are attached to glycerol through sugar 1-phosphate linkages so that the repeating units in these teichoic acids contain two phosphate groups. The D-alanyl ester residues do not show the usual high alkali lability, and so are probably attached to the N-acetylglucosamine rather than directly to glycerol. Similar teichoic acids are present in walls of various other micrococci (Oeding *et al.*, 1967; Partridge *et al.*, 1973).

3. *Phosphorylated Polysaccharides*

Walls of various bacteria contain polymers which, like the wall teichoic acid of *Staph. lactis* I3, contain sugar 1-phosphate residues as part of the repeating unit. The simplest of these polymers is the phosphorylated polysaccharide which constitutes about 40% by weight of the wall of *Staphylococcus lactis* 2102 (Archibald and Stafford, 1972). This polymer consists simply of a chain of N-acetylglucosamine 1-phosphate residues in which the adjacent acetamidosugar units are connected by phosphodiester linkages between the 1- and 6-positions (Fig. 9). Since it

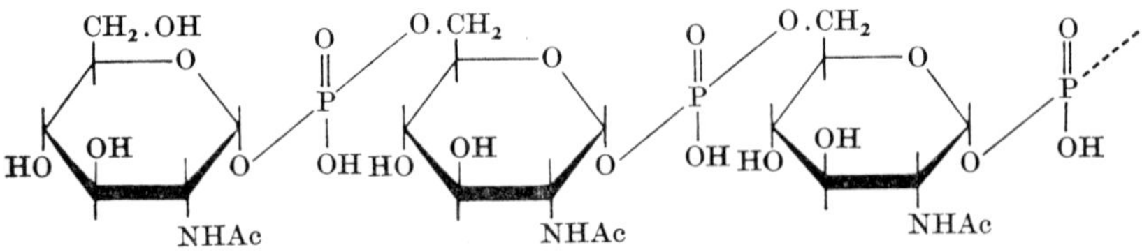

FIG. 9. Structure of the phosphorylated polysaccharide of walls of *Staphylococcus lactis* 2102.

lacks glycerol and ribitol, this polymer is not a teichoic acid but its structure and cellular location are such that it is appropriate to include it in this discussion. Walls of *Micrococcus* sp. A1 (Partridge *et al.*, 1971) contain a polymer in which adjacent units of 3-O-α-D-glucopyranosyl-*N*-acetylgalactosamine 1-phosphate are connected by phosphodiester linkage between the reducing group of *N*-acetylgalactosamine and the hydroxyl group at position 6 of glucose (Fig. 10). A similar polymer in which the repeating units are 3-O-β-D-glucopyranosyl-N-acetylgalactosamine 1-phosphate occurs, together with glycerol teichoic acid, in walls of *Bacillus subtilis* 168 (Duckworth *et al.*, 1972; unpublished observations by Dr. V. N. Shibaev). Sugar 1-phosphate moieties have been found in the wall polysaccharides of certain streptococci (Pazur *et al.*, 1971) and may also be present in the phosphorylated polysaccharide which is present, in addition to ribitol teichoic acid, in some strains of *Lactobacillus plantarum* (Anderson *et al.*, 1969; Douglas and

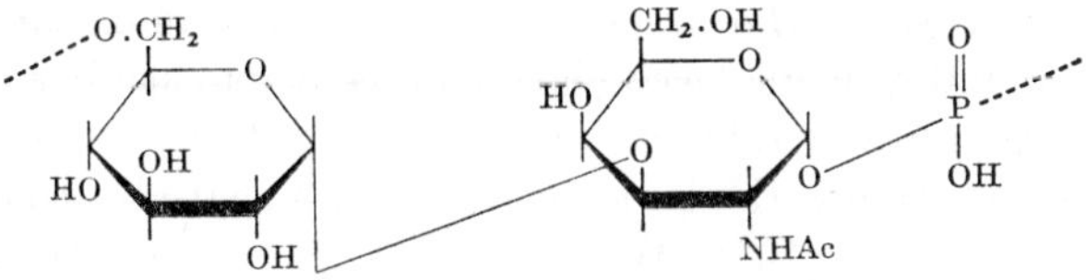

FIG. 10. Structure of the phosphorylated polysaccharide in walls of *Micrococcus* sp. Al.

Wolin, 1971). It is possible that polysaccharides and teichoic acids which contain sugar 1-phosphate linkages may be more widespread than present reports suggest, since their lability towards acid and the complexity of their hydrolysis and degradation in alkali may have resulted in their being overlooked in earlier studies.

III. Cellular Location

A. Membrane Teichoic Acids

In all Gram-positive bacteria so far examined, the "cell contents" fraction of disrupted cells contains a glycerol teichoic acid which can be isolated after extraction with trichloroacetic acid or phenol. Although the isolated teichoic acid is freely soluble in water, in its native state it is associated with a particulate component of the cell (Critchley *et al.*, 1962) which sediments on high-speed centrifugation of the "cell contents" fraction. That the teichoic acid is associated with the cytoplasmic membrane was shown directly in studies on *Streptococcus faecalis* ATCC 9790. Membranes isolated by disruption of protoplasts of this organism contained most of the glycerol teichoic acid present in the cell. This could be removed by washing with salts or, especially, water but even after extensive washing substantial proportions of teichoic acid remained in the membrane (Shockman and Slade, 1964). Other studies with *Bacillus megaterium* NCIB 7581 and various Group-D streptococci gave rather different results in that although isolated membrane contained some teichoic acid most of the "intracellular" teichoic acid was released into the supernatant solution when cells were converted into protoplasts (Hay *et al.*, 1963; Smith and Shattock, 1964). More recently it has become clear that the concentration of Mg^{2+} ions present during the formation of protoplasts largely determines the fate of membrane teichoic acid (Hughes *et al.*, 1973; unpublished observations by G. H. Pigott). When various bacteria are converted into protoplasts in the absence of Mg^{2+} ions, the membrane teichoic acid is released whereas most of it is retained in association with the membrane when protoplasts are formed in solutions containing Mg^{2+} in concentrations greater than 10m*M*. The nature of the association between teichoic acid and the cytoplasmic membrane has been clarified recently in studies which have shown that membrane teichoic acids isolated from several different bacteria by extraction with phenol are firmly associated with lipid (Coley *et al.*, 1972; Toon *et al.*, 1972; Wicken and Knox, 1970, 1971). This lipid is not removed on extraction with organic solvents or on chromatography in the presence of sodium dodecylsulphate or 6*M*-urea, and so is apparently attached covalently to the teichoic acid. All of the membrane teichoic acid is associated with lipid, and the isolated lipoteichoic acids have characteristic physical properties and form micelles which give very sharp peaks (6S to 10S) on ultracentrifugation. The nature of the association between the teichoic acid and lipid moieties has been studied in the membrane teichoic acid of *Streptococcus faecalis* NCIB 8191 (Toon *et al.*, 1972). This teichoic acid consists of a chain of 28 to 35 units of kojibiosyl glycerol phosphate, and is associated with a kojibiosyl diglyceride or, more probably, a phosphatidylkojibiosyl diglyceride. In the proposed structure, shown in Fig. 11, the teichoic acid is attached

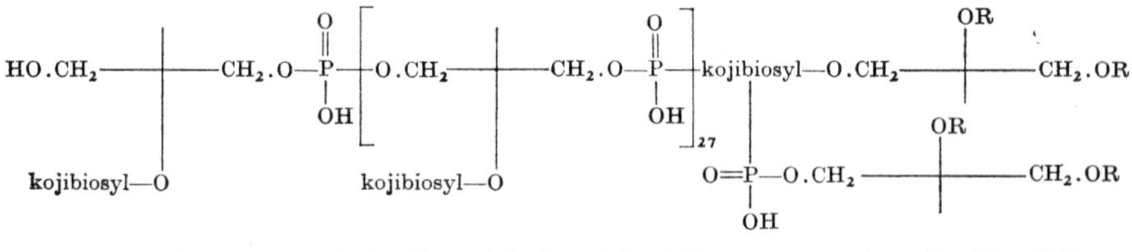

FIG. 11. Structure of the lipoteichoic acid of *Streptococcus faecalis*. R = fatty acid.

to the lipid by a phosphodiester linkage between the terminal glycerol residue of the teichoic acid chain and one of the hydroxyl groups of the kojibiosyl moiety of the glycolipid; a phosphatidyl residue is also attached to one of the hydroxyl groups of this kojibiosyl residue. The teichoic acid can thus be anchored in the membrane by the intercalation of its lipid with other phospholipids in the bilayer region of the membrane. It is unlikely that the long hydrophilic chain of the teichoic acid can be accommodated within the membrane, and this is envisaged as being located on the outer surface of the membrane extending into the space between the membrane and the wall.

B. Wall Teichoic Acids

Teichoic acids are held in the cell wall by covalent attachment to peptidoglycan, and so can be extracted only by procedures which cause fission of covalent linkages. Early studies indicated that the D-alanyl ester residues and the inter-unit phosphodiester groups of teichoic acid are not attached to peptidoglycan, and that multiple linkages do not occur (Archibald *et al.*, 1961a; Rogers and Garrett, 1965). Enzymic dissolution of walls gives complexes in which teichoic acids are attached to soluble fragments of peptidoglycan (Ghuysen and Strominger, 1963; Ghuysen *et al.*, 1962; Hughes *et al.*, 1968; Young, 1966). Because such complexes do not contain the phosphomonoester groups which are present at one end of the chain in teichoic acids which have been isolated by extraction into dilute acid, Ghuysen *et al.* (1965) suggested that teichoic acids are linked through a terminal phosphodiester to a sugar residue in the glycan chain. Direct support for this has come from the isolation of muramic acid phosphate from acid hydrolysates of walls and teichoic acid-glycan complexes from various bacteria (Grant and Wicken, 1968; Heyman *et al.*, 1967; Liu and Gottschlich, 1967; Tipper *et al.*, 1967). Muramic acid phosphate has also been shown to be involved in the linkage of polysaccharides to walls of *Lactobacillus casei* (Knox and Hall, 1965), *L. fermenti* (Knox and Holmwood, 1968) and *Bacillus licheniformis* (Hughes, 1970). It is possible that teichoic acids are linked directly by a phosphodiester bond between muramic acid and the terminal alditol moiety of the teichoic acid chain, though it has been argued (Hay *et al.*, 1965) that a direct linkage of this kind would not be sufficiently labile to account for its apparently preferential hydrolysis during extraction of walls with dilute acid or alkali. The stability of the phosphodiester linkages in teichoic acids is however substantially influenced by relatively minor stereochemical differences (Archibald *et al.*, 1971) and further study is required before any firm conclusion can be reached. Direct demonstration of the nature of the linkage has been reported only with the rather unusual (Fig. 8, p. 61) wall teichoic acid of *Staphylococcus lactis* I3 (Button *et al.*, 1966). Autolysis gives a teichoic acid-glycan complex which was treated with acid under conditions that effect selective hydrolysis of sugar 1-phosphate linkages. This left a glycerol diphosphate moiety attached to the glycan, and selective degradative procedures indicated that one of the phosphate groups was esterified to both glycerol and muramic acid. It is not known whether the muramic acid which carries the teichoic acid is specifically located in the glycan chain, but recent studies have shown that only one of the muramic acid moieties in any chain is linked to teichoic acid and all of the teichoic acid is attached to glycan (Archibald *et al.*, 1973a). Digestion of walls of

Staph. lactis I3 with a *Flavobacter* amidase gives glycan and teichoic acid-glycan fractions which can be separated from each other and from peptide. Approximately 40% of the glycan chains are attached to teichoic acid; each of these chains carries only one teichoic acid molecule, and the composition and average chain length of the glycan which is associated with teichoic acid is the same as that of the rest of the glycan in the wall. As described below, the finding that as many as 40% of the glycan chains are covalently attached to teichoic acid has implications concerning the location of the teichoic acid within the wall structure.

Although it is more than fifteen years since the discovery of teichoic acids, relatively little attention has been given to their location in the wall structure. After fixation and staining with glutaraldehyde, osmium tetroxide, uranyl acetate or lead citrate, walls of several Gram-positive bacteria have a trilaminar structure with discrete inner and outer electron-dense areas which are separated by a thicker and relatively transparent middle layer (Cole *et al.*, 1970; Higgins *et al.*, 1970). In an early electron microscope study, sections of walls from which teichoic acid had been removed by extraction into trichloroacetic acid were reported to be little different in appearance from sections of intact walls (Archibald *et al.*, 1961a). In contrast, Nermut (1967) found that treatment of walls of *Bacillus megaterium* strain M with hot formamide resulted in a substantial decrease in their thickness, apparently by removal of an outer layer which was envisaged as being composed of "bristles" of teichoic acid molecules arranged perpendicularly to the peptidoglycan

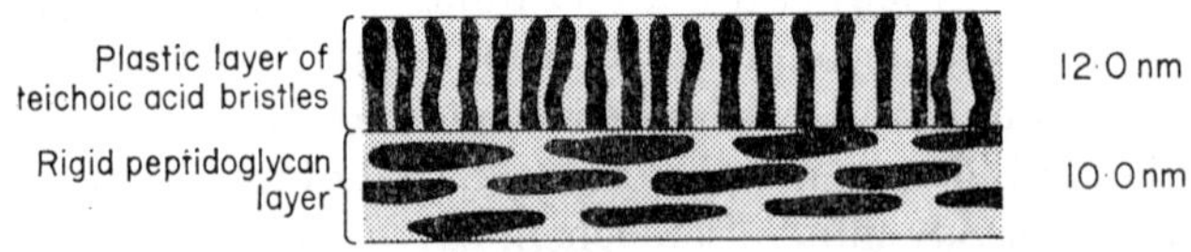

FIG. 12. Model of cell wall of *Bacillus megaterium* proposed by Nermut (1967).

(Fig. 12). Both of these studies are open to the criticism that the conditions used to extract teichoic acid were sufficiently drastic to cause damage to the structure and perhaps to the morphology of the peptidoglycan. However, the suggestion that teichoic acid is concentrated in the outer layers of the wall is given some support by the studies of Cole *et al.* (1970) on the morphology of walls of a temperature-sensitive mutant of *B. subtilis*. Walls of this mutant show a normal trilaminar structure when grown at the permissive temperature but, on growth at elevated temperatures when wall teichoic acid is not synthesized, the outer and inner dark layers are lost. These dark layers are, however, observed in another teichoic acid-deficient mutant of *B. subtilis* (Heijenoort *et al.*, 1971).

Walls of this mutant contain large quantities of a negatively charged protein and, provided this is protected by fixation in glutaraldehyde, the appearance of the walls and cells is fairly normal. Interestingly, this mutant shows fewer deviations from normal morphology and growth characteristics than does that described by Cole *et al.* (1970); a possible explanation of this might be that the negatively charged protein can fulfil some of the physiological functions for which teichoic acid is normally required (see Section V, p. 81).

It does not yet seem possible to draw a firm conclusion from these studies, but a surface location for at least some of the teichoic acid is indicated by the findings that teichoic acids contribute to the surface charge of the cell, they form part of the specific receptor sites for various bacteriophages, and both walls and whole cells are agglutinated by lectins and antibodies which are specific for the teichoic acid component of the wall.

The uptake of antibody to wall teichoic acid by whole cells and isolated walls of certain bacilli has been studied in some detail by Burger (1966). Walls of *B. subtilis* 3610 absorb very little more antibody than do whole cells. Since the antibody has unrestricted access to both the inner and outer surfaces of the isolated walls, their failure to bind appreciably more antibody suggests that little or no wall teichoic acid is accessible at the inner surface. The amount of antibody absorbed by the wall increased greatly after a brief digestion with lysozyme, and it was found that in the untreated wall only 20% of all possible antigenic sites were accessible to antibody molecules. Similar results were obtained with *B. subtilis* W23, whereas 70% of the possible antigenic sites in walls of *B. licheniformis* ATCC 9945 adsorbed antibody molecules. This could mean that most of the teichoic acid is at the surface in walls of *B. licheniformis*, whereas the polymer in the *B. subtilis* strains is buried within the wall. Alternatively, antibody molecules might be able to penetrate more deeply into walls of *B. licheniformis* but little is known about the penetration of wall by antibody or about steric factors which might limit adsorption of antibody to the wall surface.

The finding that 40% of the glycan chains in *Staph. lactis* I3 are attached to teichoic acid has some implications regarding the disposition of the teichoic acid in the wall (Archibald *et al.*, 1973a). Kelemen and Rogers (1971) considered various alternatives for the arrangement of peptidoglycan in walls. In one possible structure the glycan chains lie parallel to the surface of the wall. The thickness of cell walls is such that they would have to contain several layers of glycan and so fewer than 40% of the glycan chains could be present at the surface in this type of structure. It would follow that much of the teichoic acid in walls of *Staph. lactis* I3 would have to be attached to glycan chains which are located in the inner layers of the peptidoglycan (Fig. 13), so that teichoic acid could not be present exclusively in an outermost layer of the wall. In an alternative model (Kelemen and Rogers, 1971) the glycan chains are arranged radially so that they can extend from the inner to the outer surfaces of the peptidoglycan layer. This model could accommodate teichoic acid as an outer layer held by attachment to the outer ends of a proportion of the glycan chains (Fig. 13). Mauck and Glaser (1972a) have

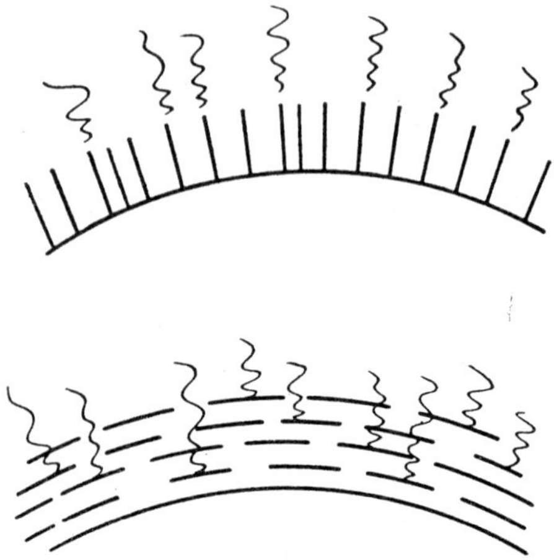

FIG. 13. Models for the structure of the wall of *Staphylococcus lactis* I3 showing attachment of teichoic acid to 40% of the glycan chains.

shown that, during growth and turnover of walls of *B. subtilis* NCTC 3610, newly synthesized teichoic acid-glycopeptide molecules are cross-linked to glycopeptides which had been synthesized previously. These authors consider that this observation can best be explained on the basis of a structure in which the glycan chains are arranged radially. However, such an arrangement need not be as highly ordered as in the Kelemen and Rogers models and an X-ray diffraction study of walls of *Staph. aureus* did not give evidence of regular structure (Balyuzi *et al.*, 1972).

In any consideration of the arrangement of teichoic acid in walls, it must be remembered that teichoic acids are fairly long and flexible molecules which are held covalently at only one end; their precise arrangement may thus be influenced by environmental conditions. James and Brewer (1968) found that teichoic acid ceased to contribute to the surface charge of *Staph. aureus* at pH values greater than 5–6. They suggested that this might be due to a change in the arrangement of the teichoic acid molecules at the cell surface brought about by pH-dependent changes in the electrostatic interactions between the various polymeric components of the wall. It is known that walls are elastic structures which expand and contract in response to changes in the external pH value and ionic strength (Marquis, 1968; Ou and Marquis, 1970). It is quite possible that interactions between teichoic acid and peptidoglycan are involved in these processes.

IV. Biosynthesis

Teichoic acids, like peptidoglycan, are synthesized by enzyme systems which are located in the cytoplasmic membrane. Procedures which have been used for the isolation of enzymically active preparations include the recovery of membranes from osmotically disrupted protoplasts and the isolation, by differential centrifugation of physically disrupted bacteria, of particulate fractions rich in fragmented membrane. The method used for the isolation of the particulate enzyme greatly influences its activity but preparations obtained after disruption of cells by grinding with alumina have generally been found to be active in the synthesis of both teichoic acid and peptidoglycan.

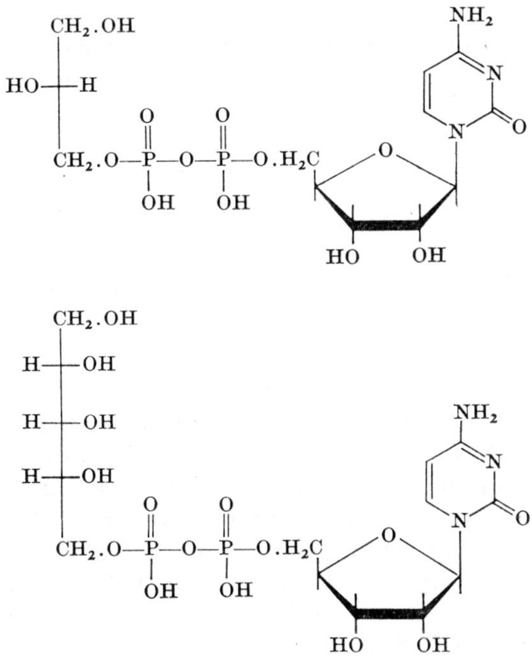

FIG. 14. Structures of cytidine diphosphate glycerol and cytidine diphosphate ribitol.

The ribitol phosphate and glycerol phosphate components of teichoic acid are derived from the nucleotides cytidine diphosphate ribitol and cytidine diphosphate glycerol (Fig. 14). The enzymes responsible (Shaw, 1962) for synthesis of cytidine diphosphate ribitol from cytidine triphosphate and D-ribitol 5-phosphate (cytidine diphosphate ribitol pyrophosphorylase) and cytidine triphosphate and D-glycerol 1-phosphate (cytidine diphosphate glycerol pyrophosphorylase) are thought to be loosely associated with the membrane, but are normally recovered in the cytoplasmic fraction of disrupted bacteria. D-Ribitol 5-phosphate (L-ribitol 1-phosphate) is formed by an $NADH_2$-dependent reduction of D-ribulose 5-phosphate (Glaser, 1963); D-glycerol 1-phosphate is presumably derived from glycolysis or by gluconeogenesis.

The glycosidic components of teichoic acid are derived from the appropriate uridine diphosphate sugars, but the origin of the D-alanyl ester residues has until recently been uncertain. Several Gram-positive bacteria are known to contain a D-alanine-activating enzyme which reacts with D-alanine, produced by the action of alanine racemase, and ATP to form inorganic pyrophosphate and an "activated alanine" thought to be an alanyl AMP-enzyme complex (Baddiley and Neuhaus, 1960). Early attempts to demonstrate transfer of alanine from this complex to teichoic acid or teichoic acid precursors or degradation products were unsuccessful, but recently an enzyme has been isolated from *Lactobacillus casei* which transfers the alanine to a membrane-bound acceptor which is believed to be the lipoteichoic acid (Reusch and Neuhaus, 1971; R. Linzer and F. C. Neuhaus, unpublished observations).

A. Teichoic Acids which Contain a Poly(alditol phosphate) Chain

Cell-free synthesis of poly(ribitol phosphate) has been achieved with particulate enzyme preparations from *Lactobacillus plantarum* ATCC 8014 (Glaser, 1964) and *Staphylococcus aureus* strain Copenhagen (Ishimoto and Strominger, 1966). The enzyme obtained after digestion of the lactobacilli with muramidase was more active than the particulate fraction recovered from mechanically disrupted cells but, on incubation with CDP-[^{14}C]ribitol in the presence of divalent cations, both preparations synthesized labelled poly(ribitol phosphate) which was characterized after extraction with phenol. The synthetase was shown to be associated with the wall-membrane fraction of the cell, but its location in the membrane was not established unequivocally because protoplasts could not be prepared. The poly(ribitol phosphate) synthetase activity from *Staph. aureus* was present in a particulate fraction which sedimented between 30,000 and 100,000 *g*. Incubation with [^{32}P]-CDP ribitol and CDP-[^{3}H]-ribitol showed that ribitol and phosphate were incorporated simultaneously and in equivalent proportions into material which was characterized as poly(ribitol phosphate). The amount of CMP formed was equivalent to the amount of poly(ribitol phosphate) synthesized, and synthesis was markedly inhibited by the addition of CMP. The overall reaction can thus be formulated as shown in Fig. 15. High concentrations (10 to 30 mM) of divalent cation were required for optimal activity, but synthesis was not stimulated by addition of potential acceptors such as poly(ribitol phosphate) or teichoic acid-peptidoglycan complexes.

$$\text{D-Ribitol 5-phosphate} + \text{CTP} \leftrightharpoons \text{CDP-ribitol} + \text{PP}_i$$
$$\text{CDP-ribitol} + (\text{ribitol phosphate})_n \rightarrow (\text{ribitol phosphate})_{n+1} + \text{CMP}$$

FIG. 15. Reactions leading to synthesis of CDP-ribitol and poly(ribitol phosphate).

Particulate enzyme systems from various strains of *Staph. aureus* also effect incorporation of N-acetylglucosamine into poly(ribitol phosphate). The reaction requires UDP-N-acetylglucosamine, Mg^{2+} and an acceptor, and is accompanied by the formation of UDP (Nathenson and Strominger, 1963). Poly(ribitol phosphate), prepared by enzymic removal of *N*-acetylglucosamine from isolated teichoic acid, can act as acceptor, but simultaneous addition of CDP-ribitol and UDP-*N*-acetylglucosamine to the particulate enzyme leads to incorporation of *N*-acetylglucosamine at a much greater rate than is observed when an equivalent amount of poly(ribitol phosphate) is added. When the enzyme was incubated first with CDP-ribitol and UDP-N-acetylglucosamine then added, incorporation of *N*-acetylglucosamine was only about half of that observed during simultaneous synthesis. Glycosylation of the growing poly(ribitol phosphate) chain is thus more efficient than glycosylation of exogenously added acceptor or of acceptor which has been preformed by the enzyme system. This suggests a close spatial relationship between the synthetase and the transferase activities so that preformed poly(ribitol phosphate) is less accessible to the transferase. The enzymes may be organized in different sites which catalyse synthesis of α-N-acetylglucosaminyl poly(ribitol phosphate) and β-N-acetylglucosaminyl poly(ribitol phosphate) respectively.

Enzyme preparations from various strains of *Staph. aureus* synthesize α and β glycosidic linkages in proportions which are similar, though not identical, to those found in the wall teichoic acid of the same organism (Nathenson *et al.*, 1966). Thus, most of the glycosidic linkages synthesized *in vitro* by the enzyme systems from strains H and Duncan had the β configuration, but α linkages were formed in appreciably greater proportions than found in the teichoic acids isolated from the walls of these organisms. The converse was observed with strain 3528. These differences may be a consequence of disorganization of the particulate enzyme during its isolation, but it is possible that in the intact cell the activities of the two transferase enzymes, or of the two distinct synthetic sites, is regulated in some way which is not yet understood.

Glucosyl transfer from UDP-glucose to poly(ribitol phosphate) has been studied with a membrane preparation from *Bacillus subtilis* W23 (Chin *et al.*, 1966). Walls of this organism contain a mixture of fully glucosylated poly(ribitol phosphate) and glucosyl-free poly(ribitol phosphate). Teichoic acid isolated by extraction of walls with trichloroacetic acid was fully glucosylated on incubation with the membrane preparation and UDP-glucose; isolated cell walls also acted as an acceptor when a "soluble" or highly fragmented membrane preparation was used. Under the latter conditions, approximately 20% of the unsubstituted ribitol residues in the wall were glucosylated (interestingly, this is similar to the proportion of antigenic sites "available" in the wall; see Section IIIB, p. 67). Since isolated walls can act as an acceptor of glucosyl units, the absence of partially glucosylated teichoic acid in native walls suggests that two distinct systems are present, one of which synthesizes poly(ribitol phosphate) while the other synthesizes glucosyl poly(ribitol phosphate). The presence of two distinct synthetic sites is supported by the finding that synthesis of glucosyl poly(ribitol phosphate) in germinating spores is little affected by chloramphenicol, while *de novo* enzyme synthesis is required before significant amounts of glucosyl-free poly(ribitol phosphate) are synthesized (Chin *et al.*, 1968).

Synthesis of poly(glycerol phosphate) has been studied using cytoplasmic membranes, or particulate fractions containing fragmented membrane, from several bacteria. Early studies (Burger and Glaser, 1964) with enzymes from *Bacillus licheniformis* ATCC 9945 and *Bacillus subtilis* NCTC 3610 established that CDP-glycerol was the sole substrate requirement, and the reaction could be formulated in a manner analogous to that shown (Fig. 15) for synthesis of ribitol teichoic acid.

It is probable that the synthesis observed with these particulate preparations resulted from an extension of preformed chains already present in the membrane particles. Thus, it was found that almost all of the glycerol phosphate incorporated into teichoic acid on brief incubation of a membrane preparation from *B. subtilis* ATCC 6051 with CDP-1(3)-[^{3}H]glycerol was present at the glycerol-terminal end of the chain and so had been incorporated into preformed teichoic acid (Kennedy and Shaw, 1968). Only the glycerol-terminal end of poly(glycerol phosphate) is susceptible to oxidation with periodate and, after this oxidation, 45% of the incorporated label was recovered as formaldehyde showing that an average of 1·1 to 1·2 glycerol phosphate units had been added to each chain. Had the existing chains been extended by addition to the phosphate-terminal end, i.e. to that end which is covalently attached to peptidoglycan in the wall or to lipid in the membrane, periodate oxidation of the product would not have given [^{3}H]formaldehyde. It follows that chain extension proceeds by addition to the glycerol terminal end (Fig. 1, p. 54), a conclusion which was confirmed by pulse-labelling experiments. The direction of extension is thus similar to that observed during biosynthesis of glycogen when glucosyl units are transferred to the non-reducing terminus of the growing polymer and different from that observed during biosynthesis of O-antigen when the growing oligomer is transferred on to the non-reducing terminus of the incoming unit (Robbins *et al.*, 1967).

Because synthesis of poly(glycerol phosphate) by particulate enzyme preparations takes place by extension of chains which are already present in the enzyme preparation, an absolute requirement for an acceptor of the glycerol phosphate units cannot be demonstrated in these systems. Recently, however, a poly(glycerol phosphate) polymerase has been solubilized by extraction of membranes of *B. subtilis* ATCC 6051 with Triton (Mauck and Glaser, 1972a). The activity of the polymerase was found to be completely dependent on the presence of a heat-stable acceptor from which the polymerase could be separated by sucrose density-gradient centrifugation. The acceptor fraction contains glycerol, phosphate, glucosamine and fatty acids, and may correspond to the phosphoglycolipid which has been found to be associated with several membrane teichoic acids. Glycerol phosphate units were transferred to the acceptor, and the chain length of the product increased continuously. Lipoteichoic acid isolated by extraction into phenol was reported to have a highly variable activity as an acceptor in the poly(glycerol phosphate) synthesis, but teichoic acid isolated from cell walls had little or no activity. This acceptor material is distinct from the lipid, thought to be undecaprenol phosphate, reported by Anderson *et al.* (1972) to act as an intermediate in synthesis of poly(glycerol phosphate) by particulate membrane preparations from *B. subtilis* NCTC 3610 and *B. licheniformis* ATCC 9945. The soluble enzyme from *B. subtilis* did not transfer glycerol phosphate to added undecaprenol phosphate nor were the exchange reactions observed which would be expected if undecaprenol phosphate participated in the system. A possible explanation of this difference could be that the soluble enzyme was engaged in synthesis of membrane teichoic acid and that this does not require a lipid intermediate, whereas the lipid intermediate in the particulate system may have been involved in synthesis of wall teichoic acid.

The particulate enzyme from *B. subtilis* NCTC 3610 which synthesized poly(glycerol phosphate) also transferred glucose from UDP-glucose to the free hydroxyl groups of the polymer to give fully glucosylated poly(glycerol phosphate) (Glaser and Burger, 1964). Exogenous poly-(glycerol phosphate) was also glucosylated, and glucose transfer, like the polymerization reaction, required high concentrations of divalent cation. Most of the glucosyl transferase activity in post-exponential cells of *B. subtilis* 168 is associated with the cytoplasmic membrane, but substantial proportions of the activity were recovered in the cytoplasmic fraction of exponentially growing cells (Brooks *et al.*, 1971). The membrane-bound and soluble forms of the enzyme did not provide any evidence that glycosyl transfer is mediated by lipid.

B. Teichoic Acids in which Sugar Residues form an Integral Part of the Polymer Chain

Although the glycosyl transferase and poly(alditol phosphate) polymerase enzymes appear to be closely associated in cells which synthesize such teichoic acids, their individual activities can be studied in cell-free systems to which only one of the nucleotide precursors has been added. Synthesis of teichoic acids in which glycosyl residues form part of the polymer chain must obviously differ in that both the sugar and the alditol phosphate nucleotides need to be present in order for polymer synthesis to take place. Walls of *B. licheniformis* ATCC 9945 contain poly(glycerol phosphate), poly(glucosyl glycerol phosphate) and poly-(galactosyl glycerol phosphate). On incubation with CDP-glycerol, particulate membrane preparations from this organism synthesize poly(glycerol phosphate); on incubation with CDP-glycerol and UDP-glucose or UDP-galactose, poly(glucosyl glycerol phosphate) or poly-(galactosyl glycerol phosphate) are also formed (Burger and Glaser, 1966). All of these polymerization reactions are believed to proceed *via* undecaprenol phosphate intermediates (Anderson *et al.*, 1972; Hancock and Baddiley, 1972). Thus, on incubation of a particulate enzyme preparation with UDP-glucose, glucose is transferred to a phospholipid to give a glucose–phosphate lipid which can then accept glycerol phosphate from CDP-glycerol to give a glycerol–phosphate–glucose–phosphate–lipid. Pulse labelling experiments show that the glycerol–phosphate–glucose unit is then transferred to the growing polymer chain according to the scheme shown in Fig. 16. The biosynthetic repeating unit is thus glycerol–phosphate–glucose rather than glucosyl-glycerol phosphate, and the unit is first assembled on a lipid intermediate before being incorporated into polymer. This incorporation is effected by a transglycosylation reaction and not by the transphosphorylation reactions involved in the synthesis of other kinds of teichoic acid. The evidence that the lipid is undecaprenol phosphate is indirect and is discussed later. The isolated intermediates do however exhibit the characteristic lability towards hydrogenolysis and gentle acid hydrolysis expected of undecaprenol phosphates.

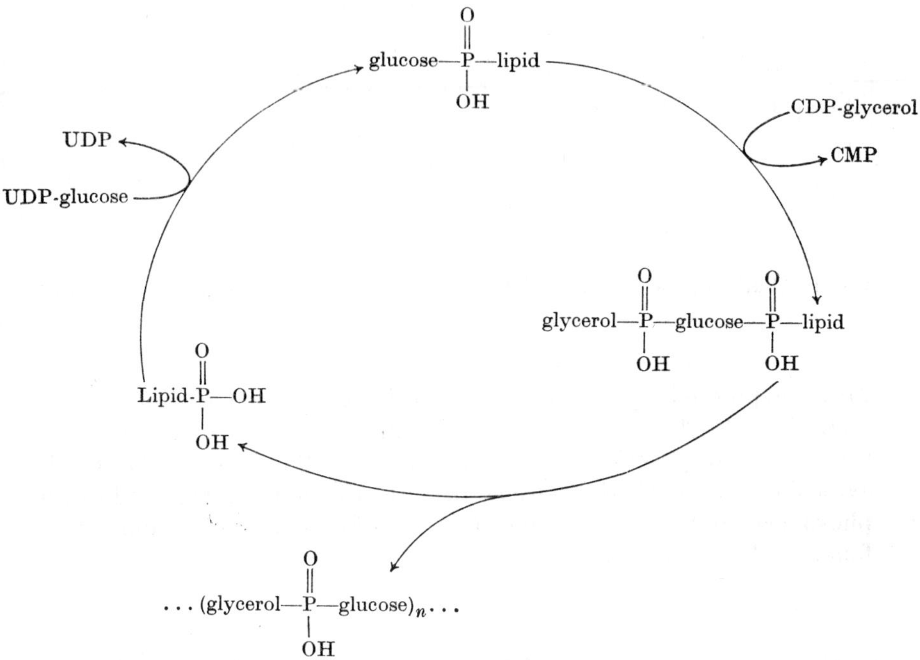

FIG. 16. Pathway for synthesis of poly(glucosyl glycerol phosphate) in *Bacillus licheniformis*.

Prior assembly of repeating unit on to a lipid intermediate has also been demonstrated in the cell-free synthesis of the wall teichoic acid of *Staphylococcus lactis* I3 (Hussey and Baddiley, 1972). Particulate enzyme preparations from this organism catalyse transfer of *N*-acetylglucosamine 1-phosphate residues from UDP-N-acetylglucosamine to a phospholipid intermediate. A glycerol phosphate residue is then transferred from CDP-glycerol to the 4-hydroxyl group of the N-acetylglucosamine to give the repeating unit. This is then transferred to the growing teichoic acid chain and lipid phosphate is released (Fig. 17). Pulse-labelling experiments show that the unit is transferred to the glycerol-terminal end of the teichoic acid chain, i.e. to the end which is remote from that which is attached to peptidoglycan in isolated walls (Hussey *et al.*, 1969). The direction of chain extension is thus the same as was found with poly(glycerol phosphate).

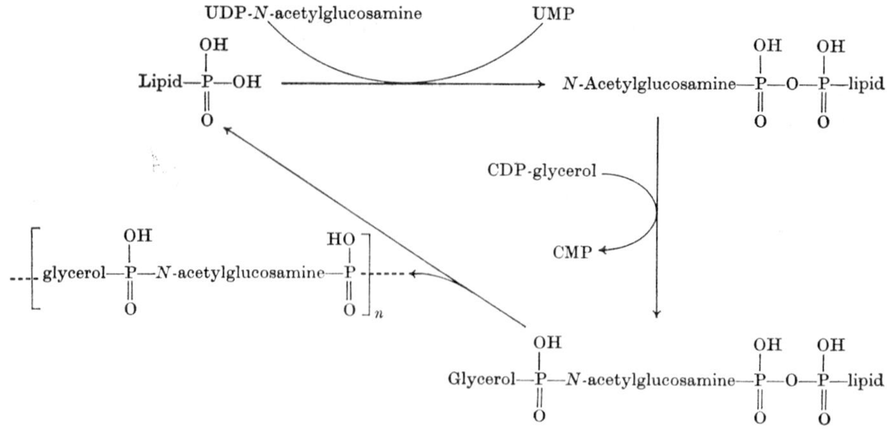

FIG. 17. Pathway for synthesis of the wall teichoic acid of *Staphylococcus lactis* I3.

The biosynthesis of the phosphorylated polysaccharide of *Staph. lactis* 2102 also proceeds by transfer of N-acetylglucosamine 1-phosphate from UDP-*N*-acetylglucosamine to a phospholipid intermediate and thence to the growing polysaccharide chain (Brooks and Baddiley, 1969a). Chain extension is brought about by addition to the non-reducing end of the phosphorylated polysaccharide, and thus is in the same direction as that found with teichoic acids.

C. NATURE OF THE LIPID INTERMEDIATE AND ITS POSSIBLE ROLE IN THE REGULATION OF CELL WALL SYNTHESIS

Undecaprenol phosphate (Fig. 18) has been isolated from several bacteria (Gough *et al.*, 1970; Higashi *et al.*, 1970; Umbreit and Strominger, 1972; Umbreit *et al.*, 1972) and has been shown to participate in the

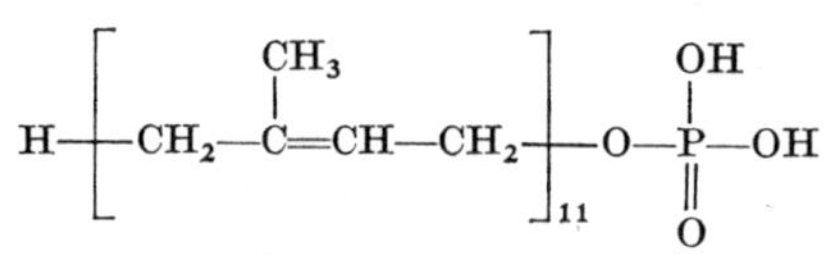

FIG. 18. Structure of undecaprenol phosphate.

biosynthesis of peptidoglycan according to the scheme shown in Fig. 19. An *N*-acetylmuramyl-peptide phosphate residue is transferred from its uridine nucleotide to undecaprenol phosphate giving the corresponding lipid pyrophosphate and uridine monophosphate. After incorporation of

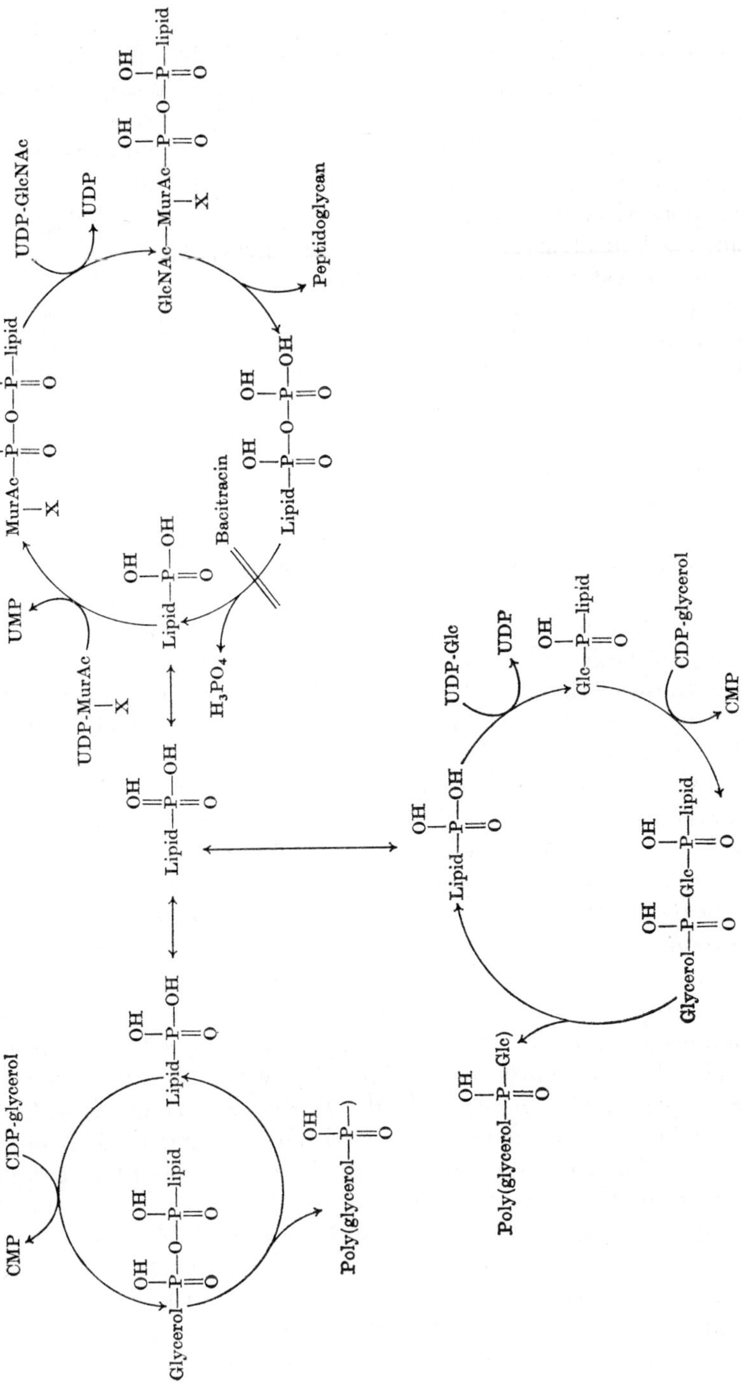

X = peptide

FIG. 19. Pathway for biosynthesis of teichoic acids and peptidoglycan in *Bacillus licheniformis* (Anderson *et al.*, 1972).

N-acetylglucosamine and modification of the peptide, the disaccharide peptide moiety is transferred to an acceptor, which is presumably the growing end of the glycan chain, and undecaprenol pyrophosphate is released. This is then hydrolysed to inorganic phosphate and undecaprenol monophosphate which can then take part in another round of synthesis (Strominger *et al.*, 1972). It is relevant to the following discussion to note that bacitracin inhibits synthesis of peptidoglycan by blocking the hydrolysis of undecaprenol pyrophosphate so that the lipid present in the system accumulates as its pyrophosphate which is inactive in the synthesis (Siewert and Strominger, 1967). Polyprenol phosphate intermediates have also been shown to participate in the biosynthesis of O-antigen (Wright *et al.*, 1967), a membrane-associated mannan (Scher *et al.*, 1968) and capsular polysaccharides (Troy *et al.*, 1971).

The solubility characteristics and observed chemical properties of the lipid intermediates concerned in the synthesis of teichoic acids are consistent with the suggestion that they too are polyprenol phosphates. Thus, the glucosyl phospholipid which participates in biosynthesis of the poly(glucosyl glycerol phosphate) teichoic acid of *B. licheniformis* ATCC 9945 gave glucose 1-phosphate on gentle acid hydrolysis or catalytic hydrogenolysis (Hancock and Baddiley, 1972). Gentle acid hydrolysis of the *N*-acetylglucosaminyl phospholipid which participates in biosynthesis of the phosphorylated polysaccharide of walls of *Staph. lactis* 2102 gave *N*-acetylglucosamine 1-pyrophosphate (Brooks and Baddiley, 1969b). Such lability is unusual and suggests the presence of an unsaturated centre in the β position to the phosphate, as found in the polyisoprenoid lipids. Although the nature of the lipid moiety has not been directly established, evidence has been obtained which indicates that the lipids which participate in biosynthesis of the teichoic acids of *Staph. lactis* I3 (Baddiley *et al.*, 1968; Watkinson *et al.*, 1971) and *B. licheniformis* ATCC 9945 (Anderson *et al.*, 1972) are the same as those which participate in biosynthesis of the peptidoglycan components of these organisms and that, at least in cell-free systems, this lipid is present in a common pool, the size of which is rate-limiting for polymer synthesis. Thus teichoic acid synthesis was inhibited when the nucleotide precursors of peptidoglycan were added to a cell-free system from *Staph. lactis* I3 which was incubated with CDP-glycerol and UDP-*N*-acetylglucosamine. This can be explained on the basis that some of the common lipid intermediate was withdrawn from the teichoic acid cycle into the peptidoglycan cycle. The diminution in the rate of teichoic acid synthesis was very much greater when bacitracin was also added, although bacitracin had almost no inhibitory effect when the nucleotide precursors of peptidoglycan were absent. The failure of bacitracin to inhibit teichoic acid synthesis can be readily understood, since lipid pyrophosphate is not formed in the teichoic acid cycle (Fig. 17, p. 76); its ability to inhibit teichoic acid synthesis when peptidoglycan precursors are present can be explained on the basis of the withdrawal of the common lipid into the peptidoglycan cycle where it accumulates as the inactive pyrophosphate (Fig. 19).

Similar results were obtained with the particulate enzyme from *B. licheniformis* ATCC 9945. The optimum conditions for synthesis of teichoic acid differed slightly from those which were optimal for peptidoglycan synthesis. Synthesis of poly(glycerol phosphate) and poly(glucosyl glycerol phosphate) was markedly inhibited by the presence of the nucleotide precursors for peptidoglycan, and peptidoglycan synthesis was similarly inhibited by the presence of the precursors for teichoic acid. In each case, this inhibition was maximal when the conditions were optimal for the competing synthesis. Again bacitracin had little effect on the rate of teichoic acid synthesis except when the peptidoglycan precursors were present. It was concluded that synthesis of all of the wall polymers in *B. licheniformis* is interrelated through the sharing (Fig. 19) of the same lipid phosphate which, by analogy with the lipid phosphate known to participate in biosynthesis of peptidoglycan in other bacteria, is probably undecaprenol phosphate. Anderson *et al.* (1972) consider that undecaprenol phosphate is unlikely to be involved in the vectorial transportation of monomer units from inside the membrane to polymer chains on the outside. They suggest that the principal function of the lipid is to control synthesis of wall components with respect to their precise structures as regularly repeating units and to control their relative amounts and location so that the wall is produced in an ordered manner. The availability of undecaprenol phosphate may be determined by the balance between phosphorylation and dephosphorylation of the lipid (Strominger *et al.*, 1972; Willoughby *et al.*, 1972) so that the rate of wall synthesis could be controlled through a single common factor, the lipid intermediate, without altering the balance of the syntheses of the different polymers present in the wall (Baddiley, 1972).

In addition to sharing a common intermediate, the syntheses of teichoic acid and peptidoglycan are interrelated in that these two polymers are covalently attached to each other. Studies using cell-free systems have not yet given any information on the process by which the polymers become linked, and it is not known whether one is built up on the other or whether attachment takes place after each has been synthesized. Studies on pulse-labelled cultures of *B. subtilis* NTCC 3610 have, however, shown that teichoic acid is attached to glycopeptide which is synthesized at the same time and not to pre-existing glycopeptide chains (Mauck and Glaser, 1972b). The biosynthetic path to peptidoglycan is such that one would expect the reducing terminal sugar in the glycan chain to be muramic acid. Approximately half of the glycan chains in walls of *Staph. lactis* I3 terminate in reducing muramic acid, while the remaining chains terminate in reducing glucosamine residues. The proportion of the two types of glycan chain present in the teichoic acid-glycan complex is the same as in the glycan which is not associated with teichoic acid. The chains which terminate in reducing glucosamine presumably arise by hydrolysis of the "biosynthesized" chains by the β-*N*-acetylglucosaminidase which is found in isolated wall-membrane preparations. The stage at which the hydrolysis takes place is not known, and it is possible that attachment of teichoic acid takes place subsequently. This could explain the presence of the same proportion of the two reducing terminal types of glycan chain in the glycan and teichoic acid–glycan fractions. It is also possible that the glycan is hydrolysed after incorporation into the wall. If teichoic acid is attached before this hydrolysis, the presence of the same proportion of the two glycan chains is most easily explained by a random location of teichoic acid along the glycan chain. However, it is difficult to envisage a process which attaches teichoic acid randomly along a glycan chain whilst giving wall material in which only one teichoic acid molecule is attached to any glycan.

Ellwood and Tempest (1969, 1972b) have shown that the amount of teichoic acid present in walls of certain bacteria is determined by the conditions under which the cells are grown, and that under certain conditions wall teichoic acid is replaced by teichuronic acid. Such changes in the composition of the wall can occur more rapidly than the rate of net wall synthesis, and it is clear that control mechanisms must exist for such regulation. The nature of these control mechanisms is unknown though Ellwood and Tempest suggest that phosphate or some phosphate-containing metabolite, possibly CDP-glycerol (or CDP-ribitol), is a key regulatory substance and that the regulation is effected by inhibition or repression (or both) of the enzymes concerned in synthesis of the nucleotide precursors of teichuronic acid. The possibility of regulation at this level is supported by the finding that the activity of CDP-glycerol pyrophosphorylase is strongly inhibited by UDP-N-acetylmuramyl peptide (R. G. Anderson, H. Hussey and J. Baddiley, unpublished observations). Also, cytidine diphosphate glycerol has been found to inhibit the activity of UDP-N-acetylglucosamine pyrophosphorylase (L. J. Douglas and J. Baddiley, unpublished observations). Glaser (1965) reported that the level of CDP-glycerol in *B. subtilis* may be controlled by the activity of CDP-glycerol pyrophosphatase, which has a very high K_m value and therefore will function only when CDP-glycerol is synthesized at a rate greater than that at which it is used for the biosynthesis of teichoic acid, and also by the activity of CDP-glycerol pyrophosphorylase which is competitively inhibited by CDP-glycerol which will thus decrease its own rate of synthesis as its concentration in the cell rises.

V. Function

A substantial proportion of the metabolic activity in many Gram-positive bacteria is directed towards synthesis of teichoic acid, and it seems reasonable to assume that these polymers have some role or function which is of value to the cell. This assumption is strengthened by the observation that the choline present in *Diplococcus pneumoniae* is found only in the wall teichoic acid (Tomasz, 1967). The organism has a dietary requirement for choline, and this is presumably indicative of a requirement for teichoic acid.

Information on the possible function(s) of teichoic acid has been gained by examining, *inter alia*, the influence of growth conditions on the amount and nature of the teichoic acid synthesized, the properties of mutants defective in synthesis of teichoic acid, the effect of agents which inhibit synthesis of teichoic acid, the effect of the presence of teichoic acid in cell-free enzyme systems, and the influence of teichoic acid on the observable physical and chemical properties of the surface layers of the cell. Not all of these approaches have yet given information which can be interpreted unambiguously, but several observations now support the idea that one important function of teichoic acid is the binding of cations and the provision of the correct cationic environment at the cell membrane. The lysis and autolysis of certain bacteria is influenced by teichoic acid, and it appears that teichoic acids may participate in the activation and regulation of autolytic enzymes during cell growth. Teichoic acids also form part of the specific receptor sites for various bacteriophages. It is, perhaps, improper to regard this as being a function of teichoic acid, However, since the nature of phage receptor sites must reflect certain aspects of the structural organization of the wall, the role of teichoic acid in adsorption of phage will be discussed briefly. The antigenic properties of teichoic acids, although of great significance in disease and in serological typing, are of little relevance to a consideration of their function and will not be discussed.

A. Role of Teichoic Acids in Cation Binding

The early suggestion that teichoic acid might participate in ion-exchange reactions and influence the passage of ionic materials through the cell surface was based largely on a consideration of its ionic nature, though it was noted that Gram-positive bacteria such as bacilli and staphylococci which tolerate relatively high concentrations of salt are characteristically rich in wall teichoic acid (Archibald *et al.*, 1961a). Magnesium ions are generally found loosely associated with the surface structures of bacteria (Strange and Shon, 1964) and the cation-binding properties of walls of certain staphylococci (Bradley and Parker, 1968; Cutinelli and Galdiero, 1967; Galdiero *et al.*, 1968) have been shown to be due almost entirely to the wall teichoic acid (Heptinstall *et al.*, 1970). Magnesium ions are known to be required for the stability of isolated cell membranes and for the activity of many membrane-bound enzymes, and it was therefore suggested that a major function of both wall- and membrane-teichoic acid might be to maintain a high concentration of magnesium ions in the region of the membrane (Heptinstall *et al.*, 1970). This suggestion has been supported by several other studies.

When grown in media of moderate ionic strength under conditions of phosphate limitation, several Gram-positive bacteria synthesize teichuronic acid instead of wall teichoic acid (Ellwood and Tempest, 1972b). Like teichoic acid, teichuronic acids are negatively charged polymers, though their charge derives from the carboxyl functions of their uronic acid components and not from phosphate ester groupings. Apparently therefore these bacteria can dispense with wall teichoic acids under such growth conditions, though they require an alternative negatively charged polymer. Cells and walls of *B. subtilis* var. *niger* grown under conditions of phosphate limitation bind Mg^{2+} much less avidly than cells and walls of the same organism grown under conditions where wall teichoic acid is synthesized (Ellwood, 1971; Meers and Tempest, 1970). When uptake of Mg^{2+} by the phosphate-limited bacteria is constrained by addition of high concentrations of Na^{+} to the medium, the bacteria respond by producing wall teichoic acid. Similarly, wall teichoic acid is synthesized when the bacteria are grown under phosphate-limiting conditions but at low pH values.

These observations strongly suggest that wall teichoic acids are functionally involved in assimilation of cations, and can be replaced by teichuronic acids only when magnesium ions are readily available (Ellwood and Tempest, 1972b). Significantly, it was found that membrane teichoic acid was still present even when bacteria were grown under conditions which led to the incorporation of teichuronic acid into the wall (Ellwood and Tempest, 1968) so that, although wall teichoic acids can be functionally replaced by other anionic polymers, the presence of membrane teichoic acid is apparently essential for proper functioning of the cell. This could explain the observation that membrane teichoic acid is present in all Gram-positive bacteria which have been examined whereas not all of these bacteria have wall teichoic acids.

It may also explain the finding that all membrane teichoic acids are of the same structural type. Possibly the poly(glycerol phosphate) teichoic acids are most suitable structurally for association with the membrane or for the provision of the correct cationic environment.

The ability of teichoic acid to maintain the optimum concentration of Mg^{2+} in the region of the membrane has been demonstrated experimentally in a cell-free system from *B. licheniformis* ATCC 9945 (Hughes *et al.*, 1973). Synthesis of poly(glycerol phosphate) and poly(glucosylglycerol phosphate) by particulate enzyme preparations from this organism exhibit a Mg^{2+} dependence which is influenced by the presence of teichoic acid. Membrane fragments isolated after disruption of cells in the absence of Mg^{2+} are devoid of membrane teichoic acid. Synthesis of wall teichoic acids by such preparations exhibits a pronounced optimum at about 12 m*M* Mg^{2+}, and synthesis falls sharply at concentrations above and below this value. Enzyme preparations which retained the membrane teichoic acid were prepared by disruption of the cells in the presence of Mg^{2+}. Synthesis of wall teichoic acids by such preparations was also stimulated by addition of low concentrations of Mg^{2+}, but the response to increasing concentrations of Mg^{2+} was considerably damped so that near-maximum activity was exhibited over a wide range of Mg^{2+} concentrations.

Even greater independence of exogenous Mg^{2+} was shown by wall-membrane preparations obtained by disruption of the bacteria in a French pressure cell. These contained both wall- and membrane-teichoic acids and, provided they were first equilibrated with Mg^{2+}, their synthetic activity was essentially independent of external Mg^{2+} up to 40 m*M*. On the basis of these and other observations, it was concluded that the Mg^{2+}-dependent components of the particulate enzyme systems interacted preferentially with the Mg^{2+} ions provided by the membrane teichoic acid rather than with those in aqueous solution. When only relatively small amounts of teichoic acid are present, as in the membrane fragments which contained lipoteichoic acid, cations must be supplied by the suspending solution to maintain the optimum concentration of bound Mg^{2+}, so that the system is dependent on added Mg^{2+} but shows little response to higher concentrations of the cation within certain limits. When wall teichoic acid is also present, as in the wall-membrane preparation, it provides a large reservoir of bound cations which can maintain the required concentration of Mg^{2+} in the membrane teichoic acid, the component which interacts directly with the membrane. The wall-membrane system thus acts as a divalent-cation buffer which maintains the optimum concentration of Mg^{2+} at the membrane so that it is almost completely independent of the concentration of Mg^{2+} in solution.

The high affinity of divalent cations for wall teichoic acid is consistent with its function in maintaining a large reservoir of bound Mg^{2+} so as to ensure the optimum cationic environment for the membrane. The amount of Mg^{2+} bound by walls of staphylococci is markedly decreased by the presence of ester alanine (Heptinstall *et al.*, 1970; Archibald *et al.*, 1973b). Interestingly, the ester-alanine content of walls of staphylococci grown in the presence of high concentrations of NaCl was very much lower and the walls had a correspondingly increased capacity for binding Mg^{2+}. It was therefore suggested that the alanyl-ester residues of the teichoic acid might function in regulating the surface charge and cation-binding properties of the wall. Some functional requirement for alanyl-ester residues is suggested by their presence in walls of *B. subtilis* var. *niger* grown at low pH values under nitrogen limitation when presumably only those nitrogen-containing components would be synthesized which are essential for growth (Ellwood and Tempest, 1972b). The alanyl-ester content of walls of both *B. subtilis* var. *niger* and *Staph. aureus* H is increased when the bacteria are grown at low pH values (Ellwood and Tempest, 1972a; Archibald *et al.*, 1973b) but the differences found in the staphylococcal walls could be accounted for by base-catalysed hydrolysis of the labile ester linkages. Since the staphylococci were grown at a low dilution rate ($0{\cdot}12\ h^{-1}$), substantial hydrolysis would have occurred even at pH 6 where the half-life of the ester linkages is several hours. Presumably, smaller differences would be found in faster growing cells, so that the physiological significance of the lability of the ester linkages is uncertain.

As well as lowering their cation-binding capacity, alanyl-ester residues cause a difference in the nature of the binding of Mg^{2+} to walls (Baddiley *et al.*, 1973). This has been shown by measurements of the binding energies of the 2S electrons in Mg^{2+} bound to cell walls containing alanine and to walls from which alanyl esters had been removed. The binding energies were determined by X-ray photo-electron spectroscopy, and the values obtained show that the alanine influences the ionic interactions of some of the Mg^{2+} ions so that they are less strongly bound than those in samples which do not contain alanine. The 2S electron-binding energy in the weakly bound Mg^{2+} corresponds to that in magnesium chloride, whereas the binding energy of the strongly bound Mg^{2+} correlates with that of the Mg^{2+} in magnesium hydrogen phosphate in which the Mg^{2+} interacts with two $P\text{-}O^-$ groups. It was therefore suggested that the strongly bound magnesium ions interact with wall teichoic acid as shown in Fig. 20a but that in the presence of alanine the NH_3^+ group of the amino acid neutralizes the charge on an adjacent phosphate thus permitting only one of the pair of phosphates to interact with Mg^{2+} as shown in Fig. 20b. Since a given alanyl residue may be capable of neutralizing any one of a number of phosphate groups, it was suggested that alternation of the amino group between a number of phosphates would produce changes in the position of the polymer chain at which Mg^{2+} could bind strongly to two phosphates, and this could produce a net movement of strongly bound Mg^{2+} on the polymer chain.

The binding of Mg^{2+} to phosphate occurs in two steps (Hammes and Levison, 1964); the first is the formation of an ion pair, and the second,

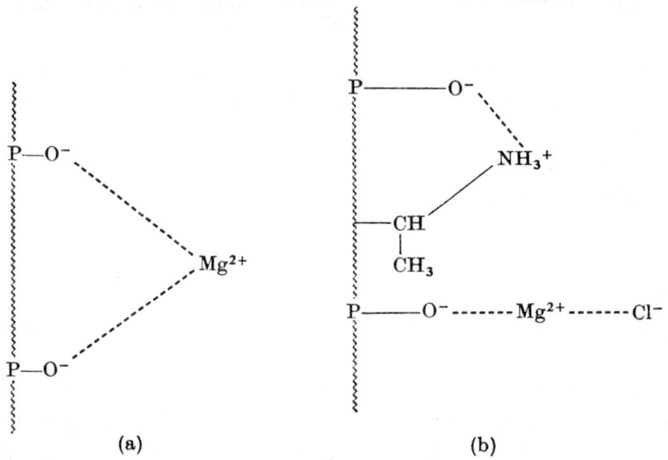

FIG. 20. Proposed explanation for the influence of alanyl esters on the binding of magnesium ions to teichoic acid (Baddiley *et al.*, 1973).

which is rate-limiting, is the loss of one or two molecules of water from the first hydration shell of the cation to produce the complex. Since the dehydration process is rate-limiting, it is possible that transfer of bivalent cations from ligand to ligand is energetically more favourable than the binding of a fully hydrated ion. The wall-membrane system may thus form an integrated array of charged groups which facilitates passage of cations through the wall and then directly on to the membrane teichoic acid and hence to the membrane.

It is interesting that the bacteriostatic action of novobiocin, which selectively inhibits the biosynthesis *in vitro* of several teichoic acids (Hughes *et al.*, 1971), appears to be due to the inhibition of several magnesium-requiring reactions which are associated with the cell membrane (Brock, 1962).

B. Influence of Teichoic Acids on Autolytic Enzymes

Autolytic enzymes are believed to play an important role during cell growth and division, and to be involved in turnover of wall material, in cell separation and in the development of competence (Forsberg and Rogers, 1971; Higgins and Shockman, 1971; Rogers, 1970). Teichoic acid binds strongly to certain autolytic enzymes, and it has been suggested that teichoic acid may, in certain cases, be involved in the localization of autolytic enzymes in the wall and in the modulation of their activity (Boylan *et al.*, 1972). Higgins *et al.* (1970) suggested that concentrations of teichoic acid in the outer surface of the nascent wall might be important in cross-wall separation both by causing electrostatic repulsion and by providing a hydrophilic environment for the lytic enzymes.

Mutants of *B. subtilis* and *Staph. aureus* have been described which lack wall teichoic acid (Boylan *et al.*, 1972; Chatterjee *et al.*, 1969, Cole *et al.*, 1970; Gilpin *et al.*, 1972). Cell-free preparations of both of these mutants contain normal levels of all of the enzymes which are known to be involved in synthesis of wall teichoic acid (Boylan *et al.*, 1972; Shaw *et al.*, 1970). The absence of wall teichoic acids in the mutants might be due to an altered organization of the biosynthetic enzymes in the cytoplasmic membrane or to the absence of a necessary enzyme, acceptor, or intermediate involved in incorporation of teichoic acid into the wall.

Both mutants show profound alterations in wall morphology and structure, in growth characteristics, in division and cell separation, and in their autolytic potentialities. The pleiotropic character of the mutants could be due primarily to damage or disorganization of the cytoplasmic membrane. It is also possible that the abnormal characteristics of the mutants are due primarily to the absence of wall teichoic acid. A consequence of this could be disorganization of the membrane because of an insufficiency of magnesium ions, and this could in turn be responsible for certain of the other abnormalities noted. Alternatively these abnormalities could be caused by the lack of a control of autolytic activity normally effected by teichoic acid. A striking example of the influence of teichoic acid on autolytic activity has come from studies on pneumococci. The dietary requirement of these organisms for choline can be satisfied by ethanolamine which then replaces choline in the wall teichoic acid (Tomasz, 1968). In media containing ethanolamine, the bacteria grow as long chains of unseparated cells; they lose their ability to be transformed, they are resistant to deoxycholate-induced lysis, and do not lyse in the presence of penicillin, D-cycloserine or phosphonomycin (Tomasz *et al.*, 1970). The walls which contain ethanolamine are not hydrolysed by autolytic enzymes isolated from pneumococci grown in the presence of either ethanolamine or in choline (Mosser and Tomasz, 1970) and this resistance to lysis can explain all of the physiological abnormalities of the cells grown in the presence of ethanolamine. These observations support the view that autolytic enzymes are involved in cell separation, in the development of competence, and in the lysis observed when antibiotics which inhibit peptidoglycan synthesis are added to growing cells. They also show clearly that the nature of the teichoic acid can profoundly affect the ability of autolytic enzymes to hydrolyse cell walls.

Teichoic acid is also involved in activation of the autolytic enzyme in pneumococci. Thus, although autolysin from cells grown in medium containing ethanolamine can lyse walls which contain choline–teichoic acid, it must first undergo a non-enzymic reaction with these walls before it becomes catalytically active (Tomasz and Westphal, 1971). Walls from which choline-teichoic acid has been removed, like those which contain ethanolamine-teichoic acid, are incapable of activating the autolysin. This activation is accompanied by conversion of the low molecular-weight (2–5 . 10^4 daltons) inactive autolysin to active material having the same molecular weight (1–3 . 10^6 daltons) as autolysin isolated from cells grown in the presence of choline. It is thought to be likely that, even in bacteria grown on choline, activation of autolysin precursors takes place only when they interact with teichoic acid at the cell surface.

Thus it seems that, in at least some Gram-positive bacteria, teichoic acids are functionally involved in the activation of autolytic enzymes, and in determining the susceptibility of the wall to autolytic activity; they may also be concerned in the localization of these enzymes in the appropriate regions of the wall.

Teichoic acids have also been shown to be responsible for the resistance of certain bacteria to a lysin produced by *Streptococcus zymogenes* (Davie and Brock, 1966; Basinger and Jackson, 1968). This is a protein, containing two distinct sub-units (Granato and Jackson, 1971), which is both haemolytic and bacteriolytic. Growing cultures of sensitive bacteria are lysed, but it is thought that the lysin acts on the cell membrane so that bacteriolysis may be effected by the cell's own autolytic enzymes. Streptococci which produce the lysin are protected from its action by an inhibitor which has been extracted from disrupted cells with phenol and shown to be an alanylated teichoic acid (Davie and Brock, 1966). The corresponding teichoic acid from sensitive streptococci lacked the alanyl-ester residues, but these were present in the teichoic acid isolated from resistant mutants of the sensitive strains. Chemical removal of the alanyl esters from teichoic acid destroyed its capacity to inhibit the lysin. The presence of alanine is therefore essential for inhibition, and the lysin is inhibited by the alanylated forms of both ribitol- and glycerol-teichoic acids. The lysin is thought to damage the membrane through an enzymic attack but the nature of the reaction(s) catalysed is unknown. Bassinger and Jackson (1968) have suggested that the lysin may normally play some role in biosynthesis of the membrane and that its action may be regulated by alanylated teichoic acid.

C. Role of Teichoic Acids in Adsorption of Bacteriophages

The first step in the infection of a bacterium by a bacteriophage is the adsorption of the phage on to the cell surface; this process requires mutual recognition between surface components of the potential host and components of the attacking bacteriophage. Wall teichoic acids and polysaccharides have been found to participate in the binding of phages by several species of Gram-positive bacteria. The involvement of teichoic acid has been most extensively studied in staphylococci and bacilli, and the present account is restricted to these organisms.

Morse (1962) showed that isolated cell walls of *Staph. aureus* NYH-6 were capable of adsorbing many different types of staphylococcal phage, but that this ability was lost on extraction of the teichoic acid. The isolated teichoic acid was also incapable of inactivating the phages, and Morse suggested that the receptor site in the intact wall is a teichoic acid–peptidoglycan complex whose stereochemical arrangement, imposed by the macromolecular structure of the wall, is involved in its specificity. This suggestion has been supported by subsequent studies.

Involvement of teichoic acid in the specific adsorption of staphylococcal phages was indicated by the observations that certain phage-resistant mutants of *Staph. aureus* had altered wall teichoic acids (Wolin *et al.*, 1966) and that enzymic removal of *N*-acetylglucosaminyl residues from the teichoic acid in isolated walls of *Staph. aureus* destroyed their ability to bind phage (Coyette and Ghuysen, 1968). Subsequently several phage-resistant mutants of *Staph. aureus* H were isolated and characterized by Chatterjee (1969). Those which did not bind phage were found to lack wall teichoic acid or to lack the N-acetylglucosaminyl substituents on the teichoic acid. Walls isolated from strains of *Staph. aureus* having exclusively α- or exclusively β-linked N-acetylglucosamine in the teichoic acid bound phages with equal efficiency, so that although the sugar is required for phage binding its anomeric configuration is unimportant. This is surprising since one might expect much more rigorous stereospecific requirements for the attachment of phage particles to the cell surface.

Teichoic acid is also a component of the receptor sites in walls of *B. subtilis* W23 for the phage-like killer particle μ (Glaser *et al.*, 1966) and in walls of *B. subtilis* 168 for representatives of five distinct groups of phage (Young, 1967). Resistant mutants of both strains lacked glucose in their wall teichoic acids. The presence of glucosylated teichoic acid is also required for adsorption of phage SP50 to walls of *B. subtilis* strains 3610 and W23 (Archibald and Coapes, 1971b, 1972; and unpublished observations). The nature of the anomeric linkage, and indeed of the alditol, seems to be unimportant since SP50 binds equally well to walls of each strain although the wall teichoic acid in strain 3610 is an α-glucosyl 1,3-poly(glycerol phosphate) whereas that in strain W23 is a β-glucosyl 1,5-poly(ribitol phosphate). This again appears to indicate a rather low degree of stereospecificity, even though it is possible that the teichoic acid is not the primary binding site for the phage but that its absence leads to the alteration or distortion of the structure which is recognized by the phage.

The importance of the peptidoglycan component of the walls in the phage-adsorption process was revealed by Young (1967) who showed that, on autolysis or digestion with lysozyme, walls of *B. subtilis* 168 lost their ability to inactivate phage. This finding suggested that the arrangement of the teichoic acid molecules in the wall is critical. The importance of this arrangement or of the size of the teichoic acid–peptidoglycan complex required may vary with different phages since μ is inactivated by the teichoic acid–glycopeptide complex obtained by digestion of walls of strain W23 with lysozyme (though it is not inactivated by teichoic acid isolated by extraction into trichloroacetic acid). Similarly Chatterjee (1969) found that the ability of wall of *Staph. aureus* strain H to inactivate phage 52A was lost when the walls were autolysed or digested with muramidase. Autolysed walls of *Staph. aureus* strain Copenhagen did not inactivate phage, but a teichoic acid–glycan complex obtained by digestion of the walls with a muramyl-L-alanine amidase did inactivate phages 3C, 71 and 77 though this inactivation was reversible and occurred with an efficiency of only 10^{-4} to 10^{-5} times than of an equivalent weight of intact walls (Coyette and Ghuysen, 1968). In a separate study of walls of strain Copenhagen, Murayama *et al.* (1968) found that teichoic acid–glycopeptide complexes, isolated by digestion of the walls with muramidase or with a glycine bridge-splitting endopeptidase, were capable of irreversibly inactivating phage 3C at about 1% of the efficiency of the intact walls but that both complexes lost their ability to inactivate the phage when they were digested with a muramyl-L-alanine amidase. Inactivation of phage 3C thus requires the presence of the three components—teichoic acid, glycan and peptide—in a single macromolecule. The low efficiency of inactivation may be a consequence of a deviation from the conformation found in the wall. Interestingly, none of the above complexes inactivated phages 3B, 42D or 53 which also bind to walls of strain Copenhagen. Possibly these phages have more definite stereochemical requirements which can be satisfied only when the teichoic acid is held in the intact walls. It is possible, also, that the peptidoglycan component of the wall is directly involved in phage binding. Thus, removal of the O-acetyl groups of the muramic acid residues of the peptidoglycan in walls of *Staph. aureus* H destroys the ability of the walls to bind phage 52A (Shaw and Chatterjee, 1971).

These studies show that teichoic acid and peptidoglycan are both important in the adsorption of phages, and that the receptor sites are organized structures whose specific stereochemical arrangement is dependent on the integrity of the wall. The nature of this stereochemical arrangement is, however, quite unknown.

References

In order to limit the number of references I have in certain cases referred to review articles or to only the most recent of a series of papers on a particular topic. I apologize to those authors whose work has not been referred to directly.

Anderson, J. C., Archibald, A. R., Baddiley, J., Davey, N. B. and Curtis, M. J. (1969). *Biochemical Journal* **113**, 183.

Anderson, R. G., Hussey, H. and Baddiley, J. (1972). *Biochemical Journal* **127**, 11.

Archibald, A. R. (1971). *Methods in Carbohydrate Chemistry* **VI**, 122.

Archibald, A. R. (1972). *In* "The Staphylococci", p. 75 (J. O. Cohen, ed.), John Wiley and Sons Inc.

Archibald, A. R. and Baddiley, J. (1966). *Advances in Carbohydrate Chemistry* **21**, 322.

Archibald, A. R. and Coapes, H. E. (1971a). *Biochemical Journal* **124**, 449.

Archibald, A. R. and Coapes, H. E. (1971b). *Biochemical Journal* **125**, 667.

Archibald, A. R. and Coapes, H. E. (1972). *Journal of General Microbiology*, **73**, 581.

Archibald, A. R. and Heptinstall, S. (1971). *Biochemical Journal* **125**, 361.

Archibald, A. R. and Stafford, G. H. (1972). *Biochemical Journal* **130**, 681.

Archibald, A. R., Armstrong, J. J., Baddiley, J. and Hay, J. B. (1961a). *Nature, London* **191**, 570.

Archibald, A. R., Baddiley, J. and Buchanan, J. G. (1961b). *Biochemical Journal* **81**, 124.

Archibald, A. R., Baddiley, J. and Blumsom, N. L. (1968a). *Advances in Enzymology* **30**, 223.

Archibald, A. R., Baddiley, J. and Button, D. (1968b). *Biochemical Journal* **110**, 543.

Archibald, A. R., Baddiley, J. and Shaukat, G. A. (1968c). *Biochemical Journal* **110**, 583.

Archibald, A. R., Baddiley, J. and Heptinstall, S. (1969a). *Biochemical Journal* **111**, 245.

Archibald, A. R., Coapes, H. E. and Stafford, G. H. (1969b). *Biochemical Journal* **113**, 899.

Archibald, A. R., Baddiley, J., Heckels, J. E. and Heptinstall, S. (1971). *Biochemical Journal* **125**, 353.

Archibald, A. R., Baddiley, J. and Heckels, J. E. (1973a). *Nature New Biology* **241**, 29.

Archibald, A. R., Baddiley, J. and Heptinstall, S. (1973b). *Biochimica et Biophysica Acta* **291**, 629.

Armstrong, J. J., Baddiley, J., Buchanan, J. G., Carss, B. and Greenberg, G. R. (1958). *Journal of the Chemical Society* 4344.

Armstrong, J. J., Baddiley, J. and Buchanan, J. G. (1960), *Biochemical Journal* **76**, 610.

Armstrong, J. J., Baddiley, J. and Buchanan, J. G. (1961). *Biochemical Journal* **80**, 254.

Baddiley, J. (1972). *Essays in Biochemistry* **8**, 35.

Baddiley, J. and Neuhaus, F. C. (1960). *Biochemical Journal* **75**, 579.

Baddiley, J., Buchanan, J. G., Hardy, F. E., Martin, R. O., RajBhandary, U. L. and Sanderson, A. R. (1961). *Biochimica et Biophysica Acta* **52**, 406.

Baddiley, J., Buchanan, J. G., Martin, R. O. and RajBhandary, U. L. (1962a). *Biochemical Journal* **85**, 49.

Baddiley, J., Buchanan, J. G., RajBhandary, U. L. and Sanderson, A. R. (1962b). *Biochemical Journal* **82**, 493.

Baddiley, J., Blumsom, N. L. and Douglas, L. J. (1968). *Biochemical Journal* **110**, 565.

Baddiley, J., Hancock, I. and Sherwood, P. M. A. (1973). *Nature. London* **243**, 43.

Balyuzi, H. H. M., Reavely, D. A. and Burge, R. E. (1972). *Nature New Biology* **235**, 252.

Basinger, S. F. and Jackson, R. W. (1968). *Journal of Bacteriology* **96**, 1895.

Bews, B. (1967). Ph.D. Thesis: University of Newcastle upon Tyne.

Bleiweis, A. S., Young, F. E. and Krause, R. M. (1967). *Journal of Bacteriology* **94** 1381.

Boylan, R. J., Mendelson, N. H., Brooks, D. and Young, F. E. (1972). *Journal of Bacteriology* **110**, 281.

Bradley, T. J. and Parker, M. S. (1968). *Experientia* **24**, 1175.

Brock, T. D. (1962). *Science, New York* **136**, 316.

Brooks, D. and Baddiley, J. (1969a). *Biochemical Journal* **113**, 635.

Brooks, D. and Baddiley, J. (1969b). *Biochemical Journal* **115**, 307.

Brooks, D., Mays, L. L., Hatefi, Y. and Young, F. E. (1971). *Journal of Bacteriology* **107**, 223.

Brundish, D. E. and Baddiley, J. (1968). *Biochemical Journal* **110**, 573.

Burger, M. M., (1966). *Proceedings of the National Academy of Sciences of the United States of America* **56**, 910.

Burger, M. M. and Glaser, L. (1964). *Journal of Biological Chemistry* **239**, 3168.

Burger, M. M. and Glaser, L. (1966). *Journal of Biological Chemistry* **241**, 494.

Button, D., Archibald, A. R. and Baddiley, J. (1966). *Biochemical Journal* **99**, 11C.

Chatterjee, A. N. (1969). *Journal of Bacteriology* **98**, 519.

Chatterjee, A. N., Mirelman, D., Singer, H. J. and Park, J. T. (1969). *Journal of Bacteriology* **100**, 846.

Chin, T., Burger, M. M. and Glaser, L. (1966). *Archives of Biochemistry and Biophysics* **116**, 358.

Chin, T., Younger, J. and Glaser, L. (1968). *Journal of Bacteriology* **95**, 2044.

Chittenden, G. F. K., Roberts, W. K., Buchanan, J. G. and Baddiley, J. (1968). *Biochemical Journal* **109**, 597.

Cole, R. M., Popkin, T. J., Boylan, R. J. and Mendelson, N. H. (1970). *Journal of Bacteriology* **103**, 793.

Coley, J., Duckworth, M. and Baddiley, J. (1972). *Journal of General Microbiology* **73**, 587.

Coyette, J. and Ghuysen, J. M. (1968). *Biochemistry, New York* **7**, 2385.

Critchley, P., Archibald, A. R. and Baddiley, J. (1962). *Biochemical Journal* **85**, 420.

Cutinelli, C. and Galdiero, F. (1967). *Journal of Bacteriology* **93**, 2022.

Davie, J. M. and Brock, T. D. (1966). *Journal of Bacteriology* **92**, 1623.

Davison, A. L. and Baddiley, J. (1963). *Journal of General Microbiology* **32**, 271.

Davison, A. L., Baddiley, J., Hofstad, T., Losnegard, N. and Oeding, P. (1964). *Nature, London* **202**, 872.

Dixon, J. T., Roberts, W. K., Mills, G. T., Buchanan, J. G. and Baddiley, J. (1968). *Carbohydrate Research* **8**, 262.

Douglas, L. J. and Wolin, M. J. (1971). *Biochemistry, New York* **10**, 1551.

Duckworth, M., Archibald, A. R. and Baddiley, J. (1972). *Biochemical Journal* **130**, 691.

Ellwood, D. C. (1971). *Biochemical Journal* **121**, 349.

Ellwood, D. C. and Tempest, D. W. (1969). *Biochemical Journal* **108**, 40P.

Ellwood, D. C. and Tempest, D. W. (1969). *Biochemical Journal* **111**, 1.

Ellwood, D. C. and Tempest, D. W. (1972a). *Journal of General Microbiology* **73**, 395.

Ellwood, D. C. and Tempest, D. W. (1972b). *Advances in Microbial Physiology*, **7** 83.

Ellwood, D. C., Kelemen, M. V. and Baddiley, J. (1963). *Biochemical Journal* **86**, 213.

Forsberg, C. and Rogers, H. J. (1971). *Nature, London* **229**, 272.

Galdiero, F., Lembo, M. and Tufano, M. A. (1968). *Experientia* **24**, 34.

Ghuysen, J. M. and Strominger, J. L. (1963). *Biochemistry, New York* **2**, 1110.

Ghuysen, J. M., Leyh-Bouille, M. and Dierickx, L. (1962). *Biochimica et Biophysica Acta* **63**, 297

Ghuysen, J. M., Tipper, D. J. and Strominger, J. L. (1965). *Biochemistry, New York* **4**, 474.

Gilpin, R. W., Chatterjee, A. N. and Young, F. E. (1972). *Journal of Bacteriology* **111**, 272.

Glaser, L. (1963). *Biochimica et Biophysica Acta* **67**, 525.

Glaser, L. (1964). *Journal of Biological Chemistry* **239**, 3168.

Glaser, L. (1965). *Biochimica et Biophysica Acta* **101**, 6.

Glaser, L. and Burger, M. M. (1964). *Journal of Biological Chemistry* **239**, 3187.

Glaser, L., Ionesco, H. and Schaeffer, P. (1966). *Biochimica et Biophysica Acta* **124**, 415.

Gough, D. P., Kirby, A. L., Richards, J. B. and Hemming, F. W. (1970). *Biochemical Journal* **118**, 167.

Granato, P. A. and Jackson, R. W. (1971). *Journal of Bacteriology* **108**, 804.

Grant, W. D. and Wicken, A. J. (1968). *Biochemical and Biophysical Research Communications* **32**, 122.

Hammes, G. G. and Levison, S. A. (1964). *Biochemistry, New York* **3**, 1504.

Hancock, I. C. and Baddiley, J. (1972). *Biochemical Journal* **127**, 27.

Hay, J. B., Archibald, A. R. and Baddiley, J. (1965). *Biochemical Journal* **97**, 723.

Hay, J. B., Wicken A. J. and Baddiley, J. (1963). *Biochimica et Biophysica Acta* **71**, 188.

Heijenoort, J. V., Menjon, D., Flouret, B., Szulmajester, J., Laporte, J. and Batelier, G. (1971). *European Journal of Biochemistry* **20**, 442.

Heptinstall, S., Archibald, A. R. and Baddiley, J. (1970). *Nature, London* **225**, 519.

Heyman, H., Maniello, J. M. and Barkulis, S. S. (1967). *Biochemical and Biophysical Research Communications* **26**, 486.

Higashi, Y., Strominger, J. L. and Sweeley, C. C. (1970). *Journal of Biological Chemistry* **245**, 3697.

Higgins, M. L., Pooley, H. M. and Shockman, G. D. (1970). *Journal of Bacteriology* **103**, 504.

Higgins, M. L. and Shockman, G. D. (1971). *Critical Reviews in Microbiology* **1**, 29.

Hughes, A. H., Stow, M., Hancock, I. C. and Baddiley, J. (1971). *Nature New Biology* **229**, 53.

Hughes, A. H., Hancock, I. C. and Baddiley, J. (1973). *Biochemical Journal* **132**, 83.
Hughes, R. C. (1964). *Biochemical Journal* **96**, 700.
Hughes, R. C. (1970). *Biochemical Journal* **117**, 431.
Hughes, R. C. and Tanner, P. J. (1968). *Biochemical and Biophysical Research Communications* **33**, 22.
Hughes, R. C., Pavlik, J. G., Rogers, H. J. and Tanner, P. J. (1968). *Nature, London* **219**, 642.
Hussey, H. and Baddiley, J. (1972). *Biochemical Journal* **127**, 39.
Hussey, H., Brooks, D. and Baddiley, J. (1969). *Nature, London* **221**, 665.
Ishimoto, N. and Strominger, J. L. (1966). *Journal of Biological Chemistry* **241**, 639.
James, A. M. and Brewer, J. E. (1968). *Biochemical Journal* **107**, 817.
Karakawa, W. W. and Kane, J. A. (1971). *Journal of Immunology* **106**, 900.
Kelemen, M. V. and Baddiley, J. (1961). *Biochemical Journal* **80**, 246.
Kelemen, M. V. and Rogers, H. J. (1971). *Proceedings of the National Academy of Sciences of the United States of America* **68**, 992.
Kennedy, D. A., Buchanan, J. G. and Baddiley, J. (1969). *Biochemical Journal* **115**, 37.
Kennedy, L. D. and Shaw, D. R. D. (1968). *Biochemical and Biophysical Research Communications* **32**, 861.
Knox, K. W. and Hall, E. A. (1965). *Biochemical Journal* **96**, 302.
Knox, K. W. and Holmwood, K. J. (1968). *Biochemical Journal* **108**, 363.
Knox, K. W. and Wicken, A. J. (1970). *Journal of General Microbiology* **63**, 237.
Knox, K. W. and Wicken, A. J. (1972). *Infection and Immunity* **6**, 43.
Liu, T. Y. and Gottschlich, E. C. (1967). *Journal of Biological Chemistry* **242**, 471.
McCarty, M. (1959). *Journal of Experimental Medicine* **109**, 361.
Marquis, R. E. (1968). *Journal of Bacteriology* **95**, 775.
Mauck, J. and Glaser, L. (1972). *Proceedings of the National Academy of Sciences of the United States of America* **69**, 2386.
Mauck, J. and Glaser, L. (1972). *Journal of Biological Chemistry* **247**, 1180.
Meers, J. L. and Tempest, D. W. (1970). *Journal of General Microbiology* **63**, 325.
Miller, F. (1969). Ph.D. Thesis: University of Newcastle upon Tyne.
Mirelman, D., Beck, B. D. and Shaw, D. R. D. (1970). *Biochemical and Biophysical Research Communications* **39**, 712.
Montague, M. D. (1964). *Biochimica et Biophysica Acta* **86**, 588.
Morse, S. I. (1962). *Journal of Experimental Medicine* **116**, 247.
Mosser, J. L. and Tomasz, A. (1970). *Journal of Biological Chemistry* **245**, 287.
Murayama, Y., Kotani, S. and Kato, K. (1968). *Biken's Journal* **11**, 269.
Nathenson, S. G. and Strominger, J. L. (1963). *Journal of Biological Chemistry* **238**, 3161.
Nathenson, S. G., Ishimoto, N., Anderson, J. S. and Strominger, J. L. (1966). *Journal of Biological Chemistry* **241**, 651.
Naumova, I. B. and Belozersky, A. N. (1966). *Biochimia* **36**, 1276.
Naumova, I. B. and Zaretskaya, M. Z. (1964a). *Doklady Academy Nauk U.S.S.R.* **156**, 1464.
Naumova, I. B. and Zaretskaya, M. Z. (1964b). *Doklady Academy Nauk U.S.S.R.* **157**, 207.
Naumova, I. B., Belozersky, A. N. and Shafikova, F. A. (1962). *Doklady Academy Nauk U.S.S.R.* **143**, 730.
Naumova, I. B., Shabarova, Z. A. and Belozersky, A. N. (1963). *Doklady Academy Nauk U.S.S.R.* **152**, 1471.

Naumova, I. B., Rogozena, C. V. and Zaretskaya, M. Z. (1969). *Doklady Academy Nauk U.S.S.R.* **188**, 710.

Nermut, M. V. (1967). *Journal of General Microbiology* **44**, 503.

Oeding, P. and Grov, A. (1972). *In* "The Staphylococci", (J. O. Cohen, ed.), p. 333. John Wiley and Sons Inc.

Oeding, P., Mykelstad, B. and Davison, A. L. (1967). *Acta Pathologica et Microbiologica Scandinavia* **69**, 458.

Ou, L. T. and Marquis, R. E. (1970). *Journal of Bacteriology* **101**, 92.

Partridge, M. D., Davison, A. L. and Baddiley, J. (1971). *Biochemical Journal* **121**, 695.

Partridge, M. D., Davison, A. L. and Baddiley, J. (1973). *Journal of General Microbiology* **74**, 169.

Pazur, J. H., Cepure, A., Kane, J. A. and Karakawa, W. W. (1971). *Biochemical and Biophysical Research Communications* **43**, 1421.

RajBhandary, U. L. and Baddiley, J. (1963). *Biochemical Journal* **87**, 429.

Rao, E. V., Buchanan, J. G. and Baddiley, J. (1966). *Biochemical Journal* **100**, 811.

Rao, E. V., Watson, M. J., Buchanan, J. G. and Baddiley, J. (1969). *Biochemical Journal* **111**, 547.

Rebers, P. A. and Heidelberger, M. (1961). *Journal of the American Chemical Society* **83**, 3056.

Reusch, V. M. and Neuhaus, F. C. (1971). *Journal of Biological Chemistry* **246**, 6136.

Robbins, P. W., Bray, D., Daukerl, M. and Wright, A. (1967). *Science, New York* **158**, 1536.

Rogers, H. J. (1970). *Bacteriological Reviews* **34**, 194.

Rogers, H. J. and Garrett, A. J. (1965). *Biochemical Journal* **96**, 231.

Sanderson, A. R., Juergens, W. G. and Strominger, J. L. (1961). *Biochemical and Biophysical Research Communications* **5**, 472.

Sanderson, A. R., Strominger, J. L., and Nathenson, S. G. (1962). *Journal of Biological Chemistry* **237**, 3603.

Scher, M., Lennarz, W. J. and Sweeley, C. C. (1968). *Proceedings of the National Academy of Sciences of the United States of America* **59**, 1313.

Shabarova, Z. A., Hughes, N. A. and Baddiley, J. (1962). *Biochemical Journal* **83**, 216.

Sharpe, E. M., Davison, A. L. and Baddiley, J. (1964). *Journal of General Microbiology* **34**, 333.

Sharpe, M. E., Brock, J. H., Knox, K. W. and Wicken, A. J. (1973). *Journal of General Microbiology* **74**, 119.

Shaw, D. R. D. (1962). *Biochemical Journal* **82**, 297.

Shaw, N. and Baddiley, J. (1964). *Biochemical Journal* **93**, 317.

Shaw, D. R. D. and Chatterjee, A. N. (1971). *Journal of Bacteriology* **108**, 584.

Shaw, D. R. D., Mirelman, D., Chatterjee, A. N. and Park, J. T. (1970). *Journal of Biological Chemistry* **245**, 5101.

Shockman, G. D. and Slade, H. D. (1964). *Journal of General Microbiology* **37**, 297.

Siewert, G. and Strominger, J. L. (1967). *Proceedings of the National Academy of Sciences of the United States of America* **57**, 767.

Smith, D. G. and Shattock, P. M. F. (1964). *Journal of General Microbiology* **34**, 165.

Strange, R. E. and Shon, M. (1964). *Journal of General Microbiology* **34**, 99.

Strominger, J. L., Higashi, Y., Sandermann, H., Stone, K. J. and Willoughby, E. (1972). *In* "Biochemistry of the Glycosidic Linkage", (R. Pinas and H. G. Pontis, eds.), PAABS Symposium, vol. 2, Academic Press Inc., New York.

Tipper, D. J., Strominger, J. L. and Ensign, J. C. (1967). *Biochemistry, New York* **6**, 906.

Tomasz, A. (1967). *Science, New York* **157**, 694.

Tomasz, A. (1968). *Proceedings of the National Academy of Sciences of the United States of America* **59**, 86.

Tomasz, A. and Westphal, M. (1971). *Proceedings of the National Academy of Sciences of the United States of America* **68**, 2627.

Tomasz, A., Albino, A. and Zanati, E. (1970). *Nature, London* **227**, 138.

Toon, P., Brown, P. E. and Baddiley, J. (1972). *Biochemical Journal* **127**, 399.

Torii, M., Kabat, E. A. and Bezer, A. E. (1964). *Journal of Experimental Medicine* **120**, 13.

Troy, F. A., Frerman, F. E. and Heath, E. C. (1971). *Journal of Biological Chemistry* **246**, 118.

Umbreit, J. N. and Strominger, J. L. (1972). *Journal of Bacteriology* **112**, 1306.

Umbreit, J. N., Stone, K. J. and Strominger, J. L. (1972). *Journal of Bacteriology* **112**, 1302.

Watkinson, R. J., Hussey, H. and Baddiley, J. (1971). *Nature New Biology* **229**, 57.

Watson, M. J., Tyler, J. M., Buchanan, J. G. and Baddiley, J. (1972). *Biochemical Journal* **130**, 45.

Wicken, A. J. (1966). *Biochemical Journal* **99**, 108.

Wicken, A. J. and Baddiley, J. (1963). *Biochemical Journal* **87**, 54.

Wicken, A. J. and Knox, K. W. (1970). *Journal of General Microbiology* **60**, 293.

Wicken, A. J. and Knox, K. W. (1971). *Journal of General Microbiology* **67**, 251.

Willoughby, E., Higashi, Y. and Strominger, J. L. (1972). *Journal of Biological Chemistry* **247**, 5113.

Wolin, M. J., Archibald, A. R. and Baddiley, J. (1966). *Nature, London* **209**, 484.

Wright, A., Dankert, M., Fennessey, P. and Robbins, P. W. (1967). *Proceedings of the National Academy of Sciences of the United States of America* **57**, 1798.

Young, F. E. (1966). *Journal of Biological Chemistry* **241**, 3462.

Young, F. E. (1967). *Proceedings of the National Academy of Sciences of the United States of America* **58**, 2377.

Zaretskaya, M. Z., Naumova, I. B. and Shabarova, Z. A. (1971). *Biochimia* **36**, 97.

Respiration and Nitrogen Fixation in *Azotobacter*

M. G. Yates

A.R.C. Unit of Nitrogen Fixation, University of Sussex, Brighton, BN1 9QJ, England

and

C. W. Jones

Department of Biochemistry, University of Leicester, University Road, Leicester, LE1 7RH, England

I. Introduction

The similar nature of respiratory enzymes in aerobically grown bacteria and cells of highly organized life forms is well established. However, bacteria often need to perform unique metabolic processes, some of which demand an anaerobic environment. Nitrogen (or dinitrogen) fixation is one such process, and the Azotobacteriaceae are the most widespread and important family of bacteria which are both genuine nitrogen-fixers and obligate aerobes. Other aerobic bacteria which fix nitrogen include some strains of mycobacteria, although their taxonomic status is doubtful (Biggins and Postgate, 1971a), some methane oxidizers and, possibly, some thiobacilli. Knowledge of nitrogen fixation, respiration, and the relation between these two processes in Azotobacter has increased considerably over the past few years. In particular, determination of the constitution and behaviour of this organism's branched respiratory chain has contributed to the understanding of how respiration controls nitrogen fixation. We here review the work done towards this aspect; towards understanding how respiration and nitrogen fixation can function simultaneously in Azotobacter. Progress in understanding the biochemistry and chemistry of nitrogen fixation has been reviewed several times in the last few years (Hardy and Knight, 1968; Hardy and Burns, 1968; Burris, 1971; Dalton and Mortenson, 1972; Benemann and Valentine, 1972; Bergersen, 1972; Streicher and Valentine, 1973) and has been the subject of a collective monograph (Postgate, 1971a). Accordingly, only those aspects of this subject which are connected with the physiology of nitrogen fixation in aerobes will be discussed.

II. The Anaerobic Nature of Nitrogen Fixation

Nitrogen fixation is more widespread amongst anaerobic bacteria than aerobes. Obligate anaerobes, such as *Clostridium pasteurianum*, *Chromatium*, *Chlorobium limicola*, *Desulfovibrio* and *Desulfotomaculum* species, all include strains which fix nitrogen (see Postgate, 1971b). Some facultative anaerobes, which grow both aerobically and anaerobically, will only fix nitrogen anaerobically; examples of this type are *Bacillus polymyxa*, *Bacillus macerans*, *Klebsiella pneumoniae* and, among the photosynthetic bacteria, *Rhodospirillum*, *Rhodomicrobium* and *Rhodopseudomonas* species. Nitrogen fixation amongst aerobes is largely confined to the Azotobacteriaceae, including species of *Azotobacter*, *Azomonas*, *Beijerinckia* and *Derxia*. *Mycobacterium flavum*, *Mycobacterium roseo-album* and *Mycobacterium azot-absorptum* (Federov and Kalininskaya, 1961; Biggins and Postgate, 1969; L'vov and Lyubimov, 1965) and a methane-oxidizing organism (Coty, 1967) also fix nitrogen aerobically. The advent of the acetylene test helped to rule out many aerobic bacteria previously thought to fix nitrogen (Hill and Postgate, 1969).

The evidence that nitrogen fixation is both an anaerobic and a reductive process is substantial and will only be outlined here.

(a) All nitrogenases from both aerobic and anaerobic organisms comprise two non-haem iron proteins which have the same requirements for enzymic activity *in vitro*, namely, an anaerobic environment, a source of ATP and a reductant (Mortenson, 1964; Bulen *et al.*, 1964, 1965). The two component proteins that make up the nitrogenase enzyme complement each other in cross reactions between nitrogenase fractions from aerobic and some anaerobic sources (see Burris, 1971), and physiological electron carriers from one class of bacteria can donate electrons to nitrogenase from the other (Bulen *et al.*, 1964; Yates, 1972b).

(b) Both component proteins from nitrogenase are oxygen-sensitive; the smaller iron-containing fraction is particularly so (Bulen and LeComte, 1966; Kelly, 1969). Proteins such as ferredoxins and flavodoxins that transfer electrons to nitrogenase are auto-oxidizable (Yoch *et al.*, 1969; Yates, 1972b).

(c) The absence of oxygen is mandatory in nearly all measurements of nitrogen fixation *in vitro*. Crude preparations which show nitrogenase activity in the presence of oxygen have been obtained from aerobic nitrogen fixers (Bergersen, 1966a, b; Bergersen and Turner, 1967; Stewart *et al.*, 1969; Yates and Daniel, 1970) but these results can be rationalized in terms of a particulate or compartmentalized nitrogenase protected from exposure to oxygen by respiration.

(d) Growth of the aerobic nitrogen-fixing bacteria, *Azotobacter chroococcum*, *Azotobacter vinelandii*, *Azomonas macrocytogenes* (Dalton and Postgate, 1969a; Drozd and Postgate, 1970a, b), *Mycobacterium flavum* 301 (Biggins and Postgate, 1969) and *Derxia gummosa* (Hill, 1972) is inhibited by excess of oxygen when the bacteria are fixing nitrogen but not when they are supplied with fixed nitrogen. Use of the acetylene test showed that nitrogenase did not function under these conditions (Yates, 1970a; Drozd and Postgate, 1970a, b; Hill, 1972). The blue-green alga, *Anabaena flos-aqua*, failed to fix nitrogen at high concentrations of molecular oxygen (Stewart, 1969). High concentrations of oxygen also inhibited nitrogen fixation by excized soybean nodules (Aprison and Burris, 1952).

(e) Nitrogenase catalyses the reduction of several substrates besides dinitrogen (see Hardy and Burns, 1968). There is no confirmed evidence of an oxidized intermediate in any of these reactions.

From the foregoing evidence it is reasonable to assume that nitrogen fixation is an anaerobic process catalysed by oxygen-sensitive proteins. How these operate and survive in an aerobic bacterium will be discussed in detail. Because oxygen uptake is the major electron-utilizing metabolic process in *Azotobacter*, the respiratory chain involving electron transport and oxidative phosphorylation will be described first.

III. Electron Transfer to Oxygen

A. Location of Respiratory Membranes

Electron micrographs of thin sections of *Azotobacter* indicate that nitrogen-fixing organisms contain an extensive intracytoplasmic membrane system which probably arises as an invagination of the peripheral membrane (Robrish and Marr, 1962; Pangborn *et al.*, 1962; Oppenheim and Marcus, 1970; Hill *et al.*, 1972). Following disruption of whole organisms, the respiratory system is readily isolated in a particulate form, the size of the particles being dependent upon the brutality of the disruption procedures and hence upon the degree of comminution of the original membranes (Alexander and Wilson, 1955; Marr and Cota-Robles, 1957; Cota-Robles *et al.*, 1958). Small surface projections are visible on some respiratory particle preparations and may, like those on the inner membranes of mitochondria, be associated with the terminal stages of oxidative phosphorylation (Jones and Redfearn, 1967b).

Disruption of whole cells by osmotic lysis yields preparations which, by electron microscopy, are seen to consist entirely of lysed cells in which the internal membrane system is still present. When sedimented, these preparations exhibit considerable respiratory activity (Robrish and Marr, 1962; Oppenheim and Marcus, 1970) but nitrogenase activity is located entirely in the supernatant (Oppenheim *et al.*, 1970). The intracytoplasmic membrane system is virtually absent from cells cultured on a source of combined nitrogen (Oppenheim and Marcus, 1970; Drozd *et al.*, 1972; Hill *et al.*, 1972) and the membrane system is almost entirely peripheral. It is therefore likely that the function of these intracytoplasmic membranes is primarily respiratory, *viz* to protect the cytoplasmic nitrogenase from the deleterious effects of molecular oxygen (see Section V, p. 124).

B. Respiratory Chain Components

Respiratory membranes isolated from *Azotobacter* grown under nitrogen-fixing conditions exhibit extremely high rates of oxygen uptake and contain a rich pattern of respiratory enzymes and carriers, *viz* several highly active dehydrogenases, a nicotinamide nucleotide transhydrogenase and a complex cytochrome system terminated by three cytochrome oxidases. A brief description of the properties of these respiratory-chain components is set out below.

i. *Dehydrogenases.* Unsupplemented respiratory membranes from *A. vinelandii* catalyse oxidation of NADH (Bruemmer *et al.*, 1957; Tissières *et al.*, 1957; Repaske and Josten, 1958), NADPH (Pandit-Hovenkamp, 1967; Ackrell *et al.*, 1972), succinate (Repaske, 1954; Alexander and Wilson, 1955), malate and lactate (Hovenkamp, 1959a). The particulate oxidation of malate and lactate is nicotinamide-nucleotide independent, but NAD^+-linked malate- and lactate dehydrogenases and an $NADP^+$-linked decarboxylating malate dehydrogenase (malic enzyme) are present in the high-speed supernatant (Jones and Redfearn, 1966; Jurtshuk and Harper, 1968; Jurtshuk *et al.*, 1969b). Of particular interest, in view of the probability that NADPH is a primary electron donor to nitrogenase (see Section IVC, p. 118) is the presence of a soluble but highly active $NADP^+$-dependent isocitrate dehydrogenase (Chung and Franzen, 1969; Barrera and Jurtshuk, 1970). Particulate preparations from *A. agilis* also catalyse oxidation of NADH, NADPH, succinate and malate (Cota-Robles *et al.*, 1958; Robrish and Marr, 1962; Pangborn *et al.*, 1962) and subcellular extracts of *A. chroococcum* oxidize NADH, NADPH, isocitrate and glucose 6-phosphate but show relatively weak activity with other substrates (Table 1). Cell-free extracts of *A. beijerinckii* readily oxidize NADH but exhibit very poor activity with NADPH (Ritchie *et al.*, 1971).

In terms of electron-transfer activity, substrate availability and flexibility of function, the two most important respiratory-chain dehydrogenases present in *A. vinelandii* are those which catalyse oxidation of NADH and NADPH.

In spite of its extremely high activity (up to 10 μmol NADH oxidized/min/mg protein were reported for respiratory membranes of *A. vinelandii* by Eilermann *et al.*, 1970), NADH dehydrogenase is nevertheless the rate-limiting component of NADH oxidase (Jones *et al.*, 1971a, b). Respiratory membranes exhibit simple hyperbolic saturation kinetics with respect to NADH (K_m value, 40 μM) and, under appropriate conditions, readily transfer reducing equivalents from NADH dehydrogenase to a variety of artificial electron acceptors (Eilermann *et al.*, 1970; Yates, 1971a; Erickson *et al.*, 1972). Following complete reduction by NADH, respiratory membranes of *A. vinelandii* exhibited electron paramagnetic resonance and visible adsorption spectra which reflected the presence of a non-haem iron protein (Beinert *et al.*, 1962; DerVartanian and Bramlett, 1970). Partially purified preparations of NADH dehydrogenase were found to contain 4 g atom each of iron and labile sulphide per mol FMN (DerVartanian, 1972). Similar preparations from cells cultured on iron-deficient medium were shown to contain 2 g atom each of molybdenum, iron and labile sulphide per mol FMN and to oxidize readily NADH, NADPH or acetaldehyde. An unusual signal at $g = 1{\cdot}95$

was observed under reducing conditions and was attributed to the presence of monomeric Mo(V), probably co-ordinated with cysteine, at the active site of the dehydrogenase (DerVartanian and Bramlett, 1970).

Jones *et al.* (1972) reported that bathophenanthroline(4,7-diphenyl-1,10-phenanthroline)sulphonate, a water-soluble metal chelator with a

TABLE 1. Dehydrogenase Activities in Ammonia-Grown and Dinitrogen-Grown *Azotobacter chroococcum* with Different Substrates

Substrate	Tetrazolium chloride reduced (μmol/mg protein/h)	
	Dinitrogen-grown cells	Ammonia-grown cells
NADH (10 μmol)	1·9	2·4
NADPH (10 μmol)	2·3	2·1
Glucose 6-phosphate	2·7	2·8
Isocitrate	1·8	1·6
6-Phosphogluconate	1·0	1·2
Citrate	0·47	0·9
Pyruvate + CoA + TPP	0·16	0·2
Succinate	0·07	0·25
α-Ketoglutarate	0·04	0·1
Malate	0·03	0·15
β-Hydroxybutyrate	0·03	0·05
Fumarate	0·01	0·02

Azotobacter chroococcum was grown in chemostat cultures under conditions (air flow, stirring rate and dilution rate) which were adjusted to give a steady-state absorbance of 0·40 ± 0·02 absorbance units/cm^{-1} in each culture at steady state. Cells were centrifuged and resuspended in 25 m*M* tris buffer (pH 7·4) before being disrupted by sonication. Dehydrogenase activities were measured anaerobically by reduction of triphenyltetrazolium chloride (Fahmy and Walsh, 1952) using 50 μmol carbon substrate and 0·5 μmol NADP or NAD. M. G. Yates and J. Easton (unpublished data).

high affinity for Fe^{2+} ions, readily inhibited oxidation of NADH, NADPH or malate by respiratory membranes from *A. vinelandii*. Kinetically, the chelator acted as a pure competitive (dead end) inhibitor with respect to the substrate, and incubation of respiratory membranes with bathophenanthroline plus substrate yielded an absorption spectrum which was identical with that obtained with spinach ferredoxin or Fe^{2+} ions in the presence of the inhibitor.

The NADH dehydrogenase of *A. chroococcum* has been extensively purified by Yates (1971a). The enzyme had a molecular weight of 450,000 daltons and transferred reducing equivalents from NADH or NADPH to

NAD^+ and a variety of artificial electron acceptors. Of particular interest was the observation that the dehydrogenase catalysed the slow reduction of acetylene by NADH in the presence of nitrogenase, benzyl viologen and ATP (see Section IVB, p. 117).

In contrast to NADH dehydrogenase, the highly active NADPH dehydrogenase of respiratory membranes from *A. vinelandii* yielded sigmoidal saturation kinetics and exhibited a relatively poor affinity for its substrate ($[S]_{0.5}$440 μM; Ackrell *et al.*, 1972). The dehydrogenase, which was the rate-limiting component of NADPH oxidase, was inhibited by adenine nucleotides (pure competitive; AMP > ADP > ATP) and by NAD^+ (partially competitive, possibly allosteric inhibition). The enzyme exhibited kinetic properties which were compatible with the presence of a catalytic site, which bound one molecule of NADPH (or adenine nucleotide) and a regulatory site which bound one molecule of either NADPH (positive homotropic effector) or NAD^+ (negative heterotropic effector); no allosteric activators, other than NADPH, were found for the enzyme. Like NADH dehydrogenase, NADPH dehydrogenase was readily inhibited by bathophenanthroline sulphonate (Jones *et al.*, 1972) and rapidly donated reducing equivalents to various artificial quinone acceptors (Erickson *et al.*, 1972).

Purification of NADPH dehydrogenase, followed by careful analysis of its electron-carrier composition and kinetic behaviour, is clearly a prime requisite to a proper understanding of the precise respiratory function of this apparently non-phosphorylating enzyme. On present evidence, the most likely function of NADPH dehydrogenase in *A. vinelandii* is to oxidize the excess NADPH which might be expected to accumulate when nitrogen fixation is inhibited, and thus to contribute towards the respiratory protection of nitrogenase (see Section VA, p. 124) and also to the control of $NADPH/NADP^+$ balance (see Section IVD), p. 120).

ii. *Nicotinamide nucleotide transhydrogenase.* A reversible nicotinamide nucleotide transhydrogenase was first reported to be present in crude extracts of *Azotobacter* by Kaplan *et al.* (1953). The enzyme appeared to comprise part of a reversible NAD(P)H-lipoatc reductase (van den Broek and Veeger, 1968) but was subsequently separated from the lipoamide dehydrogenase and substantially purified (Chung, 1970; van den Broek *et al.*, 1971a). The purified transhydrogenase catalysed reduction of $NAD(P)^+$, thio-$NAD(P)^+$, dichlorophenolindophenol and potassium ferricyanide by NADH and NADPH, although the activities with NADPH were higher than with NADH. The very high molecular-weight enzyme was shown to contain FAD and to form complexes with

$NADP^+$ which caused partial fragmentation of the enzyme's threadlike structure (van den Broek *et al.*, 1971b, c).

Anaerobic titrations indicated that four equivalents per mol of flavin were required for full reduction. A reaction scheme was proposed in which the transhydrogenase was reduced to the two-equivalent state, followed by dissociation of the oxidized nicotinamide nucleotide, internal electron transfer to an unidentified group (X), further reduction to the four-equivalent state and finally re-oxidation of the enzyme by two molecules of acceptor; this scheme was supported by detailed kinetic studies (van den Broek and Veeger, 1971a, b). The presence on the transhydrogenase of two binding sites for NADPH (one catalytic and one regulatory) together with the ability of ATP, and probably also NAD^+, to act as negative effectors on the regulatory site (van den Broek and Veeger, 1968) strongly suggest that the transhydrogenase and the NADPH dehydrogenase may share a common component.

Clearly the nicotinamide nucleotide transhydrogenase, $NADH_-$, and NADPH dehydrogenases of *A. vinelandii* serve to regulate the intracellular $NAD(P)H/NAD(P)^+$ ratio (see Fig. 1), particularly under conditions where oxidation of NADPH to provide reducing power for nitrogen fixation is either inhibited (e.g. at high partial pressures of oxygen) or impossible (e.g. in cells grown on combined nitrogen). Control of these enzymes is essential since they control electron flow to the respiratory chain and, probably, to nitrogenase. This control is intricate, involving intracellular substrate concentration, respiratory control and the energy charge (Atkinson and Walton, 1967).

iii. *Cytochromes*. Early spectroscopic investigations of whole cells of *Azotobacter* revealed the absorption bands of *b*- and *c*-type cytochromes together with cytochrome a_2 (Negelein and Gerischer, 1934). These findings were confirmed by reduced *minus* oxidized difference spectra of respiratory-particle preparations (Smith, 1954; Bruemmer *et al.*, 1957; Temperli and Wilson, 1960). Reduced + carbon monoxide *minus* reduced difference spectra further indicated that cytochromes a_1 and *o* were also present (Castor and Chance, 1959). The *c*-type cytochrome was subsequently purified to yield two major components, namely cytochromes c_4 and c_5 (Tissières, 1956; Tissières and Burris, 1956; Neumann and Burris, 1959), together with a third cytochrome designated minor c_4 (Swank and Burris, 1969b). Both major cytochromes are acidic and exhibit redox potentials in the range +0·30 to 0·32 V; cytochrome c_4 is a monomer (mol. wt. 24,000 daltons; two haems/molecule) and c_5, although isolated as a dimer (mol. wt. 24,400 daltons; two haems/molecule), readily splits into monomers under denaturing conditions. In their

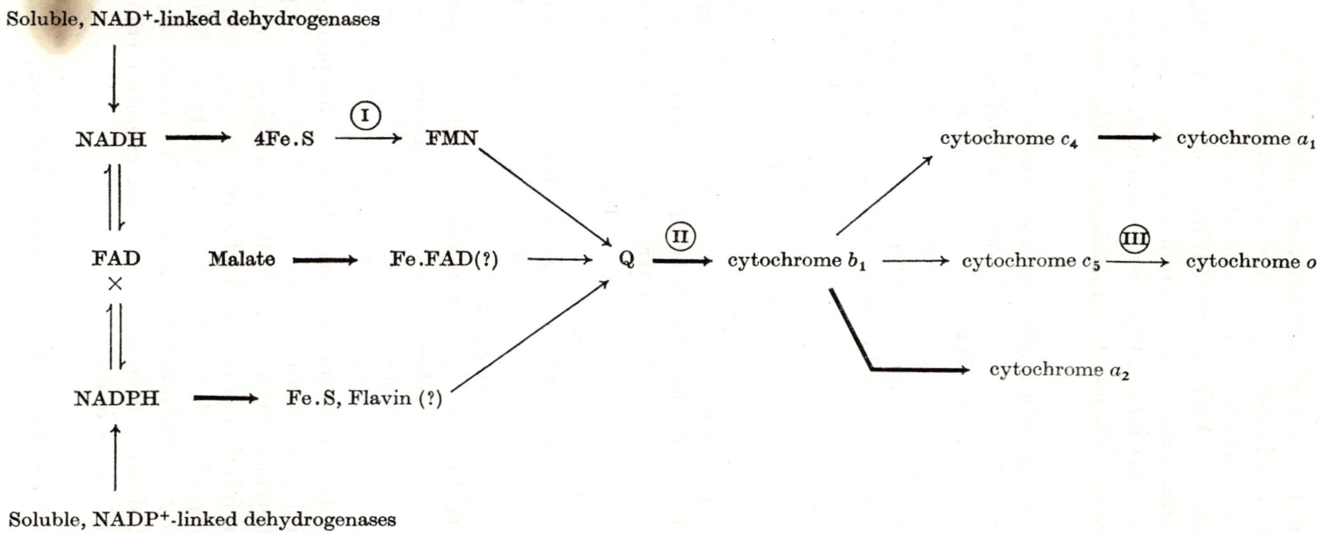

FIG. 1. The respiratory system of *Azotobacter*. Phosphorylation sites are designated I, II and III; approximate rates of electron transfer are indicated by the thickness of the arrows. Based upon Jones and Redfearn (1967a).

reduced forms, all three cytochromes are readily oxidized by respiratory membranes from *A. vinelandii* (Swank and Burris, 1969b).

The ability of cytochromes a_1, a_2 and o to combine in their reduced forms with carbon monoxide suggested that these cytochromes may function as oxidases, and this was clearly demonstrated by the measurement of photochemical action spectra of whole cells oxidizing endogenous substrates (Castor and Chance, 1959; Jones and Redfearn, 1967a). Under these conditions, cytochrome a_2 appears to be the major functional oxidase.

iv. *Quinones.* Early investigations into the nature of the quinone component of the respiratory system of *A. vinelandii* indicated the presence of ubiquinone (Temperli and Wilson, 1960; Page *et al.*, 1960), possibly accompanied by traces of an unidentified menaquinone (Lester and Crane, 1959). Later, the ubiquinone component was identified as Q-8 but both spectrophotometric and bio-assays failed to confirm the presence of menaquinone (Jones and Redfearn, 1966; Swank and Burris, 1969a).

Measurement of the redox kinetics of ubiquinone in isolated respiratory membranes, by the use of rapid chemical-extraction procedures, indicated that most of the quinone was reducible by physiological substrates, but not by ascorbate-dichlorophenolindophenol. The aerobic steady-state reduction levels of the quinone were quite compatible with it acting as a high-concentration redox carrier between the dehydrogenases and the terminal cytochrome system (Knowles and Redfearn, 1968). However, doubts were cast on the status of ubiquinone as a main-chain carrier by the observation that the rates of quinone reduction by physiological substrates were only 35–60% of the overall rates of electron transfer to oxygen.

Early attempts to investigate the function of ubiquinone in the respiratory system of *A. vinelandii via* acetone extraction–re-activation experiments were largely unsuccessful (Temperli and Wilson, 1962; Knowles and Redfearn, 1966), but later methods, which involved extraction of lyophilized respiratory membranes with pentane, proved much more useful (Swank and Burris, 1969a). In these experiments, NADH oxidase activity was lost in parallel with ubiquinone extraction, and a substantial fraction of the original activity was restored by addition of native ubiquinone (Q-8). The restored activity was fully sensitive to cyanide, and it was concluded that, in spite of the doubts expressed by Knowles and Redfearn, ubiquinone was an essential component of NADH oxidase in *A. vinelandii.*

C. Pathways of Electron Transfer

A detailed investigation of the particulate respiratory system of

A. vinelandii, mainly through the use of dual wavelength spectrophotometry and classical cytochrome oxidase inhibitors such as cyanide, azide and carbon monoxide, suggested that electron transfer in the terminal region of the respiratory chain occurred *via* a branched cytochrome system (Jones and Redfearn, 1967a). This conclusion was based upon the following observations:

(1) The aerobic steady-state reduction of cytochrome b_1 during oxidation of physiological substrates such as NADH, succinate or malate was considerably lower than that of cytochromes $c_4 + c_5$, a situation which was in direct contrast to that found in mammalian mitochondria (Chance and Williams, 1956).

(2) Oxidation of physiological substrates was much less sensitive to inhibition by cyanide and azide than was oxidation of ascorbate *via* dichlorophenolindophenol, tetramethyl-*p*-phenylenediamine (TMPD), mammalian cytochrome *c*, *A. vinelandii* cytochrome c_4 and c_5 or various other high-potential electron mediators (Jones and Redfearn, 1967a; Jurtshuk *et al.*, 1967; Jurtshuk and Old, 1968; Jurtshuk *et al.*, 1969a; Swank and Burris, 1969b). The nature of the cyanide-inhibition kinetics with respect to electrons as the variable substrate was characteristic of the different electron donors employed, *viz* uncompetitive for physiological substrates and ascorbate-TMPD (but with quite different K_i values) and non-competitive with ascorbate-dichlorophenolindophenol, which clearly implicated three different cytochrome oxidases (Table 2). The insensitivity of the oxidation of physiological substrates to cyanide

TABLE 2. Cyanide Inhibition of Cytochrome Oxidases from *Azotobacter vinelandii*

Substrate	Cytochrome oxidase	V_{max} value (μg atom O_2 min/mg protein)	Cyanide inhibition: Type	Cyanide inhibition: K_i (μM)
Ascorbate-tetramethyl-*p*-phenylenediamine	a_1	3·50	Uncompetitive	0·46
Ascorbate-dichloro-phenolindophenol	*o*	1·16	Non-competitive	0·51
Physiological (e.g. NADH, malate, NADH + malate)	a_2	$\geqslant$ 2·71	Uncompetitive	$\leqslant$ 1·15

Phosphorylating respiratory membranes from *Azotobacter vinelandii* were prepared and assayed for oxygen uptake with various substrates under State III conditions as described previously (Ackrell and Jones, 1971a; Jones *et al.*, 1971b). Cyanide was added to the reaction mixture containing membranes one minute before addition of substrate and initial oyxgen uptake rates were measured. Unpublished data of C. W. Jones.

was similar to that observed for electron transfer *via* cytochrome a_2 in *Achromobacter* strain 7 and other organisms (Arima and Oka, 1965; Oka and Arima, 1965), whereas the sensitivity of the high-potential substrates to cyanide was very similar, both qualitatively and quantitatively, to that observed for electron transfer *via* cytochrome oxidases a_1 and o in several other species of bacteria (D. J. Meyer and C. W. Jones, unpublished observations).

(3) Oxidation of physiological substrates was sensitive to inhibition by carbon monoxide, but this inhibition exhibited a marked resistance to relief by light, a property associated with cytochrome oxidase a_2 (Negelein and Gerischer, 1934; Castor and Chance, 1959). The ability of high-intensity red light, but not blue light, to relieve this inhibition substantially, and the failure of low concentrations of cyanide to abolish this relief, strongly supported the concept that cytochrome oxidase a_2 was the major physiological oxidase (Jones and Redfearn, 1967a). Conversely, the high photosensitivity of ascorbate-dichlorophenolindophenol oxidation, when inhibited by carbon monoxide, together with the ability of low concentrations of cyanide to abolish this photorelief, implicated cytochrome oxidases a_1 and/or o in oxidation of ascorbate-dichlorophenolindophenol, but a_1 could not be detected in the action spectrum (Erickson and Diehl, 1973). Similar observations were made subsequently for ascorbate-tetramethyl-*p*-phenylenediamine oxidation where an action spectrum indicated only cytochrome oxidase a_1 (Erickson and Diehl, 1973).

(4) Purified cytochromes c_4 and c_5 from *A. vinelandii* yielded significantly different P/O ratios when used as electron mediators between ascorbate and the respiratory chain (see Section IIID, p. 110) which suggested that these two cytochromes operate in parallel rather than in series (Ackrell and Jones, 1971a).

In the original formulation of the branched cytochrome system of *A. vinelandii* (Jones and Redfearn, 1967a) the branch was postulated to occur at the quinone level to yield a major terminal pathway ($b_1 \rightarrow a_2$) and a minor terminal pathway ($c_4 - c_5 \rightarrow a_1/o$). The absence of cytochrome b_1 from the minor route was deduced from the apparent inability of 2-*n*-alkyl-4-hydroxyquinoline-N-oxide (HQNO) to inhibit NADH oxidation completely. However, Jurtshuk *et al.* (1969a) subsequently showed complete inhibition of respiration by HQNO, and further analysis of the mutual depletion kinetics of this inhibition (Henderson, 1973) suggested that HQNO combined tightly with the reduced form of cytochrome b_1. A slightly modified scheme for the branched respiratory system of *Azotobacter*, which satisfies most of the experimental data, is shown in Fig. 1 (p. 105).

It should be noted that another scheme has been proposed for the respiratory system of *A. vinelandii* which involves separate phosphorylating and non-phosphorylating respiratory chains, possibly linked at the level of ubiquinone (Eilermann *et al.*, 1970, 1971). Unfortunately, the cytochrome system of this scheme, which was proposed principally to explain certain respiratory control properties of isolated membranes, has not been described and it is therefore impossible to compare the merits of the two schemes.

Although substantial purification of certain individual components of the *Azotobacter* respiratory chain has been successfully carried out (see Section IIIB, p. 100), attempts to fractionate the system into enzymically active complexes of the type extracted from mammalian systems (Green and Wharton, 1963) have met with only limited success. An NADH oxidase particle, essentially free of succinate oxidase activity, was prepared by Repaske and Josten (1958) and found to contain higher concentrations of cytochromes b_1 and a_2, relative to c_4 and c_5, than normal respiratory particles. Similarly, green respiratory particles, derived from normal respiratory particles of *A. vinelandii* by Jones and Redfearn (1967b), contained high concentrations of cytochromes b_1, a_2 and a_1; red particles were particularly rich in flavoprotein, ubiquinone, non-haem iron and cytochromes b_1, $c_4 + c_5$ and *o*, and also exhibited substantial oxidase activity with succinate or with ascorbate plus a variety of artificial electron mediators. The properties of these two types of derivative particles largely supported the concept of a branched cytochrome system in *A. vinelandii*.

Brief studies on the respiratory system of *Azotobacter* cultured on various sources of combined nitrogen have indicated that it is not strikingly different from that of nitrogen-fixing cells (Lisenkova and Khmel, 1967; Dalton and Postgate, 1969b). The lower cytochrome a_2 content of respiratory membranes from cells grown batchwise on urea, reported by Knowles and Redfearn (1969), was not confirmed by Drozd and Postgate (1970b) using continuous cultures maintained at a constant pO_2, and the original differences can probably be attributed to different ambient oxygen concentrations in the nitrogen-free and urea-containing batch cultures. Substantial repression of respiratory chain components, particularly cytochrome a_2, was observed when glucose replaced mannitol as the carbon source (Daniel and Erickson, 1969) but again no attempts were made to monitor the ambient oxygen concentration in the cultures.

D. Oxidative Phosphorylation

1. *Subcellular Preparations*

Oxidative phosphorylation by crude extracts of *Azotobacter* was first described by Hyndman *et al.* (1953). This was later extended to crude particulate preparations oxidizing succinate (Tissières and Slater, 1955; Tissières *et al.*, 1957), NADH and NADPH (Rose and Ochoa, 1956), malate and lactate (Hovenkamp, 1959a). In all of these particle preparations, the efficiency of oxidative phosphorylation was relatively low (P/O ratios with NADH as substrate very rarely exceeded 0·5) but activity was occasionally stimulated by addition of high-speed supernatant (Temperli and Wilson, 1960).

Phosphorylating activity was readily lost when the respiratory particles were suspended in weak phosphate buffer and recentrifuged (Hovenkamp, 1969b) but could be completely restored by incubating the resultant soluble fraction with the non-phosphorylating particles in the presence of high concentrations of monovalent or divalent salts (Pandit-Hovenkamp, 1966). Rather surprisingly, in view of their low P/O ratios, the respiratory particles exhibited very low ATPase activity (e.g. 0·04 μmol/min/mg protein compared with a rate of up to 3 μmol/min/mg protein for ATP synthesis) but this was increased up to 20-fold by pre-incubating the particles with trypsin to remove a natural protein inhibitor of the ATPase (Eilermann *et al.*, 1971). Latent ATPases of this type have also been detected in mitochondria and chloroplasts (Pullman and Monroy, 1963; Racker, 1963; Vambutas and Racker, 1965).

Eilermann *et al.* (1970) reported that unsupplemented respiratory membranes from *A. vinelandii* yielded consistently higher P/O ratios with NADH than with either succinate or malate as substrate, and from this deduced the presence of at least two phosphorylating sites, one at the level of NADH dehydrogenase (site I) and a second to the oxygen side of ubiquinone (site II). These findings were confirmed using respiratory membranes of considerably higher phosphorylation efficiency (P/O ratio $\not>$ 1·10 with NADH as substrate) which were prepared by extensive modification of Hovenkamp's procedure (Ackrell and Jones, 1971a). The association of site-I phosphorylation with NADH dehydrogenase was verified by the observation that ATP synthesis accompanied the anaerobic oxidation of NADH by Q-1 (Eilermann *et al.*, 1970; Erickson *et al.*, 1972), but no significant phosphorylation was detected at the level of NADPH dehydrogenase. The use of ascorbate plus a variety of high-potential electron mediators allowed detection of a third phosphorylation site located between the *c*-type cytochromes (probably c_5) and molecular oxygen, but this site exhibited a considerably lower phosphorylation efficiency than either site I or II (Ackrell and Jones, 1971a).

Energization of respiratory membranes from *A. vinelandii* as a result of electron transfer from malate to molecular oxygen was recently detected directly by measuring the quenching of the fluorescent probe atebrin (Eilermann, 1970). As expected, fluorescence was restored under anaerobic conditions or following the addition of uncoupling agents, respiratory inhibitors or ADP. In general, oxidative phosphorylation in subcellular preparations of *A. vinelandii* is sensitive to classical uncoupling agents (with the striking exception of 2,4-dinitrophenol) and phosphorylation inhibitors at concentrations similar to those which affect mitochondria (Rose and Ochoa, 1956; Pandit-Hovenkamp, 1967; Eilermann *et al.*, 1971).

a. *Effect of growth conditions*. No changes were detected in the P/O ratios with either NADH or malate when low concentrations of cyanide were used to switch electron flow completely into the major branch terminated by cytochrome oxidase a_2 (Eilermann *et al.*, 1970; Ackrell and Jones, 1971a). However, consistent differences in P/O ratios with NADH have recently been observed between respiratory membranes prepared from cells harvested at different stages of oxygen-limited batch growth and which therefore contained different relative concentrations of cytochromes oxidase o, a_1 and a_2 (Fig. 2). Highest phosphorylation efficiencies

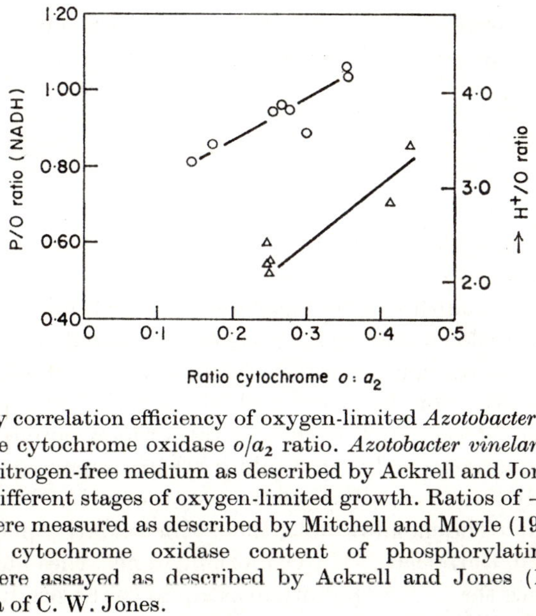

FIG. 2. Energy correlation efficiency of oxygen-limited *Azotobacter vinelandii* as a function of the cytochrome oxidase o/a_2 ratio. *Azotobacter vinelandii* was grown batchwise in nitrogen-free medium as described by Ackrell and Jones (1971a) and harvested at different stages of oxygen-limited growth. Ratios of $\rightarrow H^+/O$ for cell suspensions were measured as described by Mitchell and Moyle (1967). P/O ratios (NADH) and cytochrome oxidase content of phosphorylating respiratory membranes were assayed as described by Ackrell and Jones (1971a, b). Unpublished data of C. W. Jones.

were observed in membranes which contained high concentrations of oxidase o relative to a_2 (i.e. the terminal oxidase of the major branch appeared to be associated with respiration of low-energy conservation efficiency). Similar results were also obtained with whole cells when phosphorylation efficiencies were assayed *via* measurement of $\rightarrow H^+/O$ quotients (see Section IIID) (ii), p. 112).

Striking differences in the efficiency of energy conservation in respiratory membranes of *A. vinelandii* were observed at different stages of batch growth (Ackrell and Jones, 1971b). Ratios for P/2e at site I were lowest during very early logarithmic growth, but increased sharply as the ambient oxygen concentration decreased and reached maximum efficiency (P/2e $\ngtr$ 0·5) only under oxygen-limited growth conditions. Oxidative phosphorylation at site II was rather less susceptible to oxygen, and site III was too weakly coupled for any changes to be apparent.

As expected from the low phosphorylation efficiencies, the respiratory control index (defined as the ratio of the oxygen uptake rates in state 3 compared with 4; see Chance and Williams, 1956) with NADH as substrate was minimal in respiratory membranes prepared from early logarithmic-phase cells (Jones *et al.*, 1971a).

DerVartanian (1972) reported that respiratory membranes isolated from *A. vinelandii* grown on an iron-deficient medium exhibited lower P/O ratios with NADH than did similar membranes prepared from cells grown in iron-containing medium, but details were not given regarding possible differences in the ambient oxygen concentrations of the two cultures.

2. *Whole Cells*

Several attempts have recently been made to assay P/O ratios in whole cells of *A. vinelandii* either directly or *via* measurement of growth efficiencies with respect to oxygen.

Knowles and Smith (1970) measured rates of ATP synthesis and oxygen consumption following the addition of aerated buffer to anaerobic suspensions of stationary-phase *A. vinelandii*, and obtained P/O ratios of approximately 2 for the NAD^+-linked oxidation of β-hydroxybutyrate. Baak and Postma (1971), using a rather more sophisticated direct assay technique devised by Hempfling (1970), reported P/O ratios of approximately 3 for the oxidation of endogenous NADH.

It is implicit in the chemi-osmotic theory for oxidative phosphorylation (Mitchell, 1966, 1967) that $\rightarrow H^+/O$ quotients reflect the efficiency of ATP synthesis, since the $\rightarrow H^+/O$ quotient is equal to the product of the P/O ratio and $\rightarrow H^+/P$ ratio. A maximum $\rightarrow H^+/O$ quotient of 3·5 was observed during oxidation of endogenous substrates by *A. vinelandii* harvested during oxygen-limited growth (Fig. 2, p. 111). This was equivalent to a P/O ratio of 1·75, assuming an $\rightarrow H^+/P$ ratio of 2 as reported for mitochondria by Mitchell and Moyle (1967). This value is quite compatible with the P/O ratio of 2 for NADH observed by Knowles and Smith (since oxidation of endogenous substrates under non-growing conditions probably involves a certain amount of NADPH oxidation *via* the non-phosphorylating NADPH dehydrogenase) and both results are compatible with the concept of a respiratory system with terminal branches of different energy conservation efficiency. The P/O ratio of 3 reported by Baak and Postma (1971) is clearly at variance with this concept.

Attempts to assay P/O ratios indirectly, *via* measurement of growth efficiencies with respect to oxygen, yielded results which conflict with those obtained from direct measurements. Even under oxygen-limited conditions, when oxidative phosphorylation might be expected to operate at its highest efficiency, Y_{O_2} values $\leqslant$ 13 g cells per mol of oxygen consumed were obtained with *A. vinelandii* (Nagai *et al.*, 1969; Nagai and Aiba, 1972; A. J. Downs and C. W. Jones, unpublished data). These values are considerably lower than those obtained for several other organisms (see for example Hadjipetrou *et al.*, 1964; Schulze and Lipe, 1964; Meyer and Jones, 1973) and suggest a P/O ratio $\leqslant$ 0·6 unless, as claimed by Nagai *et al.* (1969), the value of Y_{ATP} (g cells/mol ATP consumed) for *A. vinelandii* growing on nitrogen-free minimal medium is considerably lower than the widely quoted value of approximately 10·6 (Bauchop and Elsden, 1960). This may well be because oxygen consumption in nitrogen-fixing *Azotobacter* is used both to provide energy for nitrogen fixation and to protect by the process of respiratory protection (see section VA, p. 123) the functioning nitrogenase from oxygen damage. However, preliminary results (A. J. Downs and C. W. Jones, unpublished data) indicate that *A. vinelandii* grown on a source of combined nitrogen also has low Y_{O_2} values.

E. Respiratory Control

Classical respiratory control, defined as the stimulation of respiration by ADP plus inorganic phosphate followed by a return to the controlled rate on the exhaustion of ADP (Chance and Baltscheffsky, 1958), was observed with phosphorylating respiratory membranes of *A. vinelandii* when oxidizing NADH, but not when oxidizing substrates which were incapable of site-I phosphorylation, *viz* NADPH, succinate, malate or lactate (Eilermann *et al.*, 1970, 1971; Jones *et al.*, 1971a, b). The respiratory control index with NADH was proportional to the efficiency of energy coupling at site I. Respiratory control through site II was only observed when the central Q–b_1 region of the chain was made rate-limiting either through the use of the electron-transfer inhibitor 2-*n*-alkyl-4-hydroxyquinoline-N-oxide or by employing pairs of substrates, e.g. lactate + malate (Eilermann *et al.*, 1971; Jones *et al.*, 1971a, b). Under these conditions, the respiratory control index was proportional to the efficiency of energy coupling at site II. Respiratory control was not observed with substrates such as ascorbate-dichlorophenol-indophenol which allowed only weak site-III phosphorylation. Adenosine triphosphate substantially inhibited oxidation of succinate, but not NADH, by disrupted preparations of *A. chroococcum* (Yates, 1970a).

Oxidation of NADPH, unlike NADH, is subject to several types of non-classical respiratory control. The sigmoidal saturation kinetics exhibited by the NADPH dehydrogenase (Ackrell *et al.*, 1972) ensure that small changes in the intracellular $NADPH/NADP^+$ ratio could significantly affect the rate of NADPH oxidation. Furthermore, NADPH dehydrogenase is competitively inhibited by adenine nucleotides, and thus responds to "energy charge" as defined by Atkinson and Walton (1967); minimum activity of this non-phosphorylating dehydrogenase was observed under conditions of low energy charge (Ackrell *et al.*, 1972). It is likely that, under such conditions or in the presence of the potent inhibitor NAD^+, NADPH is preferentially converted to NADH *via* the $NADPH \rightarrow NAD^+$ transhydrogenase and thence oxidized *via* the more efficiently phosphorylating NADH dehydrogenase.

IV. Electron Transfer to Nitrogen

Respiration and nitrogen fixation are the two major systems requiring electron transport in nitrogen-fixing *Azotobacter*; the latter process has been described as a form of respiration (Parker and Scutt, 1960). Since these two systems apparently share primary electron donors, and the electron carriers involved in nitrogen fixation are autoxidizable and may react with oxygen under some conditions *in vivo*, it is necessary to discuss what is known about electron transport to nitrogenase in *Azotobacter*. This discussion will be divided into five sections: (A) the electron carriers, (B) the electron-transfer sequence, and (C) the primary donors; the order being determined by the degree of uncertainty surrounding sections (B) and (C). Section D discusses the possible regulatory mechanisms for controlling NAD(P)H:NAD(P) ratios in *Azotobacter* because these electron carriers are likely electron donors to nitrogenase. Section E gives an account of the role of hydrogenase.

A. Electron Carriers

Electron transport to nitrogenase in the obligate anaerobe, *Clostridium pasteurianum*, involves a phosphoroclastic oxidation of pyruvate or oxidation of α-ketobutyrate, the electrons being transferred to ferredoxin or flavodoxin and subsequently to nitrogenase (Mortenson, 1964; D'Eustachio and Hardy, 1964; Knight and Hardy, 1966). Attempts to show cell-free nitrogen fixation with physiological donors in extracts from

Azotobacter have generally yielded low or negligible activities compared with those using extracts from *Cl. pasteurianum*. Arnon and his colleagues (Yoch and Arnon, 1969; Benemann *et al.*, 1969; Yoch *et al.*, 1969) first observed the nature of electron carriers to nitrogenase in *Azotobacter*. They found that illuminated spinach chloroplasts, which had previously been heated to remove the oxygen-producing photosystem 2, could donate electrons to nitrogenase in crude extracts from *Azotobacter vinelandii*, and reduce acetylene provided that ATP and either a flavoprotein (which they called azotoflavin) or a non-haem iron protein (Azotobacter ferredoxin) was present. These observations were confirmed by van Lin and Bothe (1972), Wong and Burris (1972), Yoch (1972) and, with *Azotobacter chroococcum*, in the A.R.C. Unit of Nitrogen Fixation in the University of Sussex, England (M. G. Yates, unpublished data). Azotoflavin, which is arguably Azotobacter flavodoxin (van Lin and Bothe, 1972), is apparently identical with the Shethna flavoprotein (Shethna *et al.*, 1964; Hinkson and Bulen, 1967; these authors did not, however, establish its biological function). The ferredoxin was identified as the non-haem iron protein III of Shethna (1970) and has been resolved into two ferredoxins of molecular weights 14,100 and 6,000 daltons, respectively. In addition to catalysing acetylene reduction in the presence of chloroplasts and crude extracts from *A. vinelandii*, both of these proteins catalyse NADP reduction with chloroplasts and pyruvate phosphoroclasm with extracts from *Cl. pasteurianum* (Yoch and Arnon, 1972).

Azotobacter flavodoxin from *A. vinelandii* or *A. chroococcum* has three oxidation states, namely fully oxidized, half reduced or semiquinone, and the fully reduced or hydroquinone state. The semiquinone is blue and unusually stable to oxygen, a property interpreted by Benemann *et al.* (1969) as a prerequisite for an electron carrier in an aerophobic system operating in an aerobic environment. Yates (1972b) showed that flavodoxin hydroquinone from *A. chroococcum* supported acetylene reduction with purified nitrogenase from *A. chroococcum*, and was oxidized to the semiquinone but no further. This evidence is summarized together with the visible spectra of the three oxidation states of Azotobacter flavodoxin in Fig. 3. This information is consistent with the reported redox potential of the flavodoxin hydroquinone/semiquinone couple from *A. vinelandii* of −495 mV (Barman and Tollin, 1972) or −465 mV (Yoch, 1972) compared with the redox potential of the semiquinone/oxidized couple which is +50 mV (Barman and Tollin, 1972) or −270 mV (Yoch, 1972) and also with the observation by Bothe and Falkenberg (1972) that either illuminated chloroplasts or the phosphoroclastic system from *Cl. pasteurianum* reduced Azotobacter flavodoxin to the hydroquinone form.

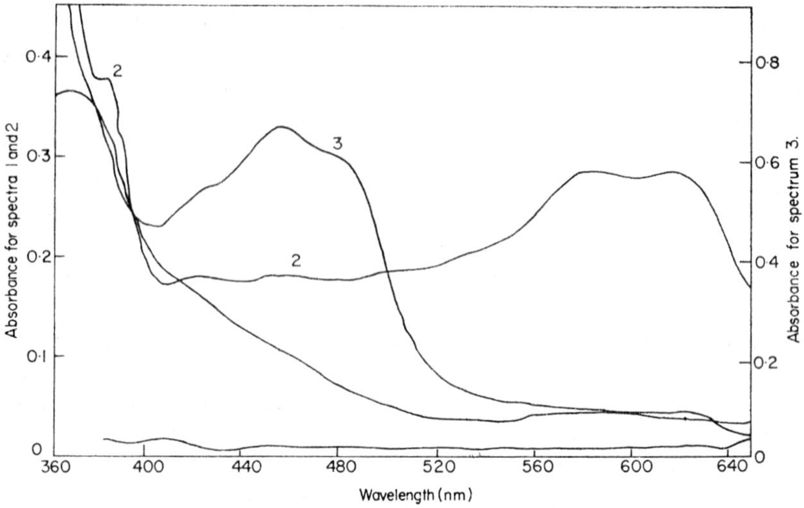

FIG. 3. Visible spectra of three oxidation states of flavodoxin from *Azotobacter chroococcum*. Curve 1 shows the spectrum of the hydroquinone state of the flavodoxin hydroquinone; curve 2, the semiquinone; and curve 3 the oxidized state. Each spectrum was measured with the same material during the course of an experiment to show reduction of the hydroquinone by purified nitrogenase fractions in the presence of ATP. The oxidized spectrum (3) was obtained by shaking the semiquinone in air in the presence of traces of potassium ferricyanide. From Yates (1972b).

B. PATHWAYS OF ELECTRON TRANSFER

Benemann *et al.* (1971) showed that flavodoxin and ferredoxin were both necessary for optimal reduction of acetylene with crude cell-free supernatants from *A. vinelandii* using NADPH or NADPH-generating carbon substrates. The activities were low and readily lost upon the ageing of the supernatant. The presence of spinach NADPH-ferredoxin reductase enhanced the rate of acetylene reduction. Wong *et al.* (1971) obtained acetylene reduction in bacteroid extracts using Azotobacter flavodoxin, bacteroid non-haem iron protein (Koch *et al.*, 1970), spinach NADPH-ferredoxin reductase and a NADPH-generating system. Yates (1971a) also obtained small but positive acetylene reduction dependent on Azotobacter flavodoxin and ferredoxin with crude extracts of *A. chroococcum*. In a review of this topic, Benemann and Valentine (1972) proposed an electron-transport pathway from NADPH → Azotobacter ferredoxin → x (unknown protein intermediate) → Azotobacter flavodoxin → nitrogenase. Their evidence was based upon three observations: (i) the requirement for Azotobacter ferredoxin, flavodoxin and a third protein factor for optimal acetylene-reducing activity; (ii) because

Azotobacter ferredoxin mediated reduction of $NADP^+$ by chloroplasts then, by analogy, in *Azotobacter* the ferredoxin was reduced by NADPH-ferredoxin reductase; (iii) reduced ferredoxin did not reduce Azotobacter flavodoxin. In part support of this scheme, Yates (1972b) gave evidence to indicate the direct transfer of electrons from flavodoxin hydroquinone to purified nitrogenase (see Fig. 3). However, van Lin and Bothe (1972) showed that Azotobacter flavodoxin catalysed $NADP^+$ reduction by chloroplasts, while Bothe and Falkenberg (1972) showed that chloroplasts reduced the flavodoxin to the hydroquinone. Both of these observations suggest that chloroplasts can reduce flavodoxin directly, which, in turn, transfers electrons to nitrogenase. Moreover, either flavodoxin or ferredoxin supports acetylene reduction with illuminated chloroplasts and extracts of *Azotobacter* (Benemann *et al.*, 1969; Yoch *et al.*, 1969). Only a small synergism was observed in the presence of both carriers (Yoch, 1972). These lines of evidence suggest that reduced ferredoxin from *Azotobacter* can transfer electrons directly to nitrogenase and, therefore, that ferredoxin and flavodoxin are alternative electron carriers to nitrogenase in *Azotobacter* as they are in *Cl. pasteurianum* (Knight *et al.*, 1966). The apparent requirement of both carriers for optimal acetylene reduction (Benemann *et al.*, 1971) and the observation that both carriers are apparently constitutive (Benemann *et al.*, 1969) are not consistent with the view that they are alternative electron carriers; this question needs further study. If these proteins are constitutive they may have functions other than transferring electrons to nitrogenase. Attempts in the A.R.C. Unit of Nitrogen Fixation to show that reduced ferredoxin from *Azotobacter* could be re-oxidized by a fully complemented nitrogen-fixing system were not unequivocally successful. From the evidence it is highly probable that these proteins operate as electron carriers to nitrogenase but, at present, we feel that it is premature to claim that they operate sequentially, simultaneously or alternatively in this respect.

A second approach in the study of electron transfer to nitrogenase is the use of redox dyes, methyl- or benzyl viologen, as substitutes for ferredoxins. For example, Yoch (1972) demonstrated that catalytic amounts of methyl viologen greatly enhanced reduction and re-oxidation of Azotobacter flavodoxin, and suggested that one-electron carriers (e.g. ferredoxins) were involved in the biological reduction of the flavodoxin. Earlier reports implicated NADH as a primary electron donor to nitrogenase in aerobic systems (Klucas and Evans, 1968; Yates and Daniel, 1970; Biggins and Postgate, 1971b). Acetylene-reducing activity was greatly enhanced by adding benzyl viologen. Yates (1971a) purified NADH dehydrogenase from *A. chroococcum*, and obtained a multi-sited preparation with a specific site to reduce benzyl viologen which, in turn, coupled to purified Azotobacter nitrogenase to reduce acetylene. However, Azotobacter ferredoxin and flavodoxin did not substitute for benzyl viologen in this reaction. Reduced benzyl- or methyl viologen also donate electrons to purified nitrogenase from *Klebsiella pneumoniae* (Ware, 1972).

In the late 1950s and early 1960s, evidence was published suggesting a role for cytochromes in electron transport to nitrogenase in *Azotobacter*. Shug *et al.* (1956) and Wilson (1958) reported that nitrogen gas could oxidize the cytochromes of *A. vinelandii*, while Ivanov and his colleagues in several publications (e.g. Ivanov *et al.*, 1965) presented evidence purported to show that electron transfer to nitrogenase occurs *via* cytochromes. This evidence has been critically reviewed elsewhere (Yates, 1971b) and, in our opinion, there is no substantial reason to believe that cytochromes transfer electrons to nitrogenase. Differences between the cytochrome composition of nitrogen-fixing and fixed nitrogen-grown *A. vinelandii* (Lisenkova and Khmel, 1967; Knowles and Redfearn, 1968) can be explained on the basis of different intracellular oxygen regimes or the need for a high Q_{O_2} value in nitrogen-fixing organisms.

C. Primary Electron Donors

The primary electron donors for respiration in *Azotobacter* have been discussed (Section IIIB, p. 101); mainly they are the tricarboxylic-acid cycle intermediates and their coenzymes NADH and NADPH. Both carriers have been implicated in electron transport to nitrogenase. Klucas and Evans (1968) showed that β-hydroxybutyric acid plus NAD^+ supported acetylene reduction by extracts from either soybean nodules or *A. vinelandii* in the presence of benzyl viologen. Benzyl viologen was later replaced by FMN, FAD or a flavin-containing factor from acetone-dried nodule extracts (Evans, 1970). Yates and Daniel (1970) and Biggins and Postgate (1971b) showed NADH-dependent acetylene reduction in particulate preparations from both *A. chroococcum* and *Mycobacterium flavum*, with or without the addition of benzyl viologen. The work of Benemann *et al.* (1971) and Wong *et al.* (1971) which was described in the last section of this review, implicates NADPH as a primary donor in extracts of *A. vinelandii* and bacteroids, in the presence of ferredoxin and flavodoxin; NADH did not support acetylene reduction in these systems.

The pyruvate phosphoroclastic reaction is the main electron-donating and energy-providing system for nitrogen fixation in the anaerobe *Cl. pasteurianum*. Components of such a system have been observed in *Azotobacter*. Rose *et al.* (1954) detected acetate kinase but not phosphotransacetylase activity in extracts of acetate-grown *A. vinelandii* strains O and S. More recently, Bresters *et al.* (1972) and Haaker *et al.* (1972) showed the presence of pyruvate dehydrogenase, phosphotransacetylase and acetokinase, and claimed anaerobic ATP production in extracts of *A. vinelandii*. They suggested that this pyruvate-dependent ATP production was an energy and electron source for nitrogen fixation in *A. vinelandii*. F. O. Campbell and M. G. Yates (unpublished data) have confirmed the presence of acetokinase in sucrose-grown *A. vinelandii* strains O and OP, but they failed to detect it in sucrose-grown *A. chroococcum* (NCIB strain 8003) or a non-gummy variant developed in their laboratory. Very low levels of acetokinase activity were observed in crude extracts of *Azomonas macrocytogenes*, *Azotobacter agilis* and *Beijerinckia indica*, but none in extracts of *Derxia gummosa*. However, not one of these organisms either grew or reduced acetylene in the absence of oxygen even after the cells had been pre-exposed to nitrogen gas for periods up to 16 h. It is unlikely, therefore, that anaerobic ATP production is a major energy source for nitrogen fixation in *Azotobacter*. It is probable that acetokinase and phosphotransacetylase, when present, operate in acetate metabolism in the reverse direction to the phosphoroclastic system.

Bresters *et al.* (1972) made an unsubstantiated suggestion that pyruvate-dependent nitrogen fixation could occur in extracts from *A. vinelandii*. It is possible, therefore, that any one of several substrates acts as primary electron donor for nitrogenase, but that NADPH and NADP-dependent carbon substrates yield the highest activities for acetylene reduction by cell-free extracts in the absence of artificial electron donors (Benemann *et al.*, 1971). Since establishing that Azotobacter flavodoxin hydroquinone or, possibly, reduced ferredoxin will link directly to purified nitrogenase, we have attempted to reduce these carriers with physiological substrates including NADH, NADPH or NADH- and NADPH-generating substrates in the presence of extracts of *Azotobacter*. These efforts have met with little success, which suggests that the responsible enzymes are inactivated during cell disruption. The experiments of Yates and Daniel (1970) showed that acetylene-reducing activity with physiological substrates was associated with large particles and cell debris after disrupting *A. chroococcum*, while Benemann *et al.* (1971) showed that such activity in crude extracts from *A. vinelandii* was very unstable on storage. Edmondson and Tollin (1971) proposed that flavodoxin semiquinone from *Azotobacter* was converted from the stable, neutral form to the reactive anionic form by losing a proton before it would either oxidize or reduce. To bring about such a loss *in vivo* may require essentially aprotic conditions; this would account for the preponderance of flavodoxin in membranes of *Azotobacter* and may also be one reason why nitrogen fixation with physiological electron donors is so difficult to reproduce *in vitro*.

Since nitrogen fixation requires a constant and plentiful supply of electrons, it is logical to suppose the presence of one or more very active dehydrogenases in *Azotobacter*. For instance, Barrera and Jurtshuk (1970) observed that $NADP^+$-dependent isocitrate dehydrogenase from acetate-grown *A. vinelandii* had a specific activity which was 10- to 100-fold greater than that of any other dehydrogenase in the cell, and they suggested that isocitrate was the primary donor of electrons for nitrogen fixation, either directly *via* NADPH or indirectly *via* NADPH-NAD^+ transhydrogenase. Experiments with continuously grown cultures of *A. chroococcum* in the A. R.C. Unit showed little difference in the activities of dehydrogenases from ammonia- and dinitrogen-grown cells after sonic disruption. Enzyme activities were tested by reduction of triphenyltetrazolium chloride. They included NADH-, NADPH-, glucose 6-phosphate-, 6-phosphogluconate-, succinate-, citrate-, isocitrate-, malate-, fumarate- and α-ketoglutarate dehydrogenases. The dehydrogenases for NADH-, glucose 6-phosphate, NADPH- and isocitrate were 5 to 10 times more active than any others (Table 1, p. 102). The first two dehydrogenases were also the most active in supporting acetylene reduction by particulate fractions from *A. chroococcum* (Yates and Daniel, 1970). As mentioned earlier, Yates (1971a) demonstrated the specific nature of the benzyl viologen (tetrazolium)-reducing site of purified NADH dehydrogenase, and that reduced benzyl viologen linked to nitrogenase.

D. Regulation of NAD(P)H:NAD(P)$^+$ Ratios

From the evidence presented in the last section, it is probable that NADH, and/or NADPH, play a significant role in electron transfer to nitrogenase. High levels of the reduced nucleotides are inhibitory, and optimal reduction of acetylene was often achieved with NAD(P)H-generating systems (Yates and Daniel, 1970; Benemann *et al.*, 1971; Wong *et al.*, 1971). Moreover, high levels of NAD(P)H inhibit key catabolic enzymes in *Azotobacter*, e.g. glucose 6-phosphate-, 6-phosphogluconate- and isocitrate dehydrogenases (Barrera and Jurtshuk, 1970; Senior and Dawes, 1971). It should, therefore, be important that *Azotobacter* exert a fine control over the internal NAD(P)H:NAD(P)$^+$ ratios. There are several enzyme systems which may do this in *Azotobacter*.

(i) Ackrell *et al.* (1972) discovered a NADPH dehydrogenase in *A. vinelandii* to which NADPH was a positive homotropic effector (Section IIIB, p. 103). This enzyme would exert a fine control over the

NADPH:NADP$^+$ ratio in *Azotobacter* particularly at high oxygen solution rates when nitrogenase is "switched off" (Dalton and Postgate, 1969a).

(ii) Flavodoxin hydroquinone from *Azotobacter* reacts very rapidly with oxygen (Edmondson and Tollin, 1971; Yates, 1972b): a 4×10^{-5} *M* solution was 50% oxidized to the semiquinone in 12 msec by oxygen-saturated tris buffer at pH 8·2 (M. G. Yates and R. N. F. Thorneley, unpublished data). If NAD(P)H is the primary electron donor for hydroquinone formation then autoxidation of the hydroquinone would help to minimize the levels of reduced carriers.

(iii) Highly active NAD(P)H-NAD(P)$^+$ transhydrogenases exist in *Azotobacter* (Chung, 1970; Senior and Dawes, 1971; Yates, 1971a). Transhydrogenase activity followed by elimination of electrons through the repiratory chain would help to maintain low NAD(P)H:NAD(P)$^+$ ratios.

(iv) A mechanism for exerting control over the NAD(P)H:NAD(P)$^+$ ratios under oxygen-limited conditions was proposed by Senior and Dawes (1971). The NADH-dependent synthesis of poly-β-hydroxybutyrate under oxygen limitation in *Azotobacter beijerinckia* was a means to relieve inhibition of the enzymes in the Entner-Doudoroff pathway such as glucose 6-phosphate- and 6-phosphogluconate dehydrogenases by high levels of NAD(P)H. This would allow the energy-producing cycles to operate and nitrogen fixation to continue.

E. The Role of Hydrogenase

Hydrogenase occurs in all organisms that fix nitrogen; in anaerobes, activity of the enzyme is apparently reversible; in *Azotobacter* it is almost completely one-directional (Hyndman *et al.*, 1953). Its role has been discussed recently, together with that of bacteroid hydrogenase by Dixon (1970), who suggested three possible functions. (i) A supplementary oxygen-scavenging effect as a protective mechanism for nitrogenase. (ii) Prevention of inhibition by molecular hydrogen of nitrogenase activity, on the presumption that the nitrogenase from *Azotobacter* evolves hydrogen *in vivo*. (iii) Enhancing the efficiency of nitrogen fixation by recycling the hydrogen evolved in (ii). On this basis, molecular hydrogen should act as an electron donor for nitrogenase and support acetylene reduction by washed cells of *Azotobacter*. The results of such an experiment have not been reported. Attempts in this laboratory to reduce acetylene by using molecular hydrogen as an electron donor with whole cells or crude extracts from *Azotobacter* have not been successful.

V. Protection of Nitrogenase Against Oxygen Damage

Several investigators have reported that growth of nitrogen-fixing *Azotobacter* is inhibited by a high pO_2 (Prazmowski, 1912; Meyerhof and Burk, 1928; Burk, 1930; Tschapek and Giambiagi, 1955; Dilworth and Parker, 1961). Meyerhof and Burk (1928) thought that this inhibition by oxygen was an effect upon nitrogen fixation but, later, Burk (1930) considered that the phenomenon was a general inhibition of growth. Schmidt-Lorenz and Rippel-Baldes (1957) proposed that oxygen inhibited nitrogen fixation by changing the redox potential of the environment. Other workers (Parker, 1954; Phillips and Johnson, 1961; Khmel *et al.*, 1965; Khmel and Gabinskaya, 1965) observed increased efficiency of nitrogen fixation in *A. vinelandii* at low pO_2 values, and Parker and Scutt (1958, 1960) calculated from data obtained by Warburg manometry that oxygen was a competitive inhibitor of nitrogen fixation. In addition, Dilworth (1962) and Dilworth and Kennedy (1963) showed that acetate kinase, malate synthase, isocitrate lyase and phosphotransacetylase activities in *Azotobacter* were damaged at a partial pressure of one atmosphere of oxygen.

Dalton and Postgate (1969a) studied the growth of *Azotobacter* in both batch and continuous culture, and found that the inhibition by oxygen was specific to nitrogen-fixing cells. They also established that the effect was one of inhibition by oxygen rather than lack of carbon dioxide which is necessary to initiate growth of many aerobes (Walker, 1932; Gladstone *et al.*, 1935). Postgate and his colleagues (Dalton and Postgate, 1969a, b; Yates, 1970a; Drozd and Postgate, 1970a, b; Hill, 1972; Hill *et al.*, 1972; Lees and Postgate, 1973) in a detailed investigation of this phenomenon using continuous cultures have established that: (i) the degree of sensitivity of nitrogen fixation towards oxygen inhibition was dependent upon the nutritional status of the bacteria: carbon-limited and phosphate-limited cells were particularly oxygen-sensitive when fixing nitrogen (Dalton and Postgate, 1969a; Hill *et al.*, 1972; Lees and Postgate, 1973). *Azotobacter* rarely showed optimal nitrogen-fixing ability at the atmospheric pO_2 value, and in "nitrogen-limited cells" (in the sense defined by Dalton and Postgate, 1969b; as a limitation of nitrogenase function) the optimal pO_2 value for acetylene reduction depended on the pO_2 value during growth and the population density (Drozd and Postgate, 1970a); (ii) nitrogen-fixing ability stopped abruptly when the oxygen solution rate was increased and restarted immediately, or after a short lag, when the oxygen solution rate was lowered again (Yates, 1970a; Drozd and Postgate, 1970a, b). The readiness with which the "switching off" occurred depended upon the population density and the nutritional status of the cells: phosphate-limited or carbon-limited cells were particularly sensitive to an increase in oxygen solution rate, whereas dense cultures of nitrogen-limited cells need very vigorous shaking to "switch off".

The interpretation of these results was formulated into a theory by Dalton and Postgate (1969a). This involved two types of protection against oxygen inhibition of nitrogen fixation. The first type, respiratory protection, operates while nitrogenase is functional and maintains an anaerobic environment. The second type, conformational protection, operates to make the nitrogenase non-functional or "switched off", and protects against damage to the enzyme by oxygen.

A. Respiratory Protection

The evidence for respiratory protection in *Azotobacter* has been discussed by Postgate (1971b, 1974) and will be re-iterated here briefly. Observations which support the main lines of evidence already mentioned are:

(i) The growth and nitrogen fixation efficiencies (yield per g of carbon substrate) of *A. chroococcum* in continuous culture are inversely proportional to the dilution rate and the pO_2. This suggests that exposure to excess oxygen, either in time or concentration, causes the organism to waste carbon substrate, presumably to protect the nitrogenase from damage by oxygen (Dalton and Postgate, 1969a, b).

(ii) Populations subjected to oxygen stress in conditions in which they cannot exert respiratory protection (phosphate-limited populations) show specific damage to their nitrogenase system (Lees and Postgate, 1973).

(iii) Nitrogen-fixing continuous cultures of *A. chroococcum* can adapt themselves to high pO_2 levels during which their Q_{O_2} values and cytochrome a_2 content increase markedly (Drozd and Postgate, 1970b). Large increases in Q_{O_2} values were also observed when batch cultures of *A. vinelandii*, subjected to restricted aeration, were exposed to highly aerobic conditions, and were accompanied by *de novo* synthesis of extra cytochrome a_2, NADH- and NADPH dehydrogenases, and by a sharp decrease in energy conservation efficiency particularly at site I (Jones *et al.*, 1973).

(iv) The concept of partly uncoupled respiration, particularly at high pO_2 levels, is supported by the high Q_{O_2} values exhibited by nitrogen-fixing continuous cultures of *A. vinelandii* (Nagai *et al.*, 1969; Nagai and Aiba, 1972).

(v) The highly branched respiratory system of *Azotobacter* (see Sections IIIC, p. 106, and D, p. 109) is clearly capable of allowing wide variation in energy conservation efficiency according to the prevailing oxygen concentration and the exact pathway of electron transfer employed. Highly aerobic conditions promote respiration of low-energy conservation efficiency *via* an uncoupled NADH dehydrogenase and the non-phosphorylating branch of the cytochrome system terminated by cytochrome a_2. At particularly high oxygen concentrations, nitrogenase is "switched off" and respiration is augmented by oxidation of NADPH *via* the non-phosphorylating NADPH dehydrogenase (Ackrell *et al.*, 1972). Since the Q_{O_2} value of an aerobic bacterial culture is inversely proportional to its efficiency of oxidative phosphorylation (Harrison and Loveless, 1971), the utilization of inefficient pathways of energy conservation by *Azotobacter* under highly aerobic conditions readily accounts for the very high respiratory activity. The involvement of respiration in the protection of nitrogenase may well be the reason for the very high Q_{O_2} values for which the Azotobacteriaceae are famous (Williams and Wilson, 1954). Thus, the need to protect functioning nitrogenase and autoxidizable ferredoxin and flavodoxin from inhibition by oxygen is met in the cell by enhanced respiration concomitant with a decrease in the efficiency of oxidative phosphorylation (see Fig. 1, p. 105).

In *A. vinelandii* cultured under oxygen-limited conditions, the activities of the "minor" branches of the cytochrome system terminated by cytochromes a_1 and o (at least one branch of which is phosphorylating) are particularly high; the efficiency of energy conservation at site I is maximum, oxidation of NADPH *via* the respiratory chain NADPH dehydrogenase is probably minimal, and whole-cell Q_{O_2} values are low (Ackrell and Jones, 1971b; Jones *et al.*, 1973). The marked increase in cytochrome content as a result of oxygen limitation probably reflects an attempt by the cell to maintain electron flow in the face of limited acceptor availability, and is a common characteristic of aerobic organisms (Sinclair and White, 1970).

In spite of recent clarification of the nature of respiratory protection of nitrogenase in *Azotobacter*, it is still not clear how, for example, the ambient oxygen concentration controls either the efficiency of energy conservation at site I or the differential synthesis of cytochrome a_2.

B. Conformational Protection

Because nitrogenase proteins are oxygen-sensitive they need to be protected during periods of oxygen stress. Dalton and Postgate (1969a) suggested that nitrogenase underwent a conformational change which made it oxygen-tolerant. Postgate (1974) has discussed the question of conformational protection in some detail. Supporting evidence for conformational protection is as follows:

(a) The nitrogenase in crude cell-free extracts from *Azotobacter* is oxygen-tolerant, whereas those in crude extracts from anaerobic nitrogen fixers are oxygen-sensitive (Kelly, 1969). The purified nitrogenase fractions from all of the azotobacters so far tested are oxygen-sensitive (Bulen and LeComte, 1966; Kelly, 1969). It follows that the nitrogenase from *Azotobacter* is protected from oxygen damage in crude extracts by association with unknown components (such as proteins, lipids, phospholipids); presumably these components assist the nitrogenase to maintain a conformation in which it is oxygen-tolerant. Yates (1970b) showed that, when nitrogenase from *Azotobacter* was separated from NADH dehydrogenase by DEAE-cellulose chromatography, it became oxygen-sensitive; adding back the NADH dehydrogenase restored oxygen tolerance. Purified NADH dehydrogenase, whether particulate or solubilized, was unable to do this.

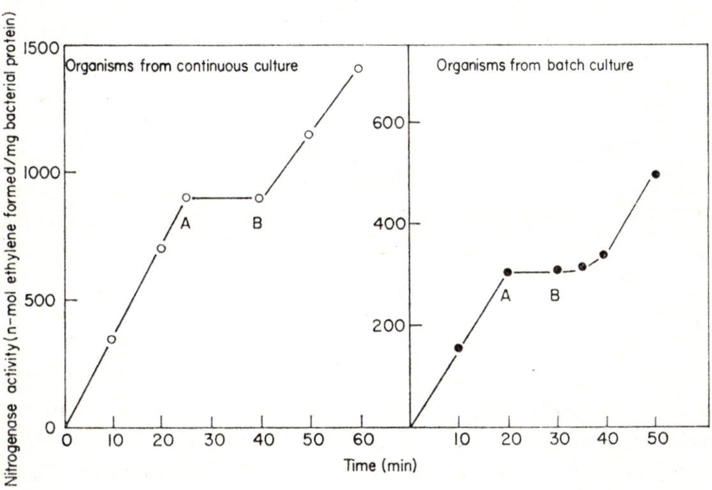

FIG. 4. "Switch off" and "switch on" of nitrogenase activity by *Azotobacter chroococcum* in response to aeration. Samples (2 ml) of growing cultures (batch or continuous) were shaken gently in argon + 0·1 atmosphere oxygen containing acetylene. Shaking rates increased at A, and returned to the original value at B. (A) Shaking rates: 75 × 1·5 cm/min to 150 × 4 cm/min and back. (B) Shaking rates: 75 × 1·5 cm/min to 250 × 1 cm/min and back. From Hill *et al.* (1972).

Presumably, nitrogenase is also protected from oxygen damage *in vivo* since acetylene-reducing ability can "switch on" immediately upon reverting from a high to a low oxygen solution rate (Yates, 1970a; Drozd and Postgate, 1970a, b; Fig. 4). This "switch on" was observed in the presence of chloramphenicol which indicated that synthesis of nitrogenase was unnecessary (Drozd and Postgate, 1970b). In some cases, "switch on" was delayed (Fig. 4) and optimal nitrogenase activity was not achieved for periods up to 40 min (Hill *et al.*, 1972). The longest "delay" recorded was 90 min (J. R. Postgate, personal communication). Such a phenomenon might be due to either: (i) a low Q_{O_2} resulting in a slow removal of excess molecular oxygen from the cell; (ii) oxygen damage to nitrogenase; or (iii) an inadequate supply of a metabolite necessary to cause the conformational change to the "switched on" form. Cells which show a "switch on" lag generally take 10–15 min to remove oxygen from solution after the change from a high to a low oxygen solution rate (Lees and Postgate, 1973). This observation correlates well with the lag time reported by Hill *et al.* (1972). In cases when the lag is longer (e.g. 90 min), addition of chloramphenicol to inhibit protein synthesis would decide whether possibility (ii) has occurred. In all cases, after "switch on" a return to the original rate of nitrogenase activity is mandatory since this activity is governed by the oxygen solution rate, which in all of the experiments reported was returned to the original low value. Possibility (iii) cannot be tested without knowing the metabolite which triggers the "switch on". The sluggish "switch off" displayed by cultures of *Derxia gummosa* was attributed to "slime" which retarded the access of oxygen to the cell (Hill, 1972).

(b) The rate of nitrogen fixation by nitrogen-limited *A. chroococcum* in continuous cultures depends upon the growth rate which, in turn, is governed by the dilution rate. This implies that nitrogenase function in slow-growing cells is controlled by the dissolved oxygen concentration (Dalton and Postgate, 1969b).

The location of nitrogenase within *Azotobacter* may be an important factor in the protection against oxygen damage: Azotobacter nitrogenase is particulate in crude extracts (Bulen *et al.*, 1964) whereas that in *Cl. pasteurianum* is soluble (Carnahan *et al.*, 1960). Russian workers (Yakovlev and Levchenko, 1964; Yakovlev *et al.*, 1965; Levchenko *et al.*, 1969) decided that nitrogenase in *Azotobacter vinelandii* is located in small cell organelles which they called "mitochondria". They used iodonitrotetrazolium indicator, and determined the sites of formazan formation in the presence of electron donors by electron microscopy. Formazan formation was inhibited by nitrogen gas, and the authors suggested that this inhibition occurred at nitrogenase-containing sites. Yates and Daniel (1970) found that acetylene reduction with physiological electron donors rather than sodium dithionite was associated with easily sedimentable membrane fractions from *A. chroococcum* and, although optimal activity was obtained under anaerobic conditions, some activity was observed in the presence of oxygen. Biggins and Postgate (1971b) obtained similar membranous active fractions from

Mycobacterium flavum. Such membrane preparations from *M. flavum* were oxygen-tolerant whereas, in particulate fractions (comparable to those obtained from crude supernatants of *Azotobacter*: Bulen *et al.*, 1965), the nitrogenase was oxygen-sensitive. On the other hand, Oppenheim and Marcus (1970) showed that nitrogenase obtained from *A. vinelandii* by osmotic shock was free from NADH dehydrogenase, less easily sedimentable, and more oxygen-sensitive than nitrogenase in crude extracts obtained by disrupting organisms in a French pressure cell. They also showed that nitrogen-fixing cells had an extensive internal membrane network which was absent from ammonia-grown organisms. This observation was confirmed by Marcus and Kaneshiro (1972) and for *A. chroococcum* by Hill *et al.* (1972). Marcus and Kaneshiro (1972) also found that, while the total lipid content in *A. vinelandii* was similar, there were apparent differences in the contents of phospholipids, anionic lipids, neutral lipids and coenzyme Q between nitrogen-fixing and ammonia-grown cells. It seems likely, therefore, that nitrogenase in *Azotobacter* and in *M. flavum* is associated with membrane fractions which serve to protect it from oxygen damage both by their respiratory activity and by physical association.

Alternative explanations can be given for the apparent "switching off" of nitrogenase under oxygen stress in *Azotobacter*. Firstly, Parker and Scutt (1960) suggested that nitrogen fixation could be regarded as a form of respiration, and that molecular oxygen competed with dinitrogen for electrons. A sudden large increase in the oxygen solution rate would change the nutrient status of the cells from that of oxygen limitation to that of apparent carbon limitation; in other words, oxidation of carbon-containing substrates would not be fast enough to produce sufficient electrons for both oxygen uptake and nitrogen fixation. From the data presented by Yates (1970a) and Drozd and Postgate (1970b), the apparent number of electrons transferred to nitrogenase during growth of *A. chroococcum* is less than 1% of the number transferred to oxygen. Therefore, in theory, a very small increase in the Q_{O_2} value would be sufficient to inhibit the electron supply to nitrogenase and cause "switch off". Figure 5 (p. 128) shows the rates of oxygen uptake and acetylene reduction in cultures of *A. chroococcum*, and that an increase in the oxygen solution rate which is sufficient to "switch off" nitrogenase does not alter the rate of oxygen uptake significantly. Secondly, the high reactivity of the flavodoxin hydroquinone from *Azotobacter* towards oxygen was mentioned earlier. Access to excess oxygen in the cell might enable the hydroquinone to transfer electrons preferentially to oxygen rather than to nitrogenase and thus cause an instant "switch off". This reactivity of the hydroquinone could explain the non-competitive nature of the inhibition of nitrogenase activity of particulate preparations from

A. vinelandii by oxygen in the presence of illuminated chloroplasts and Azotobacter flavodoxin (Wong and Burris, 1972).

Both of these explanations could account for the almost instantaneous "switch off" and "switch on" of nitrogenase activity, but they do not preclude the possibility that a conformational change occurs in the nitrogenase at the same time as "switch off" to protect it from oxygen damage. While conformational protection of nitrogenase in crude extracts of *Azotobacter* is a proven fact, the postulated regulation of this

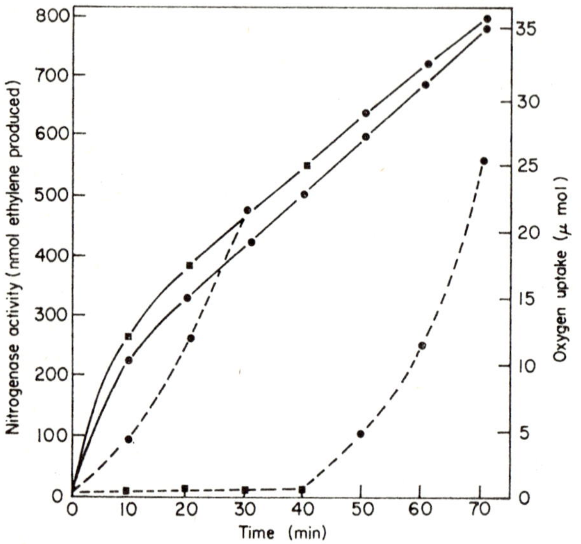

FIG. 5. Effect of changing the shaking amplitude on oxygen uptake and acetylene reduction in *Azotobacter chroococcum*. Culture (10 ml containing 0·15 mg bacterial protein/ml) was injected into a manometer flask containing glucose in the main compartment and potassium hydroxide in the centre well under argon–oxygen (4:1) containing 4% acetylene at 30°C. The shaking rate was 90 strokes/min at amplitudes of 7·9 or 4·7 cm. After 40 min the amplitude of 7·9 cm was lowered to 4·7 cm. ■——■, indicates oxygen uptake at amplitude 7·9 cm; ●——●, oxygen uptake at amplitude 4·7 cm; ■----■, ethylene produced at amplitude 7·9 cm; ●----●, ethylene produced at amplitude 4·7 cm. From Yates (1970a).

protection *in vivo*, involving a change in the conformation of nitrogenase from an oxygen-sensitive to an oxygen-tolerant state, remains an attractive theory supported only by circumstantial evidence. Supporting evidence of a physical nature is required. In other words, oxygen-tolerant and oxygen-sensitive nitrogenases should be extractable from *Azotobacter* which have been exposed to such levels of oxygen that they should possess oxygen-tolerant and oxygen-sensitive nitrogenases respectively.

Experiments designed to obtain such proteins have not been successful in this laboratory. Evidence exists, however, that nitrogenase proteins do undergo a conformational change which influences their reactivity towards oxygen. An indication that fraction 2 (Ac2) of nitrogenase from *A. chroococcum* can change its conformation to become more oxygen-sensitive in the presence of ATP was obtained by Yates (1972a). Adenosine triphosphate made purified Ac2, as well as nitrogenase activity in crude extracts, hypersensitive to oxygen. This effect was apparently related to the chelating ability of ATP and not to the ATPase activity. It was also obtained with other trinucleotides (ITP, GTP) which are inactive in nitrogenase function. Nevertheless, this may represent a real function of ATP in nitrogen fixation, namely that it associates with fraction 2 to create an "active species" to combine with fraction 1 for the subsequent reduction of substrate (Fig. 6). Evidence in support of this idea shows that ATP induces a conformational change in fraction 2 (Kp2) of *Klebsiella pneumoniae* which enhances its sensitivity towards oxygen and its reactivity towards 5,5-dithio-nitrobenzoic acid (Thorneley and Eady, 1973).

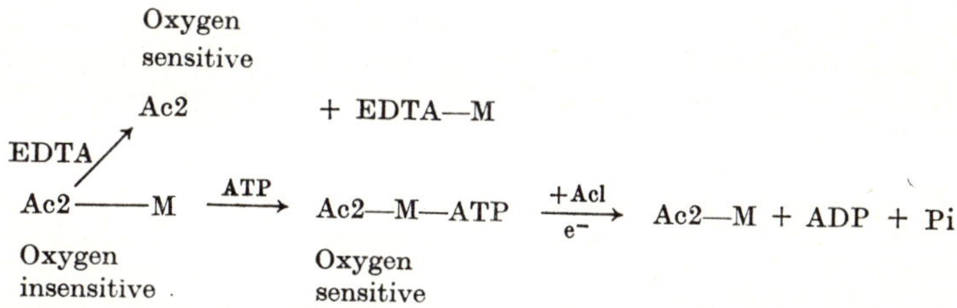

FIG. 6. Suggested mechanism whereby the iron protein (Ac2) of nitrogenase is made sensitive to oxygen by ATP or EDTA in crude supernatants from *Azotobacter chroococcum*. M = metal ion (probably Mg^{2+}, Ca^{2+} or Fe^{2+}); Ac_2 indicates the iron protein of nitrogenase. From Yates (1972a).

Finally, it is pertinent to note that the oxygen sensitivity of nitrogenase from *A. chroococcum*, either in the presence or absence of ATP, is inversely related to the protein concentration (Yates, 1970b, 1972a) and that enzyme concentrations *in vivo* are many-fold higher than is usual *in vitro*. Therefore, nitrogenase may be protected against oxygen damage in part by the concentration of protein and association with other proteins *in vivo* as well as by a specific conformational change. If specific regulation of conformational protection does occur then the mechanism can be damaged. For example, Drozd and Postgate (1970b) reported that the nitrogenase in *A. chroococcum* which had been exposed to pure oxygen for periods up to 1 h became inactive because of oxygen damage to fraction 2 (Ac2). Also, Lees and Postgate (1973) found that phosphate-limited *A. chroococcum* was non-viable on nitrogen-deficient medium after exposure to oxygen but was 100% viable in ammonia-containing medium. It follows, therefore, that in these circumstances nitrogenase is appreciably susceptible to oxygen damage even when, according to the theory of Dalton and Postgate (1969a), it ought to be in the "switched off" state. It is noteworthy that this hypersensitivity to oxygen in phosphate-limited cells (Lees and Postgate, 1973) parallels the inability of these populations to conduct respiratory protection, and implies that the regulation of conformational protection is in response to similar signals as those that initiate respiratory protection.

VI. Acknowledgement

We thank Professor J. R. Postgate for useful criticism of the manuscript.

References

Ackrell, B. A. C., Erickson, S. K. and Jones, C. W. (1972). *European Journal of Biochemistry* **26**, 387.

Ackrell, B. A. C. and Jones, C. W. (1971a). *European Journal of Biochemistry* **20**, 22.

Ackrell, B. A. C. and Jones, C. W. (1971b). *European Journal of Biochemistry* **20**, 29.

Alexander, M. and Wilson, P. W. (1955). *Proceedings of the National Academy of Sciences of the United States of America* **41**, 843.

Aprison, M. H. and Burris, R. H. (1952). *Science* **115**, 264.

Arima, K. and Oka, T. (1965). *Journal of Bacteriology* **90**, 734.

Atkinson, D. E. and Walton, G. M. (1967). *Journal of Biological Chemistry* **242**, 3239.

Baak, J. M. and Postma, P. W. (1971). *Federation of European Biochemical Societies Letters* **19**, 189.

Barman, B. G. and Tollin, G. (1972). *Biochemistry* **11**, 4755.

Barrera, C. R. and Jurtschuk, P. (1970). *Biochimica et Biophysica Acta* **220**, 416.

Bauchop, T. and Elsden, S. R. (1960). *Journal of General Microbiology* **23**, 457.

Beinert, H., Heinen, W. and Palmer, G. (1962). *Brookhaven Symposium* **15**, 229.

Benemann, J. R. and Valentine, R. C. (1972). *Advances in Microbial Physiology* **8**, 59.

Benemann, J. R., Yoch, D. C., Valentine, R. C. and Arnon, D. I. (1969). *Proceedings of the National Academy of Sciences of the United States of America* **64**, 1079.

Benemann, J. R., Yoch, D. C., Valentine, R. C. and Arnon, D. I. (1971). *Biochimica et Biophysica Acta* **226**, 205.

Bergersen, F. J. (1966a). *Biochimica et Biophysica Acta* **115**, 247.

Bergersen, F. J. (1966b). *Biochimica et Biophysica Acta* **130**, 304.

Bergersen, F. J. (1972). *Annual Review of Plant Physiology* **20**, 541.

Bergersen, F. J. and Turner, G. L. (1967). *Biochimica et Biophysica Acta* **141**, 507.

Biggins, D. R. and Postgate, J. R. (1969). *Journal of General Microbiology* **56**, 181.

Biggins, D. R. and Postgate, J. R. (1971a). *Journal of General Microbiology* **65**, 119.

Biggins, D. R. and Postgate, J. R. (1971b). *European Journal of Biochemistry* **19**, 408.

Bothe, H. and Falkenberg, B. (1972). *Zeitschrift für Naturforschung* **27b**, 1090.

Bresters, T. W., Krul, J., Scheepens, P. C. and Veeger, C. (1972). *Federation of European Biochemical Societies Letters* **22**, 305.

Bruemmer, J. H., Wilson, P. W., Glenn, J. L. and Crane, F. L. (1957). *Journal of Bacteriology* **73**, 113.

Bulen, W. A., Burns, R. C. and LeComte, J. R. (1964). *Biochemical and Biophysical Research Communications* **17**, 265.

Bulen, W. A., Burns, R. C. and LeComte, J. R. (1965). *Proceedings of the National Academy of Sciences of the United States of America* **53**, 532.

Bulen, W. A. and LeComte, J. R. (1966). *Proceedings of the National Academy of Sciences of the United States of America* **56**, 979.

Burk, D. (1930). *Journal of Physical Chemistry* **34**, 1195.

Burris, R. H. (1971). *In* "The Chemistry and Biochemistry of Nitrogen Fixation", (J. R. Postgate, ed.), p. 106. Plenum Press, London.

Carnahan, J. E., Mortenson, L. E., Mower, H. F. and Castle, J. E. (1960). *Biochimica et Biophysica Acta* **44**, 520.

Castor, L. N. and Chance, B. (1959). *Journal of Biological Chemistry* **234**, 1587.

Chance, B. and Baltscheffsky, M. (1958). *Biochemical Journal* **68**, 283.

Chance, B. and Williams, G. R. (1956). *Advances in Enzymology* **17**, 65.

Chung, A. E. (1970). *Journal of Bacteriology* **102**, 438.

Chung, A. E. and Franzen, J. S. (1969). *Biochemistry* **8**, 3175.

Cota-Robles, E. H., Marr, A. G. and Nilson E. H. (1958). *Journal of Bacteriology* **75**, 243.

Coty, V. F. (1967). *Biotechnology and Bioengineering* **9**, 25.

Dalton, H. and Mortenson, L. E. (1972). *Bacteriological Reviews* **36**, 231.

Dalton, H. and Postgate, J. R. (1969a). *Journal of General Microbiology* **54**, 463.

Dalton, H. and Postgate, J. R. (1969b). *Journal of General Microbiology* **56**, 307.

Daniel, R. M. and Erickson, S. K. (1969). *Biochimica et Biophysica Acta* **180**, 63.

DerVartanian, D. V. (1972). *Zeitschrift für Naturforshung* **27b**, 1080.

DerVartanian, D. V. and Bramlett, R. (1970). *Biochimica et Biophysica Acta* **220**, 443.

D'Eustachio, A. J. and Hardy, R. W. F. (1964). *Biochemical and Biophysical Research Communications* **15**, 319.

Dilworth, M. J. (1962). *Biochimica et Biophysica Acta* **56**, 127.

Dilworth, M. J. and Kennedy, I. R. (1963). *Biochimica et Biophysica Acta* **67**, 240.

Dilworth, M. J. and Parker (1961). *Nature* **191**, 520.

Dixon, R. O. D. (1972). *Archiv für Mikrobiologie* **85**, 195.

Drozd, J. W. and Postgate, J. R. (1970a). *Journal of General Microbiology* **60**, 427.

Drozd, J. W. and Postgate, J. R. (1970b). *Journal of General Microbiology* **63**, 63.

Drozd, J. W., Tubb, R. S. and Postgate, J. R. (1972). *Journal of General Microbiology* **73**, 221.

Edmondson, D. E. and Tollin, G. (1971). *Biochemistry* **10**, 133.

Eilermann, L. J. M. (1970). *Biochimica et Biophysica Acta* **216**, 231.

Eilermann, L. J. M., Pandit-Hovenkamp, H. G. and Kolk, A. H. J. (1970). *Biochimica et Biophysica Acta* **197**, 25.

Eilermann, L. J. M., Pandit-Hovenkamp, H. G., Van der Meer-Van Buren, M., Kolk, A. H. J. and Feenstra, M. (1971). *Biochimica et Biophysica Acta* **245**, 305.

Eilermann, L. J. M. and Slater, E. C. (1970). *Biochimica et Biophysica Acta* **216**, 226.

Erickson, S. K., Ackrell, B. A. C. and Jones, C. W. (1972). *Biochemical Journal* **127**, 73P.

Erickson, S. K. and Diehl, H. (1973). *Biochemical and Biophysical Research Communications* **50**, 321.

Evans, H. J. (1970). *In* "How Crops Grow", (J. F. Horsfall, ed.), p. 110. Centennial lecture series, Connecticut Agricultural Experimental Station Bulletin No. 708.
Fahmy, A. R. and Walsh, E. O'f. (1952). *Biochemical Journal* **51**, 55.
Federov, M. F. and Kalininskaya, T. A. (1961). *Mikrobiologiya* **30**, 9.
Gladstone, G. P., Fildes, P. and Richardson, G. M. (1935). *British Journal of Experimental Pathology* **16**, 335.
Green, D. E. and Wharton, D. C. (1963). *Biochemische Zeitschrift* **338**, 335.
Haaker, H., Bresters, T. W. and Veeger, C. (1972). *Federation of European Biochemical Societies Letters* **23**, 160.
Hadjipetrou, L. P., Gerrits, J. P., Tevlings, F. A. G. and Stouthamer, A. H. (1964). *Journal of General Microbiology* **36**, 139.
Hardy, R. W. F. and Burns, R. C. (1968). *Annual Review of Biochemistry* **37**, 331.
Hardy, R. W. F. and Knight, E. J. (1968). *In* "Progress in Phytochemistry", (L. Reinhold and Y. Liwschitz, eds.), p. 407. Wiley, London.
Harrison, D. E. F. and Loveless, J. E. (1971). *Journal of General Microbiology* **68**, 35.
Hempfling, W. P. (1970). *Biochimica et Biophysica Acta* **205**, 169.
Henderson, P. J. F. (1973). *Biochemical Journal* **135**, 101–107.
Hill, S. (1972). *Journal of General Microbiology* **67**, 77.
Hill, S., Drozd, J. W. and Postgate, J. R. (1972). *Journal of Applied Chemistry and Biotechnology* **22**, 541.
Hill, S. and Postgate, J. R. (1969). *Journal of General Microbiology* **58**, 277.
Hinkson, J. W. and Bulen, W. A. (1967). *Journal of Biological Chemistry* **242**, 3345.
Hovenkamp, H. G. (1959a). *Biochimica et Biophysica Acta* **34**, 485.
Hovenkamp, H. G. (1959b). *Nature* **184**, 471.
Hyndman, L. A., Burris, R. H. and Wilson, P. W. (1953). *Journal of Bacteriology* **65**, 522.
Ivanov, I. D., Sitonite, Yu. P. and Belov, Yu. M. (1965). *Mikrobiologiya* **34**, 193.
Jones, C. W., Ackrell, B. A. C. and Erickson, S. K. (1971a). *Federation of European Biochemical Societies Letters* **13**, 33.
Jones, C. W., Ackrell, B. A. C. and Erickson, S. K. (1971b). *Biochimica et Biophysica Acta* **245**, 54.
Jones, C. W., Ackrell, B. A. C. and Erickson, S. K. (1972). *Biochemical Journal* **127**, 74P.
Jones, C. W., Brice, J. M., Wright, V. and Ackrell, B. A. C. (1973). *Federation of European Biochemical Societies Letters* **29**, 77.
Jones, C. W. and Redfearn, E. R. (1966). *Biochimica et Biophysica Acta* **113**, 467.
Jones, C. W. and Redfearn, E. R. (1967a). *Biochimica et Biophysica Acta* **143**, 340.
Jones, C. W. and Redfearn, E. R. (1967b). *Biochimica et Biophysica Acta* **143**, 354.
Jurtshuk, P., Aston, P. R. and Old, L. (1967). *Journal of Bacteriology* **93**, 1069.
Jurtshuk, P., Bednarz, A. J., Zey, P. and Denton, C. H. (1969b). *Journal of Bacteriology* **98**, 1120.
Jurtshuk, P. and Harper, L. (1968). *Journal of Bacteriology* **96**, 678.
Jurtshuk, P., May, A. K., Pope, L. M. and Aston, P. R. (1969a). *Canadian Journal of Microbiology* **15**, 797.
Jurtshuk, P. and Old, L. (1968). *Journal of Bacteriology* **95**, 1790.
Kaplan, N. O., Colowick, S. P., Neufeld, E. F. and Ciotti, M. M. (1953). *Journal of Biological Chemistry* **205**, 17.
Kelly, M. (1969). *Biochimica et Biophysica Acta* **191**, 527.
Khmel, I. A. and Gabinskaya, K. N. (1965). *Mikrobiologiya* **34**, 661.
Khmel, I. A., Gabinskaya, K. N. and Ierusalimskii, N. D. (1965). *Mikrobiologiya* **34**, 689.

Klucas, R. and Evans, H. J. (1968). *Plant Physiology* **43**, 1458.

Koch, B., Wong, P., Russell, S. A., Howard, R. and Evans, H. J. (1970). *Biochemical Journal* **118**, 773.

Knight, E., Jr., D'Eustachio, A. J. and Hardy, R. W. F. (1966). *Biochemical and Biophysical Research Communications* **113**, 626.

Knight, E., Jr. and Hardy, R. W. F. (1966). *Journal of Biological Chemistry* **241**, 2752.

Knowles, C. J. and Redfearn, E. R. (1966). *Biochemical Journal* **99**, 33P.

Knowles, C. J. and Redfearn, E. R. (1968). *Biochimica et Biophysica Acta* **162**, 348.

Knowles, C. J. and Redfearn, E. R. (1969). *Journal of Bacteriology* **97**, 756.

Knowles, C. J. and Smith, L. (1970). *Biochimica et Biophysica Acta* **197**, 152.

Lees, H. and Postgate, J. R. (1973). *Journal of General Microbiology* **75**, 161.

Lester, R. L. and Crane, F. L. (1959). *Biochimica et Biophysica Acta* **32**, 492.

Levchenko, L. A., Ivleva, I. N. and Yakovlev, V. A. (1969). *Izvestiya Akademii Nauk S.S.S.R. Seriya Biologiya* No. 1, 165.

Lisenkova, L. L. and Khmel, I. A. (1967). *Mikrobiologiya* **36**, 905.

L'vov, N. P. and Lyubimov, V. I. (1965). *Izvestiya Akademii Nauk S.S.S.R. Seriya Biologiya* **30**, 250.

Marcus, L. and Kaneshiro, T. (1972). *Biochimica et Biophysica Acta* **288**, 296.

Marr, A. G. and Cota-Robles, E. H. (1957). *Journal of Bacteriology* **74**, 79.

Meyer, D. J. and Jones, C. W. (1973). *European Journal of Biochemistry*, **36**, 144–151.

Meyerhof, O. and Burk, D. (1928). *Zeitschrift für Physikalische Chemie* (*A*) **139**, 117.

Mitchell, P. (1966). *Biological Reviews* **41**, 445.

Mitchell, P. (1967). *Federation Proceedings. Federation of American Societies for Experimental Biology* **26**, 1370.

Mitchell, P. and Moyle, J. (1967). *Biochemical Journal* **105**, 1147.

Mortenson, L. E. (1964). *Biochimica et Biophysica Acta* **81**, 473.

Nagai, S. and Aiba, S. (1972). *Journal of General Microbiology* **73**, 531.

Nagai, S., Nishizawa, Y. and Aiba, S. (1969). *Journal of General Microbiology* **59**, 163.

Negelein, E. and Gerischer, W. (1934). *Biochemische Zeitschrift* **268**, 1.

Neumann, N. P. and Burris, R. H. (1959). *Journal of Biological Chemistry* **234**, 3286.

Oka, T. and Arima, K. (1965). *Journal of Bacteriology* **90**, 744.

Oppenheim, J., Fisher, R. J., Wilson, P. W. and Marcus, L. (1970). *Journal of Bacteriology* **101**, 292.

Oppenheim, J. and Marcus, L. (1970). *Journal of Bacteriology* **101**, 286.

Page, A. C., Gale, P., Wallick, H., Walton, R. B., McDaniel, L. E., Woodruff, H. B. and Folkers, K. (1960). *Archives of Biochemistry and Biophysics* **89**, 313.

Pandit-Hovenkamp, H. G. (1966). *Biochimica et Biophysica Acta* **118**, 645.

Pandit-Hovenkamp, H. G. (1967). *Methods in Enzymology* **10**, 152.

Pangborn, J., Marr, A. G. and Robrish, S. A. (1962). *Journal of Bacteriology* **84**, 669.

Parker, C. A. (1954). *Nature* **173**, 780.

Parker, C. A. and Scutt, P. B. (1958). *Biochimica et Biophysica Acta* **29**, 662.

Parker, C. A. and Scutt, P. B. (1960). *Biochimica et Biophysica Acta* **38**, 230.

Phillips, D. H. and Johnson, M. J. (1961). *Journal of Biochemical, Microbiological and Technical Engineering* **3**, 227.

Postgate, J. R., ed. (1971a). "The Chemistry and Biochemistry of Nitrogen Fixation", Plenum Press, London and New York.

Postgate, J. R. (ed.) (1971b). "The Chemistry and Biochemistry of Nitrogen Fixation" pp. 161. Plenum Press, London and New York.

Postgate, J. R. (1974). *In* "Biological Nitrogen Fixation", (A. Quispel, ed.). North Holland Publishing Company, Amsterdam.

Prajmowski, A. (1912). *Centralblatt für Bakteriologie Parasitenkunde und Infektionskrankheiten Abteilung II* **33**, 292.

Pullman, M. E. and Monroy, G. C. (1963). *Journal of Biological Chemistry* **238**, 3762.

Repaske, R. (1954). *Journal of Bacteriology* **68**, 555.

Repaske, R. and Josten, J. J. (1958). *Journal of Biological Chemistry* **233**, 466.

Racker, E. (1963). *Biochemical and Biophysical Research Communications* **10**, 435.

Ritchie, G. A. F., Senior, P. J. and Dawes, E. A. (1971). *Biochemical Journal* **121**, 309.

Robrish, S. A. and Marr, A. G. (1962). *Journal of Bacteriology* **83**, 158.

Rose, I. A., Gruneberg-Manago, M., Korey, S. R. and Ochoa, S. (1954). *Journal of Biological Chemistry* **211**, 737.

Rose, I. A. and Ochoa, S. (1956). *Journal of Biological Chemistry* **220**, 307.

Schmidt-Lorenz, W. and Rippel-Baldes, A. (1957). *Archiv für Mikrobiologie* **28**, 45.

Schulze, K. L. and Lipe, R. S. (1964). *Archiv für Mikrobiologie* **48**, 1.

Senior, P. J. and Dawes, E. A. (1971). *Biochemical Journal* **125**, 55.

Shethna, Y. I. (1970). *Biochimica et Biophysica Acta* **205**, 58.

Shethna, Y. I., Wilson, P. W., Hansen, R. E. and Beinert, H. (1964). *Proceedings of the National Academy of Sciences of the United States of America* **52**, 1263.

Shug, A. L., Hamilton, P. B. and Wilson, P. W. (1956). *In* "Inorganic Nitrogen Metabolism", (W. D. McElroy and B. Glass, eds.), p. 344. Johns Hopkins Press, Baltimore.

Sinclair, P. R. and White, D. C. (1970). *Journal of Bacteriology* **101**, 365.

Smith, L. (1954). *Archives of Biochemistry and Biophysics* **50**, 299.

Stewart, W. D. P. (1969). *Proceedings of the Royal Society B* **172**, 367.

Stewart, W. D. P., Haystead, A. and Pearson, H. W. (1969). *Nature* **224**, 226.

Streicher, S. L. and Valentine, R. C. (1973). *Annual Review of Biochemistry* **42**, 279.

Swank, R. T. and Burris, R. H. (1969a). *Journal of Bacteriology* **98**, 311.

Swank, R. T. and Burris, R. H. (1969b). *Biochimica et Biophysica Acta* **180**, 473.

Temperli, A. and Wilson, P. W. (1960). *Biochemische Zeitschrift* **320**, 195.

Temperli, A. and Wilson, P. W. (1962). *Nature* **193**, 171.

Thorneley, R. N. F. and Eady, R. R. (1973). *Biochemical Journal* **133**, 405.

Tissières, A. (1956). *Biochemical Journal* **64**, 582.

Tissières, A. and Burris, R. H. (1956). *Biochimica et Biophysica Acta* **20**, 436.

Tissières, A., Hovenkamp, H. G. and Slater, E. C. (1957). *Biochimica et Biophysica Acta* **25**, 336.

Tissières, A. and Slater, E. C. (1955). *Nature* **176**, 736.

Tschapek, M. and Giambiagi, N. (1955). *Archiv. für Mikrobiologie* **21**, 376.

Vambutas, V. K. and Racker, E. (1965). *Journal of Biological Chemistry* **240**, 2660.

van den Broek, H. W. J., Santema, J. S., Wassink, J. H. and Veeger, C. (1971a). *European Journal of Biochemistry* **24**, 31.

van den Broek, H. W. J., van Breemen, J. F. L., van Bruggen, E. F. J. and Veeger, C. (1971b). *European Journal of Biochemistry* **24**, 46.

van den Broek, H. W. J., Santema, J. S. and Veeger, C. (1971c). *European Journal of Biochemistry* **24**, 55.

van den Broek, H. W. J. and Veeger, C. (1968). *Federation of European Biochemical Societies Letters* **1**, 301.

van den Broek, H. W. J. and Veeger, C. (1971a). *European Journal of Biochemistry* **24**, 63.

van den Broek, H. W. J. and Veeger, C. (1971b). *European Journal of Biochemistry* **24**, 72.

van Lin, B. and Bothe, H. (1972). *Archiv für Mikrobiologie* **82**, 155.

Walker, H. H. (1932). *Science* **76**, 602.

Ware, D. A. (1972). *Biochemical Journal* **130**, 301.

Williams, A. M. and Wilson, P. W. (1954). *Journal of Bacteriology* **67**, 353.

Wilson, P. W. (1958). *In* "Encyclopaedia of Plant Physiology", (W. Ruhland, ed.), p. 9. Berlin, Germany.

Wong, P., Evans, H. J., Klucas, R. and Russell, S. A. (1971). *Plant and Soil, special volume* p. 525.

Wong, P. and Burris, R. H. (1972). *Proceedings of the National Academy of Sciences of the United States of America* **69**, 672.

Yakovlev, V. A. and Levchenko, L. A. (1964). *Doklady Akademii Nauk S.S.S.R.* **159**, 1173.

Yakovlev, V. A., Vorob'ev, L. V., Levchenko, L. A., Linde, V. R., Slepko, G. I. and Syrtsova, L. A. (1965). *Biokhimiya* **30**, 1167.

Yates, M. G. (1970a). *Journal of General Microbiology* **60**, 393.

Yates, M. G. (1970b). *Federation of European Biochemical Societies Letters* **8**, 281.

Yates, M. G. (1971a). *European Journal of Biochemistry* **24**, 347.

Yates, M. G. (1971b). *In* "The Chemistry and Biochemistry of Nitrogen Fixation", (J. R. Postgate, ed.), p. 283. Plenum Press, London and New York.

Yates, M. G. (1972a). *European Journal of Biochemistry* **29**, 386.

Yates, M. G. (1972b). *Federation of European Biochemical Societies Letters* **27**, 63.

Yates, M. G. and Daniel, R. M. (1970). *Biochimica et Biophysica Acta* **197**, 161.

Yoch, D. C. (1972). *Biochemical and Biophysical Research Communications* **49**, 335.

Yoch, D. C. and Arnon, D. I. (1969). *Federation Proceedings. Federation of American Societies for Experimental Biology* **28**, 912.

Yoch, D. C. and Arnon, D. I. (1972). *Journal of Biological Chemistry* **247**, 4514.

Yoch, D. C., Benemann, J. R., Valentine, D. C. and Arnon, D. I. (1969). *Proceedings of the National Academy of Sciences of the United States of America* **64**, 1404.

Mechanisms of Spore Heat Resistance

G. W. GOULD and G. J. DRING

Unilever Research Laboratory, Colworth House, Sharnbrook, Bedford, England.

I. Introduction

Amongst biological systems, bacterial endospores exhibit the greatest degree of thermoresistance. The heat resistance of bacterial spores may exceed that of vegetative forms by factors of 10^5 or more. However, even between spores of different species, the spread of heat resistance can be very large, ranging, for example, from the relatively heat sensitive spores of *Clostridium botulinum* type E ($D_{80°C} = < 0{\cdot}33$–$3{\cdot}3$ min; Roberts and Ingram, 1965) to the extreme resistance of some like those of *Bacillus stearothermophilus* ($D_{115°C} = 22{\cdot}6$ min; Briggs, 1966).

It is not within the scope of this review to attempt to present a detailed account of the gross structural and chemical composition of bacterial spores. Rather, we intend to consider as far as the present state of knowledge permits, the involvement of those spore components, both at the molecular and structural level, which may be related to the property of thermoresistance.

We will consider the available data selectively and try to indicate those factors for which research provides strong evidence for an association with heat resistance. We will also try to indicate factors which have at times been considered important in heat resistance, but which may in the light of more recent work, be fortuitous associations.

Finally, we will appraise those hypotheses which have, perhaps, been given insufficient consideration in the past, and briefly speculate regarding their relevance to the mechanisms of spore heat resistance.

II. Structure of the Bacterial Endospore

A. Cytology

For detailed accounts of the structure and chemistry of the spore the reader is referred to comprehensive reviews by Murrell *et al.* (1969), Murrell (1967, 1969) and, for the cytological development of the spore, to publications by Murrell (1967) and Fitz-James and Young (1969).

The typical spore structure, as revealed by the electron microscope, is shown in Fig. 1 and consists of a central protoplast or core (*c*) enclosed by the plasma membrane or inner spore membrane (*ism*) around which lies the thick cortex layer (*ctx*) and finally the multilamellar coats (*cts*). An outer spore membrane may exist between the cortex and the coat. It is inconceivable that a spore, the dimensions of which do not exceed about 1 μm, is capable of excluding heat through the existence of thermally resistant barriers. However, a number of well-defined regions do exist within the spore, and the question arises as to whether the property of heat resistance can be ascribed to any one of them. Alternatively, does heat resistance result from a combination of factors distributed throughout several or all of the regions of the spore?

B. Location of Components

1. *Calcium and Dipicolinic Acid*

From studies made relating to release of dipicolinic acid (DPA) by various solvents, detergents and mechanical disintegration, Rode and

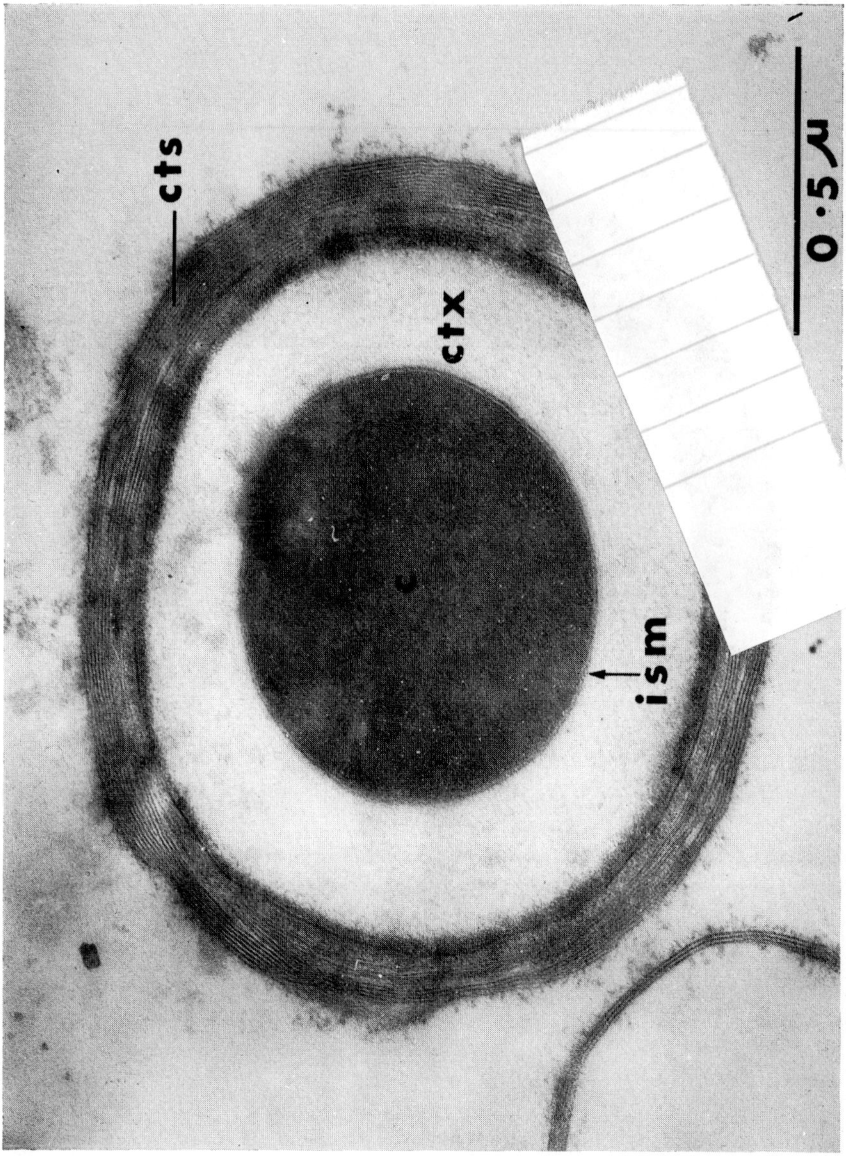

FIG. 1. Electron micrograph of a thin section of an [illegible]spore of *Clostridium sporogenes* showing the major structures—protoplast or core (*c*), plasma membrane o[illegible]nner spore membrane (*ism*), cortex (*ctx*), coats (*cts*).

Foster (1960a, b, c) reasoned that DPA was located outside the spore core. Roberts and Ingram (1965) suggested that DPA was located within the spore protoplast since autoclaving removed DPA but not the cortex. Using ultraviolet light techniques Hashimoto and Gerhardt (1960) were unable to detect DPA in the cortical region.

Evidence has accumulated suggesting that DPA and Ca^{2+} are located together within the spore. However, as shown by Nelson *et al.* (1969) other bifunctional acids such as sulpholactic acid, phosphoglyceric acid and glutamic acid occur at high concentrations in spores of some species and may share with DPA the role of metal-ion chelation. It was shown that on heating spores these compounds were released at the same time as DPA and presumably share the same location.

Prior to the isolation of the DPA-negative thermoresistant mutants (see p. 148) research proceeded on the assumption that DPA and Ca^{2+} were in all probability associated with thermoresistance. It thus appeared reasonable to expect that if the location of both DPA and Ca^{2+} within the spore could be established then their association with thermoresistance might be more clearly understood. Furthermore the nature and site of the actual mechanism of heat resistance would be resolved.

Murrell *et al.* (1969), from detailed chemical analyses, concluded that both Ca^{2+} and DPA were located within the spore cortex (i.e., in the peptidoglycan polymer that is present in this region; Warth *et al.*, 1963), and that there was evidence for Ca^{2+}, DPA and peptidoglycan being associated chemically. Their suggestions fitted well with the concept of the cortex functioning as a contractile organelle (see p. 157) and provided a suitable location for the ions needed to bring about the contraction. Considerable data exist (Slepecky, 1961; Windle and Sacks, 1963) to suggest that, wherever their precise location within the spore, Ca^{2+} and DPA are together as a calcium dipicolinate chelate. Bailey *et al.* (1965) examined DPA spectra from spores embedded in potassium bromide and obtained reuslts which supported this conclusion. Furthermore it appeared that some of the Mn^{2+} present in spores might also be complexed with DPA. However, their study did not reveal the location of the complex.

Knaysi (1965) assumed that a Ca^{2+}:DPA chelate was present in spores and used "spodography", a technique employing micro-incineration of the spore followed by microscopic examination, to argue in favour of a central core location for calcium dipicolinate.

Gerhardt *et al.* (1971) and Scherrer and Gerhardt (1972) introduced electron probe X-ray micro-analysis as a physical method for examining both the physical state and the location of electrolytes such as Ca^{2+} and DPA within the spore. Their method is of particular merit, since the examination is made directly on the intact spore, and disruption, which is known to lead to liberation and redistribution of spore components, is not required. Gerhardt *et al.* (1971) and Scherrer and Gerhardt (1972) concluded that, for *Bacillus cereus* and *B. megaterium*, Ca^{2+} was distributed throughout the spore similarly to the distribution of carbon. There was no evidence for its being present to any greater extent in the cortex, but rather that a concentration did occur within a region equivalent to the core.

In their studies on the location of DPA within the spore, Leanz and Gilvarg (1973) developed and applied the technique of beta-attenuation analysis (Leanz and Gilvarg, 1972) to whole intact spores. It has been demonstrated that the beta emission from intra-spore tritium-labelled compounds is partly absorbed before the particles escape from the spore. Thus when the label is situated deep within the spore fewer electrons will escape and the level of radiation detected by a surrounding liquid scinter will be much reduced compared with a superficially located label. With *B. megaterium* cultures, various tritium-labelled markers were introduced during sporulation at different known locations within the spore. ^{3}H-Uracil marked the core nucleic acids, ^{3}H-α-ϵ-diaminopimelic (DAP) acid, the spore cortex and ^{3}H-lysine the outer protein coats. For the unknown location of DPA, ^{3}H-DPA was added to sporulating cells of a *B. megaterium* mutant that incorporates DPA during sporulation. Attenuations in β-emission of 26, 18 and 3% were found for the uracil, DAP and lysine markers respectively. A value of 33% attenuation for the DPA marker strongly indicated a core location for this component.

Many metal ions other than Ca^{2+} are known to exist in spores, some of which may be functional in heat resistance. Other than those data cited here for Ca^{2+} and for Mn^{2+}, no evidence is available relating to their location within the spore. Gerhardt *et al.* (1971) presented evidence that electrolytes in general are probably electrostatically bound in spores and only become mobile during germination.

2. *Peptidoglycan*

Warth *et al.* (1963) studied the dissolution of disrupted spores by lysozyme and concluded that the cortex region was dependent for its integrity on peptidoglycan. The peripheral staining of mercaptoethanol-treated spores by fluorochrome-labelled lysozyme (Gould *et al.*, 1963) supported this view.

3. *Enzymes*

Sadoff *et al.* (1965) have studied the quite considerable effects of monovalent ions on the heat resistance of isolated spore enzymes (see

Section VB, p. 149). If stabilization of enzymes *in vivo* is brought about by monovalent ions concentrated in a region of the spore where, through a low water content, high solute concentration might exist, then a core location for such ion: enzyme interactions would seem most likely. It is worthwhile considering the question of the location of enzymes within the spore. Possibly, three categories of enzymes are present in the spore; (i) enzymes which act as fail-safe mechanisms during germination, (ii) enzymes which function in the germination mechanism, and (iii) enzymes required for outgrowth and vegetative growth following germination. The precise location of these enzymes has not been established, although presumably category (iii) includes most of the cells' complement of enzymes, and exist within the core; they need not function in the dormant spore. Some enzymes of category (i) are known to be intrinsically heat-stable and could therefore be located externally to any general heat protection mechanism, perhaps near the spore periphery. Category (ii) enzymes must be sufficiently active to function during germination and yet also be thermally protected. For these enzymes, like those of category (iii), a location either within the core itself or on the inner spore membrane under the cortex would seem likely.

III. Heat Resistance During Spore Formation and Germination

A. Spore Formation

Early studies by Williams (1929), Theophilis and Hammer (1938) Williams and Robertson (1954) and El-Bisi and Ordal (1956) showed that sporulation at high temperatures resulted in spores which were more heat resistant than those produced at lower temperatures. *Clostridium botulinum* spores produced at 37°C were more resistant than those grown at 41°, 24° or 29°C (Sugiyama, 1951). None of these studies precludes the possibility that selection of cells capable of producing spores of greater resistance took place.

During spore formation a number of well-defined changes occur in sequence (see Schaeffer, 1969; Mandelstam, 1969; Dawes and Hansen, 1972). The sequence, some parts of which are illustrated in Table 1, is not inviolate, depending to some extent on the medium and the organism being studied. Nevertheless, it is commonly observed that heat resistance develops late, i.e. following synthesis of dipicolinic acid, sulpholactic acid in those organisms that produce it (Bonsen *et al.*, 1969; Wood, 1971), and the incorporation of calcium, and formation of the cortex.

Heat resistance increases during a late "maturation" phase which is thought to involve dehydration of the core region, but the mechanism which may be responsible for this is still conjectural (see Section VIB, p. 157).

TABLE 1. Development of Heat Resistance and Related Changes during Spore Formation

Time of Incubation
Vegetative form
Start of spore formation ("Stage 0")
Chromatin filament formed Protease excreted ("Stage 1")
Spore septum formed ("Stage 2")
Spore protoplast engulfed Heat resistant catalase formed ("Stage 3")
Cortex synthesized Spore becomes refractile Dipicolinic acid synthesized Calcium incorporated ("Stage 4")
Spore coats formed Spores become resistant to octanol ("Stage 5")
Spores become resistant to heat ("Stage 6")
Autolysis releases spore from sporangium ("Stage 7")

The table summarises the most commonly observed sequence of events: for detailed reviews see Schaeffer (1969) and Mandelstam (1969).

TABLE 2. Loss of Heat Resistance and Related Changes during Spore Germination

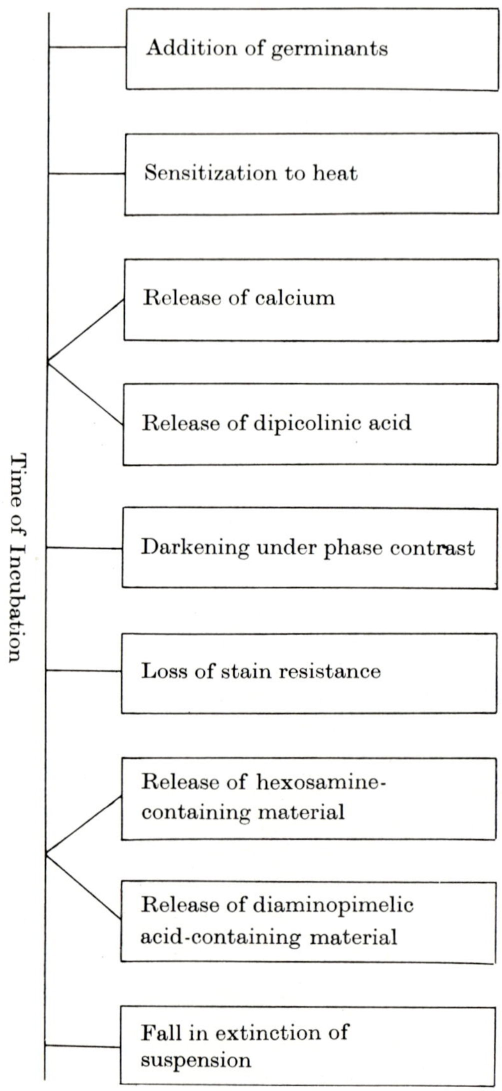

From data obtained with spores of *Bacillus cereus* (Hashimoto *et al.*, 1969; Dring and Gould (1971a); *Bacillus megaterium* (Levinson and Hyatt, 1966) and *Bacillus subtilis* (Dring and Gould, 1971b).

B. Spore Germination

When germination is initiated the characteristic properties of spores are lost in a rapid but measurable sequence. Studies with *B. megaterium* (Levinson and Hyatt, 1966), *B. cereus* T (Hashimoto *et al.*, 1969; Dring and Gould, 1971a) and *Cl. botulinum* 62A (Rowley and Feeherry, 1970) have all indicated that loss of heat resistance is one of the earliest measurable events, preceding release of calcium and dipicolinic acid, acquisition of stainability, phase-darkening, the fall in extinction of spore suspensions and the release from spores of autolysis products (Table 2). Small differences in the sequence have been reported by different researchers.

Pretreatment of spores may change markedly the relative rates of the various changes which occur during the initiation of germination; for example, Sogin *et al.* (1972) noticed that when the germination of spores of *B. cereus* T, given extended heat activation (4 h at 75°C), was initiated by the addition of a mixture of alanine and adenosine, heat resistance as measured by colony count was lost much more rapidly (99% fall in 4 min) than extinction of the spore suspension (about 5% fall in 4 min).

It is not possible from these studies to gain much insight into the mechanism of spore heat resistance, for none of the other physical or chemical changes which occur can be shown to be exactly correlated with the change in heat resistance during germination.

IV. Heat Resistance and Super-dormancy

Loss of spore viability during heating may not be analogous to the heat-induced loss of viability of vegetative forms. This was well illustrated by Cassier and Sebald (1969) who reported experiments in which spores of a strain of *Cl. perfringens* were shown to be inactivated by heat if enumerated using conventional media, and yet appeared still viable if plated in media containing lysozyme. The observations were confirmed by Duncan *et al.* (1972), and the phenomenon has also been described for spores of *Cl. botulinum* type E (Sebald and Ionesco, 1972). It seems most likely that heat normally kills these spores by inactivating enzymes necessary for the initiation of germination, whilst leaving the protoplast viable but trapped within the spore.

It is known that lysozyme will cause germination of spores of diverse species, but normally some treatment of the spores is first necessary to make the spore coat permeable to the enzyme (Gould and Hitchins, 1963). It appears that heated spores of the *Cl. perfringens* and *Cl. botulinum* type E strains become lysozyme sensitive without the necessity for this pretreatment. Lysozyme is then able to initiate germination of such spores through the by-passing of the inactivated enzyme(s) and hydrolysing cortex peptidoglycan directly.

Cassier and Ryter (1971) subsequently isolated a mutant strain of *Cl. perfringens*, the spores of which were essentially non-germinable whether or not they were heated (i.e. they were operationally "dead") unless lysozyme was present. The mutant spores could be seen by electron microscopy to have defective coats which were so altered as presumably to allow lysozyme access to the underlying cortex. At the same time the mutant spores are thought to have lost completely the enzyme(s) which normally operate to initiate germination.

The care needed in interpreting heat inactivation data for bacterial spores was further recently highlighted by Busta and Adams (1972) and Adams and Busta (1972), who found that the sensitivity of *B. subtilis* A spores to activation, and to inactivation, by heat differed according to the system chosen to initiate germination (i.e. in such a way as to suggest that heat could damage one germination system or "pathway" more than another). Previously, Busta and Ordal (1964) and Edwards *et al.* (1965) studied spores of *B. subtilis* A that had been so severely heated as to have lost the ability to grow on normal laboratory media, and would therefore conventionally be scored as dead. However, when plated on media containing calcium dipicolinate these same spores were able to grow and form colonies, and were therefore scored as viable. It is known that calcium dipicolinate will initiate spore germination.

It seems likely that it acts in this instance by by-passing the normal germination mechanism which has been inactivated by heat, although the spores' viability as judged by outgrowth and vegetative growth is not severely damaged.

The measured heat resistance of such spores can clearly depend very much on the media and methods used to enumerate the survivors. Spores may be scored as "dead" on conventional media simply because their mechanism for initiating germination has been inactivated, although we know that within the spore the protoplast remains fully viable. Spores in this situation are essentially "superdormant" (Gould, 1970). An interesting fundamental question that arises is: "What is the true heat resistance of such spores?" A related practically important question is: "What heat resistance curves should one use, for instance, as a guide to the inactivation of spores like those of *Clostridium perfringens* in foods?"

V. Spore Components and Heat Resistance

A. Dipicolinic Acid

Biologically, dipicolinic acid (pyridine-2,6-dicarboxylic acid, or DPA) is of restricted occurrence. It was first recognized by Udo (1936) as a product of microbial metabolism in the fermented soybean product, "Natto". The identification of DPA as a component of bacterial spores resulted from the studies of Powell and Strange (1953). Spores of *Bacillus megaterium* and *B. subtilis* were found to shed some 30% of their dry weight during germination, and Powell (1953) identified DPA as the material present, together with calcium, various amino acids (Strange and Thorne, 1957), peptides and peptide-linked hexosamine (Strange and Powell, 1954; Strange, 1956) in exudates. A calcium dipicolinate chelate comprised 50–60% of the germination exudate from *B. megaterium* spores. These studies were paralleled almost simultaneously by those of Perry and Foster (1955) and Foster (1956), who also suggested that DPA might be related to spore heat resistance.

Dipicolinic acid has been isolated from the spores of all *Bacillus* and *Clostridium* species so far examined and from spores of *Sarcina ureae* (Thompson and Leadbetter, 1963), *S. ventriculi* and *S. maxima* (Knoll and Horschak, 1971) and the endospores of some thermophilic actinomycetes (*Thermoactinomyes vulgaris, Th. sacchari* and *Actinobifida dichotomica*; Cross *et al.*, 1968; Kalakoutskii *et al.*, 1969; Lacey and Vince, 1971). Analyses of bacterial spores of different species show that the DPA content ranges from 5 to 15% of the dry weight. It has not proved possible, through spore fractionation, to locate DPA since mechanical disruption results in its release. It can also be extracted by boiling spores in dilute mineral acid (Perry and Foster, 1955), and by water when spores are autoclaved (Janssen *et al.*, 1958). Harrell and Mantini (1957) induced release of DPA from *B. cereus* without causing germination or killing the spores by heating the spores in phosphate buffer (10^{-1} M) at 65°C for 60 mins. Dipicolinic acid is released from spores in small amounts at those temperatures employed for heat activation in germination studies (Keynan *et al.*, 1961), *B. cereus* T shedding only 10% of its available DPA when optimally activated. Rowley and Levinson (1967) showed that spores of *B. megaterium* QM B1551 remained phase bright but lost DPA and viability and were rendered sensitive to lysozyme when heated at 55°C in acid thioglycollate. Flemming (1964) considered that since DPA was not extracted from spores with water at room temperature it was probably bound as a non-dissociable complex or that its exit was prevented by a permeability barrier.

Woese and Morowitz (1958) studying germination of spores of *B. subtilis* with physiological germinants demonstrated that DPA loss practically co-incided with the decrease in extinction and suggested that both parameters were manifestations of similar processes taking place in the spore. More recently, studies with *B. cereus* T (Hashimoto *et al.*,

1969; Dring and Gould, 1971a) showed that release of DPA was preceded by loss of refractility and co-incided with release of calcium.

The observations of Powell (1950, 1951, 1957) that germination of spores is accompanied by loss of heat resistance, together with those of Powell and Strange (1953) and Powell (1953) characterizing the material lost from spores during germination, were foundations for subsequent studies which have attempted to establish whether or not DPA is directly associated with the mechanism of thermoresistance.

Evidence relating DPA content of spores to their heat resistance is conflicting. Church and Halvorson (1959) showed that a decrease of DPA in *B. cereus* spores was brought about by inclusion of DL or L-phenylalanine in the culture medium and that heat resistance of the spores was also reduced. In contrast Bryne *et al.* (1960) were unable to lower the thermoresistance of spores of *Cl. roseum* grown in medium supplemented with L-phenylalanine. However, increasing the concentration of L-alanine in the culture medium did result in a lowering of the spores' DPA content, but was unexpectedly accompanied by an increase in spore heat resistance. Grecz and Tang (1970) did not find any correlation for heat resistance and DPA for a variety of strains of *Cl. botulinum* spores whose DPA content ranged from 7·4 to 13·4% of the dry weight.

Recently, however, the dogma of the last twenty years concerning the role of DPA in spore heat resistance has been challenged. The challenge developed from the isolation by Wise *et al.* (1967) of a mutant of *B. cereus* T, the spores of which contained no DPA. As expected the mutant spores were found to be heat-sensitive when compared with wild-type spores. However, subsequent studies by Curry *et al.* (1971) and Hanson *et al.* (1972) resulted in the isolation of a revertant, *the spores of which were DPA-negative and yet were fully as heat-resistant as the DPA-positive wild type spores*. Further DPA-negative mutant spores are now available and confirm the fact that DPA is not essential for full heat-resistance (Zytkovicz and Halvorson, 1972).

The studies of Curry *et al.* (1971) and Hanson *et al.* (1972) showed that freshly harvested spores of the *B. cereus* DPA-negative revertant, which possessed normal thermoresistance, contained only one-tenth to one-twentieth of the wild-type calcium concentration. The DPA-negative spores differed from the DPA-positive spores in a number of ways. Of particular importance was the observation that their heat-resistance was not well maintained; e.g. it was lost following lyophilization, after two weeks storage at 4°C, or after cleaning in a two-phase polyethyleneglycol–phosphate buffer. Resistance was not lowered by treating the spores with acid (0·033 *N*-HCl). Interestingly the DPA-negative thermosensitive parent spores (Halvorson and Swanson, 1969) do not readily germinate in physiological germinants, such as L-alanine, either by itself or in combination with adenosine.

It has been generally observed that the DPA-negative spores so far studied are reluctant to germinate under most conditions that initiate germination of spores of the wild type, although they will give a full colony count on agar-solidified media. These observations have led Halvorson and Swanson (1969) and Hanson *et al.* (1972) to suggest that, rather than being responsible *a priori* for heat-resistance, DPA might help to stabilize established resistance and that its major functions are probably in the maintenance of dormancy and the initiation of germination.

B. Metal Ions

Spores contain a wide variety of inorganic elements (Curran *et al.*, 1943; Thomas, 1964; Murrell, 1967). Excellent comprehensive reviews concerning the physiological roles of ions in spores and in particular relating ionic composition to thermoresistance have recently been published (Murrell, 1967, 1969; Lewis, 1969; Roberts and Hitchins, 1969).

One can speculate as to the reasons why spores contain metal ions. They may participate directly or indirectly in conferring properties of dormancy and heat resistance. They may be prerequisite for the germination event, perhaps as activators of lytic enzymes, Strange and Dark (1957a, b) showing that Co^{2+}, Cu^{2+}, Ni^{2+} and Mn^{2+} ions would activate isolated spore lytic enzymes. Alternatively some ions within spores might have resulted from activities during sporulation, K^{+} being required during protein synthesis (Lubin and Ennis, 1965), or Mg^{2+} stabilizing both the physical integrity of membranes and membrane-ribosome associations during protein synthesis (Coleman, 1969), roles which the ions concerned might to some extent re-assume following germination. Some ions may be removed from spores using mild treatments. It is known that spores, akin to ion exchange resins, can retain ions in association with electronegative groups within their peripheral structures (Slepecky and Foster, 1959; Alderton and Snell, 1963). The removal of these spurious ions accumulated from the sporulation medium does not affect the normal thermal properties of the spores.

Progress in the elucidation of the functions of metal ions in spores has largely resulted from the following lines of investigation. (a) Gross ion analyses have been made of various spore species, either in the mature state or at various stages during sporulation, grown in rich or defined media using batch, endotrophic or replacement procedures, and the thermal properties of the spores examined. With defined media the ionic composition, for example in terms of Ca^{2+}, Mg^{2+} and Mn^{2+}, can be experimentally varied. (b) A variety of spore enzymes have been isolated, purified and their thermostability determined in the types of ionic environment which might prevail within the intact spore. (c) Attempts have been made both to locate ions in spores and establish by physical techniques how they may be associated with spore components and thus relate structure to function. (d) Spore components other than enzymes have been isolated and their association affinities with various typical spore ions determined. The complexes so derived may have similar associations as regards thermoresistance as in the intact spore. (e) Mutant strains of spores have been isolated whose ionic composition is different from that of the wild type and their thermal properties have been examined. (f) By treatment of mature spores with mineral acid, cations can be removed. Reloading of the spores with other metal cations can then be carried out and thermal resistance examined.

The notion that metal ions might be involved in spore thermoresistance originated from Williams (1929), who observed that the composition of the sporulation medium, particularly in terms of iron, calcium, magnesium and phosphate, influences the heat resistance of spores. Until very recently the general pattern established has been that, in addition to twenty or so other ionic species found within spores, significantly higher levels of the divalent cations Ca^{2+}, Mg^{2+} and Mn^{2+} are present than in the corresponding vegetative cells. Calcium ions occur in spores at levels of 1–3% of the dry weight and a level of 6·4% has been reported for *Cl. bifermentans* (Bailey *et al.*, 1965). From their observations, Murrell and Warth (1965) concluded that, for various strains of *Bacillus* exhibiting a 700-fold range in thermal resistance, the level of calcium ion was significantly correlated with thermoresistance. For the Mg^{2+} content an inverse correlation was found. The Mg:Ca ratio decreased significantly with increase in heat resistance. Similar observations were made by Walker *et al.* (1961). Slepecky and Foster (1959) showed that when spores were grown in medium containing calcium (1·8 μg/ml) together with low and high levels of other divalent cations (Zn^{2+}, Ni^{2+}, Cu^{2+}, Co^{2+}, and Mn^{2+}) the resultant calcium levels in the spores obtained were decreased in the presence of high concentrations of the other ions which themselves exhibited enhanced levels in the spores. However, analyses confirmed that these spores contained but 1% Ca^{2+} and did not survive heating 10 min at 60°C. Other replacement studies (Black *et al.*, 1960; Halvorson and Howitt, 1961; Pelcher *et al.*, 1963; Foerster and Foster, 1966b) indicated that heat-resistant spores could be grown using Sr^{2+} and Ba^{2+} as replacers for Ca^{2+}, but that for maximum heat resistance calcium was essential. Vinter (1960, 1962) showed that the developing spore acquires Ca^{2+} after refractility has been established and at about the same time as heat resistance develops.

Thus it would appear that as regards spore thermoresistance calcium exhibits a specificity which cannot be replaced by other divalent ions examined. The concept has developed that calcium has a central role in the property of thermoresistance which it was thought likely to share in some as yet unresolved manner with dipicolinic acid (Aronson *et al.*, 1967; Black *et al.*, 1960; Church and Halvorson, 1959; Levinson *et al.*, 1961; Murrell and Warth, 1965). It is possible that both Ca^{2+} and Mn^{2+} are associated with DPA in the spore, the molar ratio of the cation content to that of DPA usually being near to unity. Windle and Sacks (1963), from electron paramagnetic resonance studies, showed that Mn(II) probably exists in spores in a chelated state.

Reference has been made elsewhere (Section VA, p. 146) to the mutant spores *Bacillus cereus* T isolated recently (Hanson *et al.*, 1972) which, whilst containing no detectable levels of DPA, are nevertheless thermoresistant. It has also been determined that both the Ca^{2+} and Mn^{2+} levels of these spores are low. Furthermore, it has been reported that the spores can be successfully produced in media essentially free of added Ca^{2+}. To what extent adequate levels of Ca^{2+} might be supplied as contamination from the apparatus used or other medium ingredients which would facilitate adequate sporulation of what are, albeit low, Ca^{2+} spores is not reported.

Aoki and Slepecky (1973) found for *Bacillus fastidiosus* the presence or absence of manganese in the sporulation medium affected the heat resistance of the spores obtained. $D_{85°C}$ values of 6·5 min and 17·0 min were recorded for spores grown without added Mn^{2+} and in the presence of 10^{-5} *M* Mn^{2+} respectively. Very little change in Ca^{2+} and DPA levels occurred in the Mn^{2+}-grown spores. Bailey *et al.* (1965) found evidence for the presence of a Mn^{2+} : DPA chelate in spores, and it has been reported that *B. coagulans* var. *thermoacidurans* spores were most heat resistant when obtained from media supplemented with Mn^{2+} and Ca^{2+} (Amaha and Ordal, 1957). These observations lead Aoki and Slepecky (1973) to suggest that Mn^{2+} may be related directly itself with thermoresistance or that it may interact in some manner with Ca^{2+}. Sugiyama (1951) pointed out that Ca^{2+} might associate with neighbouring electronegatively charged groups on folded peptide chains and thus confer resistance to thermal unfolding of the chains and hence greater resistance to thermal denaturation. Calcium ions would be more effective than Na^{+} in this respect (Hober, 1945).

Hitherto, emphasis has been centred on the possible functions of divalent cations in thermoresistance. However, note should be taken of the high levels of potassium ions, approaching the levels found for Ca^{2+}, that have been reported for certain spores. Bailey *et al.* (1965) found K^{+} levels of 1·6% for *Cl. bifermentans*, 1·4% for *B. macerans*, 3·8% for *B. megaterium* and 1·8% for *B. subtilis*. However, in *B. cereus* T and

B. coagulans the K^+ contents were low, being but 0·05% and 0·65% of the dry weight. Dring and Gould (1971b) reported levels of K^+ up to 1·7% for *B. subtilis* and found that between one-third and one-half of the K^+ was shed from the spore early during germination. Such levels of K^+ would appear excessive if the ion is derived solely from protein synthesis activities during sporulation. There is no direct evidence implicating monovalent ions in the thermal properties of intact spores, Murrell (1967) reporting that two resistant spore strains had low K^+ content whilst four others had high K^+ levels.

C. Enzymes

Although highly heat-resistant within the spore, most enzymes and other spore components are not intrinsically heat-stable and lose resistance when extracted. In some instances, extracted spore enzymes have been shown to maintain resistance to heat whilst in particulate form, but to lose resistance on solubilization (alanine racemase (Stewart and Halvorson, 1954) and purine ribosidase (Nakata, 1957) from *B. cereus*). Furthermore, many of the spore enzymes that have been studied in detail have properties that are very similar or identical to those of their vegetative counterparts.

Spore and vegetative inorganic pyrophosphatases from *B. subtilis* (Spudich *et al.*, 1966; Tono and Kornberg, 1967a) and from *B. megaterium* (Tono and Kornberg, 1967b) were indistinguishable by polyacrylamide gel electrophoresis, molecular weight, amino-acid composition, kinetic constants, activation by metal ions and inhibition by various inhibitors. Similarly, spore and vegetative adenylate kinases were identical in all respects studied (Spudich and Kornberg, 1969). Deoxyribonucleic acid polymerases had the same heat sensitivities and other properties once extracted from vegetative cells or from spores (Falaschi and Kornberg, 1966) and no differences were detected in alkaline phosphatases from vegetative and sporulating *B. subtilis* cells by Glenn and Mandelstam (1971).

Ribosomes isolated from lysozyme-lysed spores of *B. megaterium* were basically similar to vegetative cell ribosomes (Chambon *et al.*, 1968), and the overall protein synthetic machinery of *B. megaterium* spores was also essentially intact and similar to that of the vegetative form (Deutscher *et al.*, 1968).

In contrast to the above examples, Sadoff and his colleagues have described a number of enzymes, the spore and vegetative forms of which have different properties. Nicotinamide adenine dinucleotide oxidase from spores of *Cl. botulinum* had a molecular weight about double that of the vegetative form, was more heat stable and was antigenically distinct (Green and Sadoff, 1965). Spore and vegetative cell fructose 1, 6-diphosphate aldolases from *B. cereus* had similar heat resistances, antigenic properties, pH optima and K_m values, but differed in other respects (Sadoff *et al.*, 1969). The addition of calcium ions increased the thermal stability of the spore aldolase whilst decreasing the resistance of the vegetative enzyme. The enzyme also differed in electrophoretic mobility in polyacrylamide gels and in Stokes' radii. The vegetative aldolase consisted of two species in equilibrium with molecular weights of 115,000 and 79,000 daltons, whilst spore aldolase was smaller, with a molecular weight of 44,000 daltons. It seemed likely that syntheses of the spore and vegetative enzymes were not directed by separate genomic units, but that vegetative-type enzyme was converted to spore-type enzyme during sporulation. Sadoff *et al.* (1970) later showed that protease formed during sporulation would bring about the vegetative-to-spore modification of aldolase, whereas non-spore proteases (e.g. trypsin, pronase) did not bring about the modification. Sadoff's discovery of enzyme modification during sporulation received further support from a study of spore and vegetative purine nucleoside phosphorylases from *B. cereus* T. Gardner and Kornberg (1967) found these enzymes to have identical sensitivities to heat, and indistinguishable physical and kinetic properties; Gilpin and Sadoff (1971) further found the enzymes to have very similar subunit molecular weights and catalytic properties (e.g. K_m values for inosine and for phosphate); however, the enzymes responded differently to phosphate. In the absence of phosphate the spore enzyme was a dimer of molecular weight 47,000 daltons, whilst the vegetative enzyme remained a tetramer with a molecular weight of about 92,000 daltons. In high concentrations of phosphate both enzymes were tetrameric. The nucleoside phosphorylase was thus modified during sporulation, but without the extensive proteolysis that accompanied modification of the aldolase.

Certain spore enzymes therefore differ from their vegetative counterparts in being modified after synthesis. In some instances it has been shown that the spore form of the enzyme, although not inherently heat resistant, can be made more resistant than the vegetative form by modification of the environment: e.g. by decrease in phosphate concentration (*B. cereus* nucleoside phosphorylase), by change in pH value and by increase in concentration of group IA cations, which together can increase the thermal resistance of *B. cereus* spore glucose dehydrogenase about 10^6-fold (Sadoff *et al.*, 1965). It was also shown that thermal stability of the enzyme was dependent upon hydrogen ion concentration. Thus at pH 6·5, enzyme half-life was 780 min, whereas at pH 7·5 it was only 3 min. The enzyme in the intact spore is some 5,000-fold more stable than the isolated enzyme at pH 6·5, and studies suggested that a dialysable factor in the spore might be responsible for the difference.

Dialysates, when recovered and added to the extract, restored thermal-stability as did sodium chloride (0·5 *M*) to an extent similar to that achieved by the factor. It was concluded that intramolecular forces might account for induced heat resistance of the glucose dehydrogenase. Alternatively, for fructose 1,6-diphosphate aldolase from *B. cereus* increase in calcium ion level was found to protect the enzyme.

The spore forms of these enzymes are generally less aggregated and smaller than the vegetative forms (Sadoff, 1970), a situation that may well contribute to their stabilization within the spore by a low general free ion level (Carstensen *et al.*, 1971) and possibly additionally by spore calcium.

Hachisuka *et al.* (1967) and Tochikubo *et al.* (1968) showed that the heat resistance of isolated vegetative *B. subtilis* glucose dehydrogenase could be increased about three-fold by dipicolinic acid (0·17 *M*) and 25-fold or so by some of its analogues (e.g. lutidinic acid). Sodium and calcium dipicolinate also delayed the heat denaturation of human serum albumin and other proteins (Mishiro and Ochi, 1966), but the relevance of these small protective effects of dipicolinic acid and its analogues to *in vivo* heat resistance of spores is not certain. Hachisuka and Tochikubo (1971) found that even ethylene diamine tetracetic acid re-activated glucose dehydrogenase from heat-inactivated *B. subtilis* spores.

D. Water

A low water activity in the spore protoplast may be inferred to be important in heat resistance from the studies by Murrell and Scott (1966) of the heat resistance of spores held at different water activities. The high refractive index of spores, which contributes to their phase brightness and to the high extinction of spore suspensions, is thought to indicate a low water content in at least a part of the spore, and yet DPA-negative spore mutants of *B. subtilis* have recently been isolated which are heat resistant but phase dark (Zytkovicz and Halvorson, 1972).

Murrell and Warth (1965) conducted a comprehensive study of the composition of spores of a number of species with heat resistance covering a wide range. They measured levels of calcium, magnesium, diaminopimelic acid, hexosamine and dipicolinic acid, and attempted to correlate these with heat resistance. One of the most statistically significant correlations that they found was that the diaminopimelic acid (DAP) content of the spores increased with increase in heat resistance. This positive correlation of DAP level strongly suggests the involvement of the peptidoglycan cortex structure and indicates that the electronegative function of this polymer could be an important factor in determining heat resistance. The negative correlation with magnesium content suggests that an excess of these cations may interfere with full thermal resistance.

The possible role of peptidoglycan in maintaining a low core water content is discussed on p. 157.

VI. Ion Exchange and Heat Resistance

A. Ion Exchange Properties of Spores

It has been noted (Slepecky and Foster, 1959) that, if spores are treated with mineral acid (0·03–0·05 *M*-HCl) at temperatures less than 5°C to facilitate removal of metals and metal phosphate precipitates held within the spores' peripheral structure by electronegative groups, then little effect on thermoresistance is incurred. Murrell and Warth (1965) reported that D_{100} values for a selection of spores treated thus were, prior to and following acid treatment, 0·99 and 1·00 min, 35·2 and 34·6 min and 2·38 and 2·17 min for three unidentified *Bacillus* strains and 270 and 232 min for *B. coagulans*. For *B. stearothermophilus*, washing with 0·3 *M*-HCl at 5°C resulted after 20 min in a loss of less than 10% Ca^{2+} whereas 55% Mg^{2+} was removed after 1–2 min. Similarly for *B. subtilis* only marginal loss of Ca^{2+} occurred in 0·033 *M*-HCl whilst between 70 and 80% Mn^{2+} and Mg^{2+} were shed in 1–2 min. Removal of Mn^{2+} and Mg^{2+} probably represented removal of spurious metal precipitates. However, when spores were lyophilized and then acid treated (0·03 *M*-HCl) they continued to lose Ca^{2+} but not DPA.

It has been demonstrated that in other circumstances the treatment of spores with acid does result in removal of Ca^{2+} and that thermoresistance is adversely affected. Alderton and Snell (1963, 1964) have extensively examined the effect of acid treatment on spore thermal resistance. Alderton and Snell (1963) demonstrated that in particular the spores of *B. megaterium* exhibit a considerable capacity for cation exchange, and that this property permits compositional changes to be brought about experimentally on the mature spores without causing their death. This facilitates control of the level and nature of the metal-ion content of the spore more precisely than is possible through modification of the ionic composition of the sporulation medium. Thus for lyophilized spores held at 25°C and pH 4·0 (nitric acid), and by monitoring cation removal in a pH-stat over a 4–5 hr period, so-called "acid stripped" or "H-(hydrogen) spores" were obtained. Such spores could then be reloaded with metal ions of choice and their thermal properties examined. Large effects relating cation load to heat resistance were found. It was also found that heat resistance could be re-instated to various extents by exposing the H-spores to various metal hydroxide solutions. Titration of spores with calcium hydroxide to a final pH of 9·5 was most efficacious in restoring thermoresistance.

In an extension to these studies (Alderton and Snell, 1964) it was demonstrated that, if acid-stripped spores were rapidly exposed to lethal temperatures in 20 m*M* calcium acetate buffer, classical logarithmic order of death rate did not ensue, but instead a decelerating death rate was obtained. Similarly, when exposure of H-spores to calcium buffer was made at lower than lethal temperatures (50°C) or through the increasing of warm-up time at the lethal temperature it was possible to modify the initial slope of the survivor, the result being that the rate of death was progressively reduced. Thus it was shown that re-adjustment within the spore, perhaps in relation to its calcium complement was taking place. It was further established that restoration of the calcium content occurred in two stages; firstly a rapid uptake not correlated with acquisition of heat resistance, and which probably equates with the non-specific binding of Ca^{2+} (Slepecky and Foster, 1959; Murrell and Warth, 1965), followed by a slow phase of uptake, found to be both temperature- and pH-dependent and correlated with restoration of heat resistance.

In quantitative studies (Rode and Foster, 1966) the amounts of calcium removed from spores during the acid-stripping process were found to be only some 5% of the spores' total calcium content. Thus, since an acid-cleaning step (Slepecky and Foster, 1959) was not used by Alderton and Snell (1963), the amounts of metal ions removed by their acid-stripping technique, and which were of obvious significance in the thermoresistance of the spores, could only be equal, at the most, in terms of calcium, to 5% of the spores' total calcium. Of this 5% it was shown that only a portion was correlated with the spores' thermal properties, the remainder presumably being accounted for by non-specific ion binding.

Whilst this type of approach shows that spore heat resistance can be modified and restored, it provides no evidence either for the location of calcium within the spore nor of the nature of the heat-resistance mechanism. Rode and Foster (1966) showed that isolated coats of H-spores suspended in ^{45}Ca-acetate almost instantaneously accumulated Ca^{2+} to about 0·5% of their dry weight. This Ca^{2+} may represent the non-specifically bound fraction not important in heat resistance, and which gains greater accessibility to sites in the isolated coat than are available in the intact spore. The precise location of that calcium fraction important in the maintenance of thermoresistance is not known. If it is associated with the cortex peptidoglycan, resulting in the formation of a contractile organelle (see p. 157) which induces and maintains thermoresistance, then removal of any of the cross-linking calcium ions would result in relaxation of the polymer and the associated loss of some or all of the spores thermoresistance.

B. Pressure and Maintenance of Heat Resistance

1. *Contractile Cortex*

Cohn (1876) and Lewith (1890) first proposed that the characteristic spore properties of refractility and heat resistance might result from a low content of water in spore cytoplasm. Since that time, however, little data on the true water content of the spore protoplast have been obtained. Measurements of the total water content of spores (Black and Gerhardt, 1962; Marshall and Murrell, 1970) have resulted in figures ranging from 50 to 80% w/w depending on the assumptions made, but it has not been possible to determine the distribution of this water within the spore.

The observations of Murrell and Scott (1966) lent strong circumstantial support to the hypothesis that there is a low water content in the spore interior. They found that lowering the environmental equilibrium relative humidity (ERH) caused an increase in the heat resistance of spores. This increase was greatest for those spores which are normally most heat sensitive; i.e. at high ERH values. The overall result was that when equilibrated at ERH values of 20–30% even spores of *Cl. botulinum* type E became nearly as heat resistant as those of *B. stearothermophilus*, whereas at ERH near 100% spores of *B. stearothermophilus* were about 10^5 times as resistant as those of *Cl. botulinum*. The clear inference was that an ERH value of 20–30% represented the situation approached within a highly heat resistant spore (e.g. *B. stearothermophilus*), whilst less heat-resistant spores could not normally maintain such a low ERH value; nevertheless, if the water level in these spores was sufficiently lowered by adjusting the environmental ERH, then heat resistance rose dramatically.

Lewis *et al.* (1960) suggested a means whereby a relatively dehydrated protoplast within a spore could be produced and maintained. They reasoned that a truly water-impermeable coat was inconsistent with the known permeabilities of organic materials; indeed, the water permeability of spores is known to be high and the available evidence strongly suggests that spore water is mostly freely exchangeable with external deuterated water (Black and Gerhardt, 1962; Marshall and Murrell, 1970). Lewis *et al.* (1960) originated the concept that a low water content in the spore protoplast could arise through compressive contraction of the surrounding cortex during sporulation.

The hypothesis has received support from the known contractile properties of anionic polymers like the peptidoglycan which contributes to cortex structure. Peptidoglycan in isolated vegetative cell walls can be caused to contract by electrostatic interactions with salts (Marquis, 1968; Ou and Marquis, 1970). Hitchins and Gould (1964) noticed that isolated spore cores with residual cortex attached became contracted and increased in refractility as the pH value was lowered or as the multivalent cation level was increased. It seemed likely that the contraction resulted from neutralization or cross-linking of electronegative groups in polymers around the core. Čáslavská *et al.* (1970) found that a variety of basic compounds added to germinated spores caused increases in refractility, presumably by cross-linking and contracting residual electronegative peptidoglycan fragments within the spore.

All these observations supported the hypothesis that spore cortex peptidoglycan could be contracted, for instance by spore calcium or calcium dipicolinate, or by other cations in spores in excess of those equimolar to DPA (Murrell *et al.*, 1969), and exert a dehydrating pressure on the spore protoplast. Pearce and Fitz-James (1971), however, studied a mutant of *B. cereus* that produced cortex-less spores and, most interestingly, the spores of this organism became phase bright during sporulation, suggesting that cortex peptidoglycan was not essential, at least for the development of spore phase brightness and refractility. Further study suggested that phase brightness of the spore protoplasts depended upon calcium binding to the protoplast membrane phospholipids. The possibility that dipicolinic acid could act to remove this bound calcium during germination is supported by the extreme dormancy of the dipicolinic acid-negative mutant spores so far reported (Wise *et al.*, 1967; Hanson *et al.*, 1972).

2. *Expanded Cortex*

Two observations suggest that the hypothesis of an expanded cortex as discussed by Alderton and Snell (1963) might have more validity than the original hypothesis of a contractile cortex (Lewis *et al.*, 1960). Firstly, the total amount of peptidoglycan in spores, measured as hexosamine or diaminopimelic acid, has not been observed to be much greater than that in vegetative forms (Murrell, 1967, 1969). Secondly, the volume occupied by the cortex in spores is large, even accounting for 30–60% of the spore, which is a much greater fraction of the cell volume than that occupied by the cell wall of the vegetative form (Table 3). The contractile cortex hypothesis would predict the opposite; i.e. that spores should contain a high level of peptidoglycan, and that it should be highly contracted and compressed and therefore occupy a very small volume.

Data available on the amount of peptidoglycan in spores, and the space it seems to occupy, are more compatible with the cortex peptidoglycan being highly expanded. Such an expanded structure can help to explain many properties of the spore. For instance in the expanded state the water content of the cortex would be high and thus compatible with the observed high water content of spores (Black and Gerhardt, 1962) whilst still allowing maintenance, by expansive pressure, of a relatively low water content in the protoplast or core. An expanded electronegative cortex would demand the *absence* of neutralizing cations, rather than their

TABLE 3. Estimated Volumes of Spore and Vegetative Cell Structures

Organism	Volume (%) of whole cell*		
	Coat	Cortex	Protoplast
Bacillus cereus spores (fixed and sectioned)	16	49	35
Bacillus cereus spores (unfixed, freeze-etched)	20	60	20
Bacillus coagulans spores (fixed and sectioned)	47	36	17
	Cell wall		Protoplast
Bacillus cereus vegetative cells (fixed and sectioned)	23		77

* Volumes were estimated from measurements of electron micrographs.

presence, in order to maintain expansion and pressure. Indeed, injection of cations or other basic molecules, or high concentrations of salts, into the cortex region would be expected to cause its collapse. It is significant that spores of some organisms can be caused to lose resistance rapidly (i.e. germinate) by exposure solely to non-specific ions (Foerster and Foster, 1966a). The recent work, mentioned on p. 141, concerning the probable location of calcium dipicolinate in the spore protoplast, has indicated that the cortex does not contain the cross-linking agents that would be needed for contraction, but is more likely to be devoid of them.

Furthermore, analysis has indicated (Warth and Strominger, 1969, 1971) that spore peptidoglycan is generally more electronegative, and the amino-sugar backbone structure is more loosely cross-linked with peptide side chains, than that in vegetative peptidoglycan. One would expect such a loose anionic polymer to be highly expandable Finally, Murrell and Warth (1965) found that the level of the electronegative peptidoglycan amino-acid diaminopimelic acid was significantly higher in heat-resistant than in heat-sensitive spores.

C. Possible Role of Calcium Dipicolinate as a Metal Ion Buffer

As detailed above, the general occurrence of dipicolinate acid in spores led to the widespread supposition that this component, perhaps complexed with calcium and other spore components, was somehow directly concerned with conferring heat resistance on spores. However, we have pointed out that this view is no longer tenable following the isolation of heat-resistant DPA-negative spore mutants.

Fitz-James (1971) and Pearce and Fitz-James (1971) presented strong evidence to support the hypothesis that calcium was necessary to maintain the integrity of the spore "core" or "protoplast", perhaps bound to phosphate groups on membrane phospholipids. These observations, and the observed instability of DPA-negative spores on storage, have led us to consider a new hypothesis, namely that a major role of calcium dipicolinate in spores is that of a calcium buffer. By this we infer a system which maintains for long periods of time within the spore protoplast a constant low pool concentration of calcium which has a physiological role in dormancy and perhaps also in heat resistance. Calcium lost from spores by slow leakage would be replenished by dissociation of the calcium-DPA complex, for which there is good evidence of a core location (e.g. for DPA (Leanz and Gilvarg, 1973) and for calcium (Scherrer and Gerhardt, 1972)). The mutant spores lacking dipicolinic acid could have internal calcium levels sufficient to satisfy the requirements for dormancy and resistance, and indeed some of the DPA-negative spore mutants now known are certainly as dormant and heat resistant as wild type spores. However, lacking the calcium buffer and the resultant large potential pool of calcium, such spores would be expected to have difficulty in maintaining sufficiently high calcium levels for long periods of time. It is therefore significant that the DPA-negative mutant spores have, in fact, been shown to be generally unstable and to lose their heat resistance and their dormancy rapidly on storage (Hanson *et al.*, 1972).

The last few years have witnessed radical changes relating to the mechanisms of spore heat resistance. Amongst the most profound changes has been the challenge to the dogma that dipicolinic acid in spores is essential for full heat resistance. Evidently a new role must be allocated to DPA. One possibility is its involvement as a metal ion buffer, as suggested above. Location studies, particularly relating to Ca^{2+} and DPA, have begun to yield the kind of data which will be necessary in order that complete resolution of spore thermoresistance can be made. Evidence is now available strongly implicating the cortex in heat resistance, either through the "contractile cortex" mechanism or the hitherto neglected alternative the "expanded cortex". In our view the latter hypothesis has much to support it. Furthermore an expanded electronegative cortex devoid of cations would allow a rational explanation of Ca^{2+}:DPA-induced germination; the neutral chelate would function to carry cations into the cortex. Such a hypothesis may certainly be tested experimentally.

VII. Acknowledgements

Our thanks are due to Prof. R. J. P. Williams for valuable discussion, particularly relating to the concept of dipicolinic acid and metal ion buffers, and to Mr. J. M. Stubbs for the electronmicrograph of *Clostridium sporogenes* shown in Fig. 1.

References

Adams, D. M. and Busta, F. F. (1972). *In* "Spores V", (H. O. Halvorson, R. Hanson and L. L. Campbell, eds.), pp. 368–377. American Society for Microbiology, Washington, D.C.

Alderton, G. and Snell, N. (1963). *Biochemical and Biophysical Research Communications* **10**, 139.

Alderton, G. and Snell, N. (1964). *Science, New York* **143**, 141.

Amaha, M. and Ordal, Z. J. (1957). *Journal of Bacteriology* **74**, 596.

Aoki, H. and Slepecky, R. A. (1973). *Journal of Bacteriology* **114**, 137.

Aronson, A., Henderson, E. and Tincher, A. (1967). *Biochemical and Biophysical Research Communications* **26**, 454.

Bailey, G. F., Karp, S. and Sacks, L. E. (1965). *Journal of Bacteriology* **89**, 984.

Black, S. M. and Gerhardt, P. (1962). *Journal of Bacteriology* **83**, 960.

Black, S. H., Hashimoto, T. and Gerhardt, P. (1960). *Canadian Journal of Microbiology* **6**, 213.

Briggs, A. (1966). *Journal of Applied Bacteriology* **29**, 490.

Bonsen, P. P. M., Spudich, J. A., Nelson, D. L. and Kornberg, A. (1969). *Journal of Bacteriology* **98**, 62.

Busta, F. F. and Adams, D. M. (1972). *Applied Microbiology* **24**, 412.

Busta, F. F. and Ordal, Z. J. (1964). *Applied Microbiology* **12**, 106.

Bryne, A. F., Burton, T. H. and Koch, R. B. (1960). *Journal of Bacteriology* **80**, 139.

Carstensen, E. L., Marquis, R. E. and Gerhardt, P. (1971). *Journal of Bacteriology* **107**, 106.

Čáslavská, J., Šťastná, J. and Vinter, V. (1970). *Folia Microbiolica Praha* **15**, 197.

Cassier, M. and Ryter, A. (1971). *Annales Institute Pasteur, Paris* **121**, 717.

Cassier, M. and Sebald, M. (1969). *Annales Institute Pasteur, Paris* **117**, 312.

Chambon, P., Duetscher, M. P. and Kornberg, A. (1968). *Journal of Biological Chemistry* **243**, 5110.

Church, B. D. and Halvorson, H. (1959). *Nature, London* **183**, 124.

Cohn, F. (1876). *Beitrage zur Biologie der Pflanzen* **2**, 249.

Coleman, G. (1969). *Biochemical Journal* **112**, 533.

Cross, T., Walker, P. D. and Gould, G. W. (1968). *Nature, London* **220**, 352.

Curran, H. R., Brunstetter, B. C. and Myers, A. T. (1943). *Journal of Bacteriology* **45**, 485.

Curry, M., Patt, T. and Hanson, R. S. (1971). *Bacteriological Proceedings* G240.

Dawes, I. W. and Hansen, J. N. (1972). *Critical Reviews in Microbiology* **1**, 479.
Deutscher, M. P., Chambon, P. and Kornberg, A. (1968). *Journal of Biological Chemistry* **243**, 5117.
Dring, G. J. and Gould, G. W. (1971a). *Journal of General Microbiology* **65**, 101.
Dring, G. J. and Gould, G. W. (1971*b*). *In* "Spore Research 1971", (A. N. Barker, G. W. Gould and J. Wolf, eds.), pp. 133–143. Academic Press, London.
Duncan, C. L., Labbe, R. G. and Reich, R. R. (1972). *Journal of Bacteriology* **109**, 550.
Edwards, J. L., Busta, F. F. and Speck, M. L. (1965). *Applied Microbiology* **13**, 851.
El-Bisi, H. M. and Ordal, Z. J. (1956). *Journal of Bacteriology* **71**, 10.
Falaschi, A. and Kornberg, A. (1966). *Journal of Biological Chemistry* **241**, 1478.
Fitz-James, P. C. (1971). *Journal of Bacteriology* **105**, 1119.
Fitz-James, P. and Young, E. (1969). *In* "The Bacterial Spore", (G. W. Gould and A. Hurst, eds.). Academic Press, London.
Flemming, H. P. (1964). Ph.D. Thesis: University of Illinois, Urbana, Illinois.
Foerster, H. F. and Foster, J. W. (1966a). *Journal of Bacteriology* **91**, 1168.
Foerster, H. F. and Foster, J. W. (1966b). *Journal of Bacteriology* **91**, 1333.
Foster, J. W. (1956). *Quarterly Reviews of Biology* **31**, 102.
Gardner, R. and Kornberg, A. (1967). *Journal of Biological Chemistry* **242**, 2383.
Gerhardt, P., Scherrer, R., Carstensen, E. L. and Marquis, R. E. (1971). *In* "Spore Research 1971", (A. N. Barker, G. W. Gould and J. Wolf, eds.), p. 341. Academic Press, London.
Gilpin, R. W. and Sadoff, H. L. (1971). *Journal of Biological Chemistry* **246**, 1475.
Glenn, A. R. and Mandelstam, J. (1971). *Biochemical Journal* **123**, 29.
Gould, G. W. (1970). *Journal of Applied Bacteriology* **33**, 34.
Gould, G. W., Georgala, D. L. and Hitchins, A. D. (1963). *Nature, London* **200**, 385.
Gould, G. W. and Hitchins, A. D. (1963). *Journal of General Microbiology* **33**, 413.
Grecz, N. and Tang, T. (1970). *Journal of General Microbiology* **63**, 303.
Green, J. H. and Sadoff, H. L. (1965). *Journal of Bacteriology* **89**, 1499.
Hachisuka, Y. and Tochikubo, K. (1971). *Journal of Bacteriology* **107**, 442.
Hachisuka, Y., Tochikubo, K., Yokoi, Y. and Murachi, T. (1967). *Journal of Biochemistry, Tokyo* **61**, 659.
Halvorson, H. and Howitt, C. (1961). *In* "Spores II", (H. O. Halvorson, ed.), 149–179. Burgess Publishing Co., Minneapolis.
Halvorson, H. O. and Swanson, A. (1969). *In* "Spores IV", (L. L. Campbell, ed.), pp. 121–132. American Society for Microbiology, Bethesda, Maryland.
Hanson, R. S., Curry, M. V., Garner, J. V. and Halvorson, H. O. (1972). *Canadian Journal of Microbiology* **18**, 1139.
Harrell, W. K. and Mantini, E. (1957). *Canadian Journal of Microbiology* **3**, 735.
Hashimoto, T. and Gerhardt, P. (1960). *Journal of Biophysical and Biochemical Cytology* **7**, 195.
Hashimoto, T., Frieben, W. R. and Conti, S. F. (1969). *Journal of Bacteriology* **100**, 1385.
Hitchins, A. D. and Gould, G. W. (1964). *Nature, London* **203**, 895.
Hober, R. (1945). "Physical Chemistry of Cells and Tissues". The Blakiston Company, Philadelphia, Pa.
Janssen, F. W., Lund, A. J. and Anderson, L. E. (1958). *Science, New York* **127**, 26.
Kalakoutskii, L. V., Agre, N. S. and Anslanjan, R. R. (1969). *Doklady AN SSS* **184**, 1214.
Keynan, A., Murrell, W. G. and Halvorson, H. O. (1961). *Nature, London* **192**, 1211.
Knaysi, G. (1965). *Journal of Bacteriology* **90**, 453.
Knoll, H. and Horschak, R. (1971). *Monatsberichte* **13**, 222.

Lacey, J. and Vince, D. A. (1971). *In* "Spore Research 1971", (A. N. Barker, G. W. Gould and J. Wolf, eds.), pp. 181–188. Academic Press, London.

Leanz, G. F. and Gilvarg, C. (1972). *In* "Spores V", (H. O. Halvorson, R. Hanson and L. L. Campbell, eds.), American Society for Microbiology, Washington, D.C.

Leanz, G. and Gilvarg, C. (1973). *Journal of Bacteriology* **114**, 455.

Levinson, H. S. and Hyatt, M. T. (1966). *Journal of Bacteriology* **91**, 1811.

Levinson, H. S., Hyatt, M. T. and Moore, F. E. (1961). *Biochemical and Biophysical Research Communications* **5**, 417.

Lewis, J. C., Snell, N. S. and Burr, H. K. (1960). *Science, New York* **132**, 544.

Lewis, J. C. (1969). *In* "The Bacterial Spore", (G. W. Gould and A. Hurst, eds.). pp. 301–352. Academic Press, London.

Lewith, S. (1890). *Archiv für Experimentelle Pathologie und Pharmakologie* **26**, 341.

Lubin, M. and Ennis, H. L. (1965). *Biochimica et Biophysica Acta* **80**, 614.

Mandelstam, J. (1969). *Symposim of the Society for General Microbiology* **19**, 377.

Marquis, R. E. (1968). *Journal of Bacteriology* **95**, 775.

Marshall, B. J. and Murrell, W. (1970). *Journal of Applied Bacteriology* **33**, 103.

Mishiro, Y. and Ochi, M. (1966). *Nature, London* **211**, 1190.

Murrell, W. G. (1967). *Advances in Microbial Physiology* **1**, 133.

Murrell, W. G. (1969). *In* "The Bacterial Spore", (G. W. Gould and A. Hurst, eds.), pp. 214–273. Academic Press, London.

Murrell, W. G., Ohye, D. F. and Gordon, R. A. (1969). *In* "Spores IV", (L. L. Campbell, ed.), pp. 1–19. American Society for Microbiology, Bethesda, Maryland.

Murrell, W. G. and Scott, W. J. (1966). *Journal of General Microbiology* **43**, 411.

Murrell, W. G. and Warth, A. D. (1965). *In* "Spores III", (L. L. Campbell and H. O. Halvorson, eds.), pp. 1–24. American Society for Microbiology, Ann Arbor, Michigan.

Nakata, H. (1957). *In* "Spores", (H. O. Halvorson, ed.). American Institute of Biological Sciences, Washington, D.C.

Nelson, D., Spudich, J., Bonsen, P., Bertsch, L. and Kornberg, A. (1969). *In* "Spores IV", (L. L. Campbell, ed.). American Society for Microbiology, Ann Arbor, Michigan.

Ou, L-T. and Marquis, R. E. (1970). *Journal of Bacteriology* **101**, 92.

Pearce, J. M. and Fitz-James, P. C. (1971). *Journal of Bacteriology* **107**, 337.

Pelcher, E. A., Flemming, H. P. and Ordal, Z. J. (1963). *Canadian Journal of Microbiology* **9**, 251.

Perry, J. J. and Foster, J. W. (1955). *Journal of Bacteriology* **69**, 337.

Powell, J. F. (1950). *Journal of General Microbiology* **4**, 330.

Powell, J. F. (1951). *Journal of General Microbiology* **5**, 993.

Powell, J. F. (1953). *Biochemical Journal* **54**, 210.

Powell, J. F. (1957). *Journal of Applied Bacteriology* **20**, 349.

Powell, J. F. and Strange, R. E. (1953). *Biochemical Journal* **54**, 205.

Rode, L. J. and Foster, J. W. (1960a). *Proceedings of the National Academy of Sciences, Washington* **46**, 118.

Rode, L. J. and Foster, J. W. (1960b). *Journal of Bacteriology* **79**, 650.

Rode, L. J. and Foster, J. W. (1960c). *Archives fur Mikrobiologie* **36**, 67.

Rode, L. J. and Foster, J. W. (1966). *Journal of Bacteriology* **91**, 1589.

Roberts, T. A. and Hitchins, A. D. (1969). *In* "The Bacterial Spore", (G. W. Gould and A. Hurst, eds.), pp. 611–670. Academic Press, London and New York.

Roberts, T. A. and Ingram, M. (1965). *Journal of Applied Bacteriology* **28**, 125.

Rowley, D. B. and Feeherry, F. (1970). *Journal of Bacteriology* **104**, 1151.

Rowley, D. B. and Levinson, H. S. (1967). *Journal of Bacteriology* **93**, 1017.

Sadoff, H. L. (1970). *Journal of Applied Bacteriology* **33**, 130.

Sadoff, H. L., Bach, J. A. and Kools, J. W. (1965). *In* "Spores III", (L. L. Campbell and H. O. Halvorson, eds.), pp. 97–110. American Society for Microbiology, Ann Arbor, Michigan.
Sadoff, H. L., Celikkol, E. and Engelbrecht, H. L. (1970). *Proceedings of the National Academy of Sciences of the U.S.A.* **66**, 844.
Sadoff, H. L., Hitchins, A. D. and Celikkol, E. (1969). *Journal of Bacteriology* **98**, 1208.
Schaeffer, P. (1969). *Bacteriology Reviews* **33**, 48.
Scherrer, R. and Gerhardt, P. (1972). *Journal of Bacteriology* **112**, 559.
Sebald, M. and Ionesco, H. (1972). *Comptes Rendu Hebdomandaire des Seances de l'Académie des Sciences, Paris*, **275**, 2175.
Slepecky, R. A. (1961). *In* "Spores II", (H. O. Halvorson, ed.). Burgess Publishing Co., Minneapolis.
Slepecky, R. A. and Foster, J. W. (1959). *Journal of Bacteriology* **78**, 117.
Sogin, M. L., McCall, W. A. and Ordal, Z. J. (1972). *In* "Spores V", (H. O. Halvorson, R. Hanson and L. L. Campbell, eds.), pp. 363–367. American Society for Microbiology, Washington, C.D.
Spudich, J. A. and Kornberg, A. (1969). *Journal of Bacteriology* **98**, 69.
Spudich, J. A., Tono, H. and Kornberg, A. (1966). *Federation Proceedings* **25**, 276.
Stewart, B. T. and Halvorson, H. O. (1954). *Archives of Biochemistry and Biophysics* **49**, 168.
Strange, R. E. (1956). *Biochemical Journal* **64**, 23P.
Strange, R. E. and Dark, F. A. (1957a). *Journal of General Microbiology* **16**, 236.
Strange, R. E. and Dark, F. A. (1957b). *Journal of General Microbiology* **17**, 525.
Strange, R. E. and Powell, J. F. (1954). *Biochemical Journal* **58**, 80.
Strange, R. E. and Thorne, C. R. (1957). *Biochimica et Biophysica Acta* **24**, 199.
Sugiyama, H. (1951). *Journal of Bacteriology* **62**, 81.
Theophilis, D. R. and Hammer, B. W. (1938). *Journal of Hygiene, Cambridge* **32**, 535.
Thomas, R. S. (1964). *Journal of Cell Biology* **23**, 113.
Thompson, R. S. and Leadbetter, E. R. (1963), *Archives für Mikrobiologie* **45**, 27.
Tochikubo, K., Hachisuka, Y. and Murachi, T. (1968). *Japanese Journal of Microbiology* **12**, 435.
Tono, H. and Kornberg, A. (1967*a*). *Journal of Biological Chemistry* **242**, 2375.
Tono, H. and Kornberg, A. (1967*b*). *Journal of Bacteriology* **93**, 1819.
Udo, S. (1936). *Journal of the Agricultural Chemical Society, Japan* **12**, 386.
Vinter, V. (1960). *Folia Microbiologica, Praha* **5**, 217.
Vinter, V. (1962). *Folia Microbiologica, Praha* **7**, 115.
Walker, H. W., Matches, J. R. and Ayres, J. C. (1961). *Journal of Bacteriology* **82**, 960.
Warth, A. D., Ohye, D. F. and Murrell, W. G. (1963). *Journal of Cell Biology* **16**, 593.
Warth, A. D. and Strominger, J. L. (1969). *Proceedings of the National Academy of Sciences, U.S.A.*, **64**, 528.
Warth, A. D. and Strominger, J. L. (1971). *Biochemistry* **10**, 4349.
Williams, O. B. (1929). *Journal of Infectious Diseases* **44**, 421.
Williams, O. B. and Robertson, W. J. (1954). *Journal of Bacteriology* **67**, 377.
Windle, J. J. and Sacks, L. E. (1963). *Biochimica et Biophysica Acta* **66**, 173.
Wise, J., Swanson, A. and Halvorson, H. O. (1967). *Journal of Bacteriology* **94**, 2075.
Woese, C. R., and Morowitz, H. J. (1958). *Journal of Bacteriology* **76**, 81.
Wood, D. A. (1971). *Biochemical Journal* **123**, 601.
Zytkovicz, T. M. and Halvorson, H. O. (1972). *In* "Spores V", (H. O. Halvorson, R. Hanson and L. L. Campbell, eds.), pp. 49–52. American Society for Microbiology, Washington, D.C.

Experimental Bacterial Ecology Studied in Continuous Culture

H. W. JANNASCH

Woods Hole Oceanographic Institution, Woods Hole, Mass., 02543, U.S.A.

and

R. I. MATELES

Laboratory of Applied Microbiology, The Hebrew University, Jerusalem, Israel

I. Introduction

More than two decades ago, in a symposium on the Chemistry and Physiology of Growth, van Niel (1949) emphasized that "Growth is the expression *par excellence* of the dynamic nature of living organisms. Among the general methods available for the scientific investigation of dynamic phenomena, the most useful ones are those which deal with kinetic aspects", and "Kinetic investigations on cultures of micro-organisms are eminently suited for establishing relations between growth and environmental factors, especially the nature and amount of nutrients". As a most fitting motto for the present article, these statements endorsed the usefulness of continuous culture techniques in microbial ecology even before the concepts of steady-state cultures and of the chemostat were formulated simultaneously by Monod (1950) and Novick and Szilard (1950a). Growth and multiplication of microorganisms are the decisive parameters for successful competition in nature as well as in continuous culture. This concept bases soundly on Beijerinck's (1921–1940) and Winogradski's (1949) ecological principles of microbial niches and selective conditions in the natural environment.

The principles and applications of the continuous culture of microorganisms have enormously contributed to the advancement of three areas of microbiology: metabolic regulation and genetics, technological and applied microbiology, and microbial ecology. This fact, as far as the first two areas are concerned, has been accounted for and acknowledged in the proper literature. The present article brings up to date, and extends, an earlier attempt by Jannasch (1965a) to complete a similar task for microbial ecology. An exhaustive review of the particularly widespread literature is not intended here, and redundancy is avoided especially with respect to the following articles.

Bungay and Bungay (1968) reviewed studies on microbial interactions in continuous culture and focused on their definitions. Therefore, terms such as commensalism, mutualism and neutralism, will not receive major emphasis nor be used for classifying purposes. Some ecological implications and reflections were made by Veldkamp and Jannasch (1972) in an article on the behaviour of mixed populations in the chemostat. Veldkamp and Kuenen (1973) discussed continuous culture techniques with respect to microbial ecology. A recent review by Meers (1973) deals with growth of mixed microbial cultures in general and includes a chapter on continuous-culture studies. The latest International Symposium on the Continuous Culture of Microorganisms (Dean *et al.*, 1972) contains a number of related topics. The individual contributions will be referred to the place of their original publication in the *Journal of Applied Chemistry and Biotechnology*. In view of this extensive literature, the justification for the present review is its exclusive emphasis on microbial ecology.

Difficulties in assigning proper limits to the field of microbial ecology, and its status in general microbiology, draw one more precursory comment. Criticism of incompatible approaches with respect to other disciplines of microbiology are often not unjustified. Some undeniable shortcomings can be ascribed to a misconception of the difficulties that are inherent in dealing with systems of such enormous complexity and unpredictable reactivity as natural microbial populations. While it is perfectly legitimate for a microbial physiologist or biochemist to simplify a complex system by working with synthetic media, high substrate concentrations, cell extracts, and so on, for the sake of well-defined and reproducible experimental conditions, these helpful tools are of limited use to the ecologist because the very complexity is the ultimate object of his research. As a corollary, compromises are necessary, and the art of microbial ecology appears to reside in operating in that narrow area where those two seemingly incongruous aims accord: the scientific soundness of data (reproducibility, statistical significance, methodical compatability, etc.) and their ecological relevance.

In this situation, continuous-culture approaches have contributed decisively to microbial ecology. A number of applications, including hetero-continuous flow systems, have been profitably used in studying unusual prokaryotes that were never successfully grown in batch culture. But most significant is the technical possibility of measuring microbial activities in steady-state systems, eliminating the artificial lag and stationary growth phase phenomena characteristic for closed culture systems. While steady-state systems are certainly not reproductions of natural conditions, they ascertain the reproducibility of data and offer a possibility of studying one or a few environmental factors at a time with the aim of reconstructing more complex systems from known elements. Applications of continuous flow systems in technical and industrial microbiology are of ecological interest, especially with respect to the behaviour of mixed bacterial populations and the mode of microbial breakdown and conversion of complex substrates such as crude hydrocarbons, pesticides, or sewage.

The general scope and the selection of details in this article attempt to combine the interests of the ecologically oriented microbiologist with those of the environmental biologist. The titles of the sections are not intended to reflect a systematic subdivision of the general topic but follow a logical order of studies actually done.

II. Pure Culture Studies

A. Steady-State Kinetics

The present plethora of theoretical model building calls for a compensating emphasis on practical experimental approaches. We are therefore limiting these discussions to only few theoretical derivations for cases of special ecological significance. One of these is the formulation of steady state.

The description of the chemostat and the definition of steady state of growing microbial cultures (Monod, 1950; Novick and Szilard, 1950a;

Perret, 1960) represented the important step by which microbial growth kinetics became a most useful tool in studying metabolic, genetic, and ecological phenomena. Borrowed originally from chemical reaction kinetics (Hinshelwood, 1946), the term steady state was widely used in biology in a rather undefined sense until its applicability was up-graded in continuous culture studies. As a result, some confusion can still be encountered in general ecology. Herbert *et al.* (1956) presented a most straightforward definition, and their terminology will be used throughout. A more recent comprehensive treatise on the theory of microbial growth in the chemostat was given by Tempest (1970).

If in a bacterial population, for a given time, the immediate environment of all cells contains the nutrients required for growth, and if the genetic composition remains unchanged, exponential growth will result:

$$X = X_0 e^{\mu(t-t_0)} \quad (1)$$

where X is the final population density when the time t has elapsed, and X_0 and t_0 are the initial population density and starting time respectively. The constant μ is the specific growth rate.

If one assumes a constant flow of sterile medium through the well-mixed suspension of non-multiplying cells, the population would be diluted following:

$$X = X_0 e^{-D(t-t_0)} \quad (2)$$

where $-D$ is the dilution rate. If growth and dilution occurs at the same time, the population density would either increase or decrease according to the net rate resulting from:

$$X = X_0 e^{(\mu-D)(t-t_0)} \quad (3)$$

The intrinsic advantage of the chemostat lies in the fact that the rate of dilution also controls the rate of growth via the concentration of the growth-limiting substrate contained in the medium. As long as the dilution rate is lower than the maximum attainable growth rate, this relationship will result in the self-adjustment of a steady state where μ and D become numerically equal. This state is characterized by the constancy of all growth parameters as long as the external physico-chemical conditions remain unaltered and no genetic change occurs. Although the importance of classical and current genetic studies in this area (Novick and Szilard, 1950b; Novick, 1958a; Kubitschek, 1970) are relevant to all experimental chemostat work, only some technological studies on the establishment of mutants (see p. 199) are directly applicable to ecological considerations.

The physiological response of a chemostat population to a change of external conditions will upset a steady state and result in the establishment of another. The new constant values for population density and concentration of limiting substrate in the culture are accurate and sensitive indicators of subtle physiological responses and, therefore, of utmost importance for ecological studies. This assumes, of course, the technical quality of complete mixing and the constancy of culture conditions including medium composition, flow rate, culture volume, temperature, and pH value.

Monod (1942) described the relationship between growth rate and concentration of limiting substrate as a data-fitting function:

$$\mu = \mu_m \frac{s}{K_s + s} \tag{4}$$

where K_s is the saturation constant (numerically equal to the limiting substrate concentration at which half of the maximum specific growth rate is reached), and μ_m the maximum specific growth rate. Using it as a first approximation of an obviously more complex process, Herbert *et al.* (1956) derived the mathematical expression of the steady state value for the population density x:

$$x = Y(S_R - s) \tag{5}$$

and for the concentration of the limiting substrate s:

$$s = K_s \frac{D}{\mu_m - D} \tag{6}$$

where S_R is the concentration of the limiting substrate in the sterile medium (reservoir), and Y the yield coefficient. This rather simple mathematical treatment has been the subject of many theoretical and experimental studies (summarized by Pirt, 1972) to show that, and under what conditions, Monod's equation does not appear satisfactory. Teissier's (1942) expression assumes a diminishing effect of the concentration of the limiting substrate with increasing growth rate. Moser (1958) introduces a higher flexibility of the model by replacing the saturation constant with an arbitrary factor. Starting with Herbert's (1958) consideration of the type of limiting substrate, i.e., its role as an energy source or essential nutrient, most deviations from theory have been expressed as changes of the yield coefficient (van Uden, 1968), in other words, changes of the proportion between anabolism and catabolism. An excellent discussion of this central topic in continuous-culture work has been given by Powell (1967, 1972) who includes the consideration of cell size, permeability and substrate diffusion rates. In most ecological applications, the required level of resolution makes Monod's original relation the preferable and most practical tool. In fact, much of the early physiological work on continuous-culture populations with interesting ecological implications originated from a variety of inconsistencies observed when Monod's model was applied.

The earlier mentioned confusion in applying open flow systems for quantitative studies in ecology relates to the fact that the time-independence of steady state conditions is not always fully understood and that exponential growth is a necessary condition. Proposed steady states of complex mixed or natural populations in the presence of diurnal or seasonal environmental changes constitute confusing misconceptions. Fairly simple mathematical relationships between growth parameters do apply only when a true steady state has established that they are not affected by fluctuations of external conditions. Obviously, transient state kinetics and their mathematical treatment (Schaezler *et al.*, 1971; Sinclair *et al.*, 1971) are far more complex than steady-state kinetics.

A substantial source of difficulties in reaching steady states experimentally results from a combination of all those effects that prevent the constancy of physico-chemical conditions. Wall growth is the most difficult to deal with; it impedes the strict requirement of complete mixing, thus imposing batch culture characteristics. The apparent effect of wall growth on the specific maximum growth rate when measured by approaching wash-out of a continuous culture has been described by Larsen and Dimmick (1964) and by Topiwala and Hamer (1971). Detailed discussions on "apparatus effects" on steady-state kinetics also have been given by Malek and Fencl (1966).

Steady states have often been proposed for mixed culture, mixed substrate, or dispersed phase (e.g. aqueous-oil or aqueous-solid) systems. As to be discussed later in detail, such systems cannot *a priori* be expected to attain steady states; on the contrary, they rarely will. It is necessary to mention this, since a considerable number of ecologically oriented continuous culture studies (especially on algae) have been based on steady-state kinetics; in these studies the type of population and mode of cultivation made the establishment of a true steady state unlikely or impossible.

There are many specific environments that are essentially heterogeneous open flow systems and resemble continuous cultures, the rumen of ruminant mammals being the most conspicuous example. But they are not "chemostats", and steady states (in the sense used here) are never achieved. Growth of microbial populations in natural environments will reflect characteristics of both systems, batch and continuous culture, at the same time.

Due to its apparent environmental constancy, the ocean, especially the deep-sea, has been suggested to possess chemostat qualities or to be in steady state. Since no complete mixing can be assumed, constant growth of a microbial population in such a highly dilute system can be visualized only in the unlikely case of being matched by the diffusion rate of nutrients.

The fact that many micro-organisms do naturally occur in situations that strongly select against steady-state conditions supports the notion that the value of the well-defined steady state lies in its potential as an experimental tool rather than in the apparent possibility of reproducing natural conditions. While the changing environmental conditions are most important for the types of interrelationships and successions of microbial species in nature, the artificially induced constancy in the chemostat, and the elimination of time as a factor, makes intricate processes amenable to analytical studies.

B. Substrate-Limited Growth

The theoretical description of continuous culture requires that growth is limited by only *one* factor or substrate throughout the experiment. At steady state, the concentration of the limiting substrate is independent of its concentration in the reservoir, but dependent on the dilution rate [equation (6)]. At the same time, the population density becomes a function of the substrate concentration in the reservoir [equation (5)]. This principle of substrate-limited growth in continuous culture is experimentally easier to realize when the population density is kept at a low value and, therefore, secondary effects, such as partial growth limitation by oxygen or a metabolic product, are unlikely to occur.

The main reason, however, to limit the discussion in this chapter to continuous-culture studies with *low* substrate concentrations and dilution rates concerns the actual ecological situation. With few exceptions, microbial growth and turnover of organic matter in natural environments, or *in vivo*, proceed at extremely slow rates compared with those obtained in artificial culture media commonly prepared to produce high yields of organisms and/or product. While microbial ecology has profited from some physiological studies, e.g. on spore formation, starvation phenomena, and certain aspects of metabolic regulation, the primary interest in employing relatively and "unnaturally" rich culture media in general microbiology limits the usefulness of most of the available information for environmental microbiology.

During the decomposition of organic matter, largely taking place in soil and water under the influence of an enormous variety of metabolic types of organisms, the complex flow of energy through the "food chain" (the product of one organism's metabolism being rapidly used by the next) has the effect of dampening fluctuation of substrate availability and of keeping the concentrations of substrates at a low level. Measuring these "natural" rates of growth and chemical transformations below certain critical levels of substrate meets with traditional difficulties. In batch culture the exponential and measurable rate of growth is trimmed on both sides by growth phases that are characteristic for closed systems: lag and stationary phase. At a decreasing initial concentration of the limiting substrate (S_R), these growth phases expand and finally appear to overlap rendering measurements of an exponential growth rate more

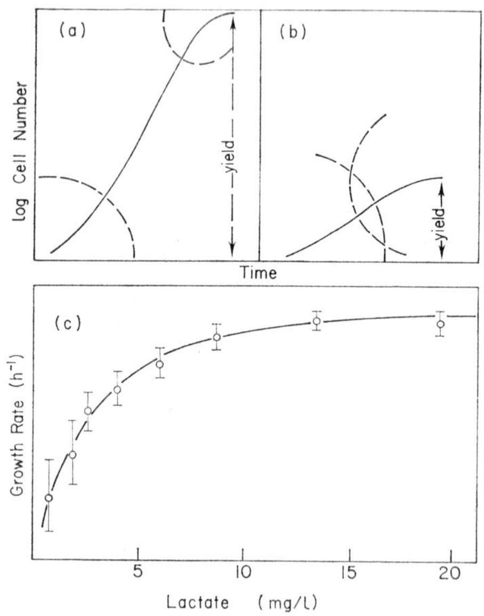

FIG. 1. (a) and (b) Range of exponential growth phase compared with the ranges of the lag and stationary phases at high and low concentrations of the limiting substrate in batch culture (schematic); (c) change of variance of growth rate data with initial concentration of limiting substrate in batch culture (data from Jannasch, 1963).

and more difficult (Fig. 1a, b). As a corollary, the statistical significance of such measurements decreases with decreasing initial concentrations of the limiting substrate below the saturation level (Fig. 1c). These values, however, are of particular importance for determining the saturation constant, K_s, the index of substrate affinity, e.g. graphically in reciprocal plots. In addition, the direct relationship between low initial substrate concentration and growth rate in closed culture systems may be obscured by a phenomenon described as "cryptic growth" (Postgate and Hunter,

1963), i.e. one part of the population growing on the autolytic products of another. In short, before kinetic studies in continuous culture became possible, the complexity of natural microbial populations as well as the limited capabilities of closed culture systems precluded experimental studies on growth and transformations at substrate concentrations comparable to natural levels.

Considering this state of affairs, a number of approaches resorted to extrapolating from data obtained at convenient substrate levels. Work on metabolic regulation (e.g. Maaløe and Kjeldgaard, 1966) has shown, however, that biosynthetic abilities, as well as minimum requirements of the bacterial cell, are strongly affected by changes of the growth rate, or by the concentrations of the limiting substrate. This renders extrapolations highly hazardous.

In a well-defined continuous-culture system, the artificial growth phases occurring in batch culture are eliminated, and it appears conceivable that steady-state growth could be achieved at any low substrate concentration and dilution rate. When this was tried experimentally, a number of phenomena were discovered which contain interesting implications with respect to natural rates of growth and metabolic transformations.

Herbert (1961a) and Tempest *et al.* (1967) point out that studies on the effect of the growth rate on macromolecular composition of cells and on the metabolic activity suggest the existence of a finite minimal growth rate. At steady state, this growth rate would correspond to a threshold concentration of the limiting substrate in the reservoir below which no growth will occur. As discussed earlier, the observed decrease of the yield coefficient with decreasing growth rate in carbon-limited cultures (Herbert, 1958) has been related to the increasing proportion of maintenance energy requirements (Marr *et al.*, 1963; McGrew and Malette 1962). While Novick (1955) summarized the facts known at that time by stating that *Escherichia coli* was "forced into lag" at imposed doubling times (ln 2/dilution rate) of more than 15 h, Tempest *et al.* (1967) were able to attain steady states at retention times (reciprocal of dilution rate) of 180 h in glycerol-limited cultures of *Aerobacter aerogenes*. However, in these cultures the percentage of non-viable cells increased with decreasing dilution rates (Fig. 2), so that the actual maximum doubling time of the viable cells was calculated as 80 h (equivalent to a minimum growth rate of about 0·009 h^{-1}).

The fact that essentially similar results were found with nitrogen (NH_4)-limited cultures indicates that energy requirements do not appear to be the critical factor. Yield and respiration decreased in proportion to the growth rate. In steady-state cultures at retention times above 66 h, the RNA and carbohydrate concentration of the cells and their morphology varied inconsistently. Emphasizing this departure from growth behaviour in the presence of higher substrate concentrations, Pirt (1972) discussed Tempest's data in detail. The interesting aspect of using morphological characteristics (Macura and Kotkova, 1953; Dean and Rogers, 1967; Luscombe and Gray, 1971) of certain suitable micro-organisms as an indicator for growth activities under complex natural

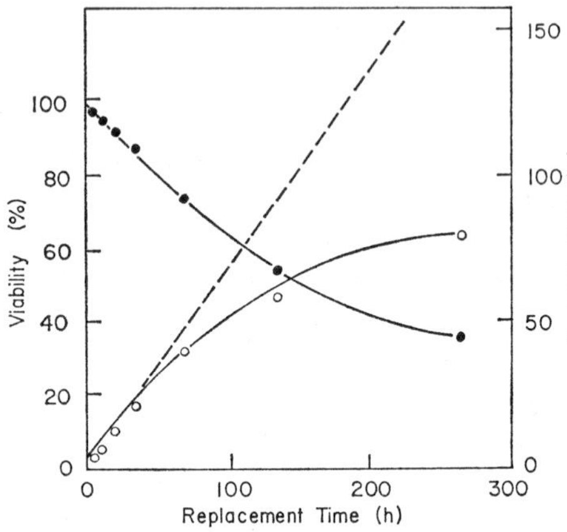

FIG. 2. Changes in the "steady state" viability (●) and doubling time of viable cells (○) in a glycerol-limited culture of *Aerobacter aerogenes*. The broken line represents the theoretical doubling time of the culture ($\ln 2/D$) (from Tempest *et al.*, 1967).

conditions has not yet been tapped. On the other hand, a thorough study by Shehata and Marr (1971) showed that the mean cell volume in *E. coli* is not uniquely determined by the specific growth rate. In studies on spore formation in *Bacillus subtilis*, as affected by the growth rate and the type and concentration of the limiting substrate, Dawes and Mandelstam (1970) found a multiplicity of interacting factors de-emphasizing starvation as the principal initiating effect.

In the natural habitat, microbial growth will alternate with prolonged periods of starvation of non-growing cells. During slow growth, with decreasing rates, the viability also decreases dependent upon the type of growth limitation (Postgate and Hunter, 1962, 1963), and small concentrations of limiting substrate added may lead to "substrate accelerated death" of the cells. This particular research has been reviewed by Postgate (1967). The practical consequences of those processes determining the survival of micro-organisms during storage of cultures (such as measurements of maintenance metabolism) apply particularly to natural populations. The commonly high proportion of microscopic counts to viable counts of bacteria in seawater and most freshwaters (Jannasch and Jones, 1959) may indicate that large parts of the microbial populations are moribund under starving conditions.

Stress conditions of a natural environment may be indicated by minimum growth rates, minimum population densities, or threshold concentrations of the limiting substrate that are much higher than those found in the studies of Tempest *et al.* (1967). When in lactate-limited chemostat cultures of *Spirillum serpens* and other aquatic spirilla, S_R (the lactate concentration in the reservoir) was lowered stepwise at a fixed dilution rate, the resulting steady state population densities did not correspond to the theoretical values calculated according to equation (5). The deficit of x (Fig. 3) increased with decreasing values of S_R until complete wash-out occurred in the presence of a sizeable amount of lactate.

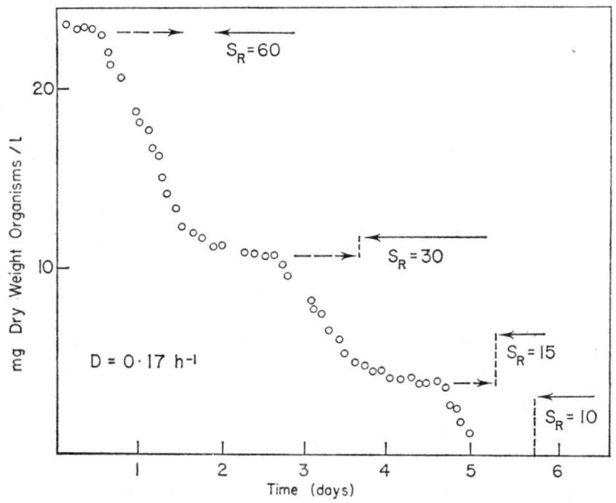

FIG. 3. Decrease of population density with a stepwise lowering of the limiting substrate concentration in the reservoir (S_R, in mg lactate/l). Expected and actually obtained steady state population densities are indicated by solid and broken arrows respectively (data from Jannasch, 1963).

This threshold concentration of S_R varied with the dilution rate. In Fig. 4 the actual steady state population densities are given, indicating that *S. serpens* under the particular culture conditions could only be grown within a certain range of dilution rates dependent on the concentration of lactate in the reservoir. The premature decrease of x can be expressed as a sharp decline of the yield coefficient (Jannasch, 1963), however, the levels of S_R at which the yield is so drastically affected are too high by far to be related to energy requirements of maintenance metabolism or endogenous respiration (Herbert, 1958; Marr *et al.*, 1963; Pirt, 1965). In addition, as wash-out occurs, the substrate concentration in the chemostat increases (*s* in Fig. 4). This clearly indicates that growth limitation has shifted to another factor that, in a relative sense, acts like an added inhibitor.

Searching for an explanation, it was found that these aquatic spirilla, grown in vigorously aerated media, were actually micro-aerophilic. They seemed to overcome, however, the inhibitory effect of oxygen saturation

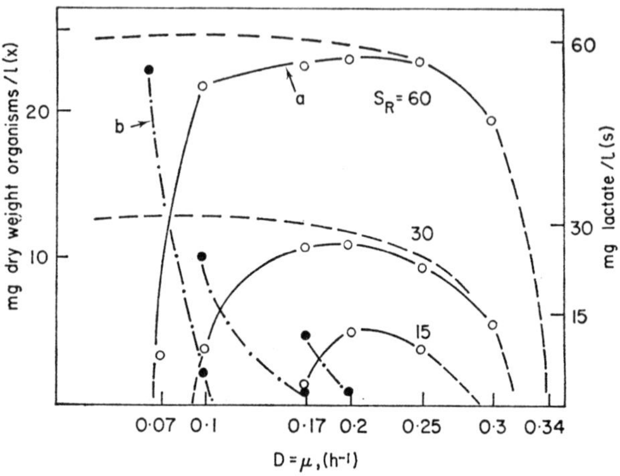

FIG. 4. Steady state population density (x, line a) and concentration of limiting substrate at steady state (s, line b) plotted against the dilution rate (D) for three concentrations of the limiting substrate in the reservoir (S_R) (data from Jannasch, 1963).

at high growth rates or high population densities by affecting the rH of the medium (Harrison, 1972). This effect is a function of both metabolic activity and population density. An addition of ascorbic acid lowered considerably the apparent "threshold concentration" of the growth-limiting substrate which, thereby, can be explained as a corollary of cultural or, ecologically speaking, environmental conditions. A mathematical description of the phenomenon was based on the linear response of the inhibitory effect at different steady-state growth rates (Jannasch, 1965b). Similar observations indicating a micro-aerophilic response or oxygen sensitive enzymes has been reported for nitrogen fixing bacteria grown in continuous culture by Hill *et al.* (1972).

It may be of more general importance that possible inhibitory constituents of an apparently "optimal" medium are masked at high growth rates and/or high population densities of a culture. The possibility of lowering both parameters in steady-state cultures without changing the composition of the medium will reveal such a growth inhibitory effect by resulting in "minimum population densities" and "threshold concentration" of the limiting substrate. The occurrence of certain substrates in seawater that are readily available for microbial utilization but obviously not attacked under the given conditions has been explained on this basis (Jannasch, 1967a).

In search for a technique whereby microbial growth rates in natural waters could be measured, filter-sterilized seawater was run through a

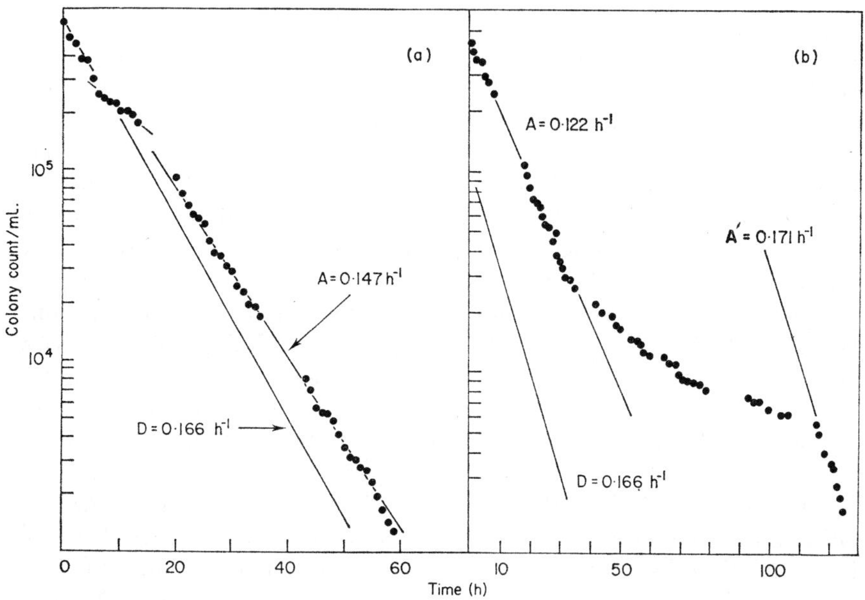

FIG. 5.(a) Wash-out rate (A) of *Achromobacter* sp. in filter-sterilized seawater as compared to the dilution rate (D). The calculated growth rate is 0·019 h^{-1} (from Jannasch, 1969). (b) Washout rate (A, A') of *Pseudomonas* sp. in autoclaved seawater as compared to the dilution rate (D). The calculated growth rate is 0·044 h^{-1}. A' corresponds to a death rate of –0·005 h^{-1} (from Jannasch, 1969).

chemostat at low but still practicable dilution rates (0·25 h^{-1}) and inoculated with a cell suspension of a marine bacteria isolate (Jannasch, 1969). In the presence of an unknown growth-limiting substrate, it was hoped that a steady state would be established indicating the potential growth rate of the test organism in the particular seawater. Instead, wash-out occurred invariably (Fig. 5) i.e., $\mu_m < D$, except with some highly enriched inshore waters. Nevertheless, from the constant washout rates, growth rates could be calculated as the difference between the set dilution rate and the recorded washout rate following the re-written equation (3)

$$\mu = D + \frac{1}{t} \ln (x/x_0) \tag{7}$$

where the washout rate is:

$$-A = \frac{1}{t} \ln (x/x_0) \tag{8}$$

Figure 5a shows data that led to the calculation of a mean generation time (reciprocal of growth rate) of 53 h.

In Fig. 5b the colony counts appear to approach a steady state. But at a population density below 8×10^4 viable cells/ml, the washout rate increases abruptly indicating a sudden cessation of growth. Since in this system growth occurs in the presence of a decreasing population density, it may be possible to explain the sudden stoppage of growth in terms of a minimum population density, as discussed in the preceding paragraphs.

Apart from studies on minimum growth of micro-organisms, continuous culture has been used for measuring rates of specific activities such as nitrogen fixation, nitrification, and sulphate reduction under characteristic environmental conditions. Some of this work has given first ideas on the magnitude and range of activities that may occur under natural conditions.

Munson and Burris (1969) showed that nitrogen fixation by *Rhodospirillum rubrum* in the chemostat was stimulated by nitrogen deficiency in the medium but not by high levels of molecular nitrogen. Hill *et al.* (1972) review continuous culture work on nitrogen-fixing bacteria, adding their own observation of micro-aerophilic behaviour of *Azotobacter croococcum* which is similar to that discussed above for *S. serpens*, with the exception that even at relatively high population densities growth was inhibited by high oxygen concentrations (Dalton and Postgate, 1969a, b). They also describe characteristic differences in N_2-limited and carbon-limited growth based on energy requirements.

If oxygen is growth limiting in chemostat cultures of *A. chroococcum* the production of storage products is initiated. Lees and Postgate (1973) summarize their results: "Oxygen-limited chemostat cultures of nitrogen-fixing *Azotobacter chroococcum* showed an inverse relation between biomass and dilution rate, accounted for largely by increased polysaccharide and polyhydroxybutyrate content. Abrupt increase in pO_2 led to immediate increase in carbon dioxide output followed later by increase in biomass and transition to N_2-limitation; viability on ammonia-free and ammonia-containing media remained at 80–100% during oxygen stress. Phosphate-limited populations showed no respiratory response to oxygen stress, viability dropped rapidly on nitrogen-free medium though the populations were 100% viable on ammonia-containing medium. These findings support the view that respiration in these bacteria has, in part, a protective function for nitrogenase."

Van Uden (1968) demonstrated the feasibility of making accurate measurements of energy requirements in chemostat cultures growing at substrate levels that are of real ecological significance. In an example (Watson, 1970), the yield-lowering effect of sodium chloride in *Saccharomyces cerevisiae* was found and linked to an increased energy requirement for the maintenance of the intracellular concentration of the salt.

C. Product-Limited Growth

From the discussion in the preceding section it is clear that some irregular responses of cultures to substrate-limited growth can be explained by a change in the nature of the growth limitation as a consequence of growth. Here the borderline between substrate- and product-limitation is often diffuse. Depending on the growth conditions and the metabolic type of organism studied, the requirements for substrate-limited growth may be difficult to realize experimentally. As substrate is utilized in the growth vessel, the composition of the medium changes. The degree of this change depends on the dilution rate, and may or may not lead to conditions at which growth is affected. If, for instance, the population of an aerobic culture is allowed to rise sufficient to decrease the oxygen concentration below a critical level, most micro-organisms will excrete significant amounts of partly oxidized metabolites such as acetoin, ethanol, lactic acid and acetic acid. The accumulation of products of incomplete oxidation can be prevented by working at very low population densities. In a similar fashion, growth may also be affected by a drop of the pH in inadequately buffered media. The effect of the dilution rate in pH and/or the concentration of metabolic products was observed by Brooks and Sikyta (1967) and Pokorna *et al.* (1967).

Pirt and Callow (1958) emphasized early the importance of adequate oxygen supply. Its ecological significance is evident, and the shift from substrate-limited to product-limited growth in natural environments may be common or even characteristic. Experimental work in this area has been confined to technical microbiology, but a discussion of the results may be useful in the present context.

The theory of microbial growth in the chemostat also applies to product-limited cultures, but with the added complication that the maximum specific growth rate and the saturation constant, which are defined by the limiting factor, may change with the dilution rate as the nature of growth limitation changes. This was found in the experiments of Zines and Rogers (1971), who observed an apparent effect of ethanol on the two growth constants of *Saccharomyces cerevisiae* growing under ammonia-limited conditions.

The production of acetate has been observed by Chian and Mateles (1968) and Mateles and Chian (1969) in glucose-limited mixed cultures (see p. 206) even though oxygen was present in excess in the growth vessels. This was confirmed in pure culture studies with *E. coli* and *Aerobacter aerogenes* by Masurekar (1968). In his glucose-limited cultures, at dilution rates near the maximum growth rate of the organism, the concentration of non-glucose carbon was 20–30% of the glucose carbon in the reservoir. At low dilution rates, the non-glucose carbon (principally acetate) amounted to 5–20%. Similar results have been reported by Harrison and Loveless (1971). The phenomenon indicates an uncoupling between catabolism and the utilization of the resulting energy for biosynthetic purposes. This effect has not yet been reported with obligate aerobic organisms. It appears desirable to confirm the assumption of complete oxidation in substrate-limited cultures experimentally, particularly when calculations of cell yield constants are concerned.

Natural populations of organisms that change their environment to their own disadvantage obviously will be highly dependent on compensating biological or chemical reactions. Gould and Lees (1960) studied the accumulation of nitrate, the change of pH, and inadequate oxygen supply in continuous cultures of *Nitrobacter*. Calculated from their data, natural populations of *Nitrobacter* may be limited by any one of these factors and thus profit highly from simultaneous growth of other microorganisms utilizing nitrate or buffering the effect of other chemical changes.

Results of continuous culture studies on the repression of nitrogenase by ammonium ions in nitrogen fixing bacteria have direct ecological implications (Hill *et al.*, 1972). In steady state cultures it was possible to examine the effect of population density, and it was found that effective repression was proportional to dry weight measurements, "an observation which may be of relevance to the natural micro-environment, where population densities are often low and thus repressible by low concentrations of free ammonium ions". On the other hand, derepression mechanisms may also be ecologically significant when enzyme synthesis and activity may both increase at low concentrations of the limiting substrate and repressor. Since the early work in this area reported by Magasanik (1957), Magasanik *et al.* (1959) and Gorini (1960), no work with a specific aim on ecological problems has been done.

A case of catabolite repression by succinate of amidase synthesis accompanied by incomplete utilization of acetamide and its metabolites was observed in continuous cultures of *Pseudomonos pyocyanea* (Boddy *et al.*, 1967) in connection with oscillations in the population density. Similarly Gilley and Bungay (1967) and Regan and Roper (1971) found damped oscillations to follow changes of the dilution rate or substrate concentration in chemostat cultures of *Saccharomyces cerevisiae* in ammonia-limited but not in glucose-limited media. Bergter *et al.* (1968) and Knorre (1968) discuss cases of sustained oscillations on the basis of regulatory metabolic processes. In *E. coli* oscillation in the cellular concentration of constitutive β-galactosidase was controlled by catabolite repression.

As to be shown in a later section, damped and sustained oscillations are a much more common feature of mixed populations in the chemostat than of pure cultures.

Product limitation of growth in continuous culture is closely related to the controlled formation of desirable metabolic intermediates and has been exploited in the technical production of lactic acid (Luedeking and Piret, 1959), acetone–butanol (Dyr *et al.*, 1958) and ethanol (Holzberg *et al.*, 1967; Zines and Rogers, 1971). Most of these substrates have been shown to occur in natural, highly productive environments with seasonally or locally distinct oxygen deficiencies. Studies on the role of individual micro-organisms in such complex food chains can not only be done at the multi-substrate level but also with multi-organism systems (see p. 191).

D. Multisubstrate-Limited Growth

In natural conditions, growth limitation of a particular organism by a single substrate, for a certain length of time, is a theoretically correct assumption. It is the valid basis for pure culture single-substrate experiments, but it does not preclude studies that deal with growth in the presence of a multiplicity of substrates such as is commonly found in natural environments.

The sequential utilization of two carbon sources has been observed in batch culture and was described by the diauxie phenomenon by Monod (1942, 1947). It cannot occur in continuous culture because of the imposed steady-state conditions. The behaviour of pure cultures in mixed substrate media was studied by Mateles *et al.* (1967). It was found that growth at the expense of the substrate utilized at the highest efficiency led to establishment of steady states accompanied by an incomplete utilization of other substrates present in the feed medium. Figure 6, a–c, shows examples of substrate utilization in cultures of *E. coli* and *Pseudomonas fluorescens* growing on different pairs of substrates. Effects

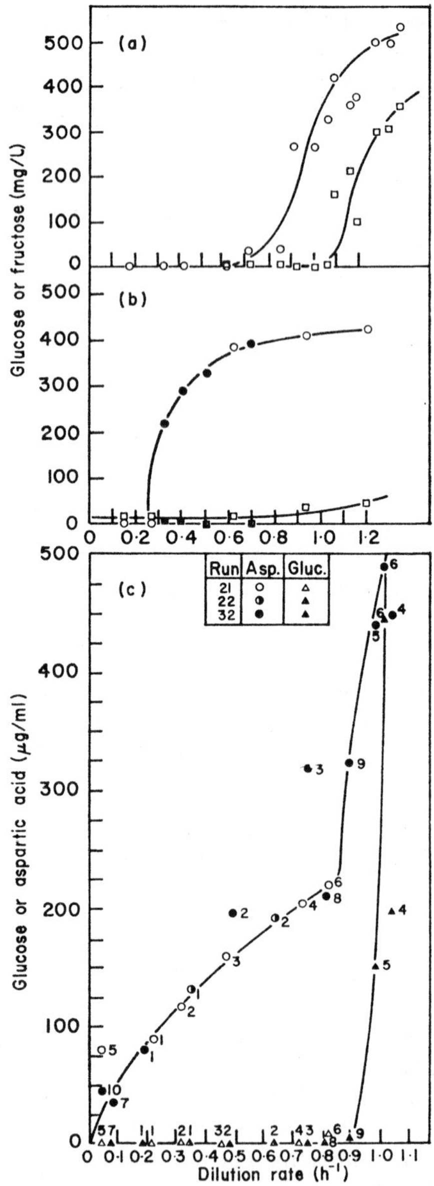

FIG. 6(a). Substrate utilization curves for continuous cultures of *Escherichia coli* with a mixed feed of glucose and fructose. The carbon sources were the growth-limiting factor. ○, fructose; □, glucose (from Mateles *et al.*, 1967).
(b). Substrate utilization curves for continuous cultures of *Pseudomonas fluorescens* with a mixed feed of glucose and fructose. The carbon sources were the growth-limiting factor. Data of two experiments: ○ and ●, fructose; □ and ■, glucose (from Mateles *et al.*, 1967).
(c). Substrate utilization curves for continuous cultures of *Escherichia coli* with a mixed feed of glucose and aspartic acid. The carbon sources were the growth-limiting factor (from Silver and Mateles, 1969).

of catabolite repression (Magasanik, 1961) and catabolite inhibition (Zwaig and Lin, 1966) appear to restrict the uptake of the "poorer" carbon source increasingly with increasing dilution rates. Quantitative differences among substrate pairs are large. This effect has been widely observed with micro-organisms although some results reported recently by Standing *et al.* (1972), with *E. coli* growing in a glucose–galactose or a glucose–xylose medium, are an exception to the general rule.

E. Multistage Culture Systems

After an early theoretical description by Maxon (1955), a large variety of technically possible multistage open flow systems have been classified by Herbert (1961b, 1964). In a number of treatises by Czechoslovakian scientists (Fencl, 1966; Řičica *et al.*, 1967; Fencl *et al.*, 1972) the point has been made that dual-stage chemostats are advantageous in producing and in collecting extracellular metabolites of commercial interest.

Apart from technical developments, the potentials of multistage chemostats operated with pure cultures have been little exploited for ecological purposes. According to equation (5), when substituted by equation (6):

$$x = Y\left(S_R - \frac{K_s D}{\mu_m - D}\right) \tag{9}$$

the degree of substrate utilization is determined by the dilution rate. The lower the dilution rate (i.e. the larger the volume of the growth vessel at a fixed flow rate), the better the utilization of the limiting substrate (i.e. the lower the substrate concentration in the growth vessel). If the volume of the second vessel in a dual stage chemostat is the same as that of the first, the population (x) and the remaining substrate concentration (s) in the second vessel will not be different from a single stage chemostat with the volumes of stage one and two combined. If, however, the volume of the second vessel is larger than that of the first, substrate utilization will be increased in proportion to the decrease of the dilution rate. As a characteristic of the chemostat, the indiscriminate dilution of the culture at a constant rate is an entirely artificial parameter, not reflecting any phenomenon occurring in natural environments. Using a chain of culture vessels, several dilution rates can be applied at the same time for studying the utilization of a particular substrate. Experiments in this area have only been done in mixed culture studies (see p. 198).

Novick (1958b) discussed theoretically a dual-stage chemostat that would facilitate measurements of maximum growth rates in the absence of possible wash-out of the culture. The outlet of the first growth vessel, run at near-maximum growth rates, is connected to a second vessel furnished with a separate inlet for fresh medium ("dual or multistream" system according to Herbert, 1964). While the combined dilution rate in the second vessel is higher than the maximum growth rate, no wash-out can occur as cells are continuously fed into the culture. The steady-state population density in the second stage related to the combined dilution rate will indicate the maximum growth rate of the population.

Experimenting with this system, Mian *et al.* (1970) found a phenomenon that was later termed by Powell (1972) "hypertrophic" growth. The maximum growth rate attained in the second vessel of a dual chemostat of the above type overshot the predicted value considerably. The explanation of this effect resides in a phenomenon that has been discussed before in the sections on substrate and product limitation and appears to be of fundamental significance in microbial ecology; an actively metabolizing population of micro-organisms may excrete one or more important growth factors into the medium enhancing growth in a feed-back fashion, thereby invalidating the theoretical relationship between growth rate and concentration of limiting substrate.

Basically the same phenomenon, relating back to Rahn's (1932) studies on the effect of culture filtrates on microbial growth, is encountered in experimental work on species succession and competition. Since this is a subject of mixed-culture studies, it will be dealt with more appropriately in a later section (see p. 188). In pure culture studies in the chemostat, the failure of metabolite production by a certain minimum population density has been found to express itself as an apparent threshold concentration of the growth-limiting substrate (see p. 175). While the effect of the environment on synthesis and excretion of metabolites has been studied exhaustively for practical reasons (Demain, 1972; Bull, 1972), the intricate role of these "adjuvant substances" (Powell, 1972) on growth, competition and survival of species under natural or laboratory conditions has received much less attention. Demain (1972), in discussing environmental factors that affect the formation of exudates, infers the ecological principle that competition in a natural population will select against metabolically "wasteful" organisms. Dual-stage chemostats are the ideal tool for studying this situation. By simply feeding the effluent of a chemostat into a second culture vessel with a larger volume, the population will continue to grow in this second stage either on the residual substrate not utilized in the first stage, or on some intermediate product that becomes utilizable at the lower dilution rate.

F. Temperature-Related Studies

A recent article by Hunter and Rose (1972) has provided a comprehensive review of the effect of temperature on the physiology and the chemical composition of cells growing in continuous culture. Therefore, we will restrict the present discussion to the question of transient behaviour and its possible relation to ecological problems.

Working with *E. coli* growing in continuous culture under ammonia-limited conditions, Ryu and Mateles (1968) found transient periods in the growth rate response ranging from 1·6 to 11 h in step-wise shifts of 5–10°C upwards and downwards in the range of 17–37°C. Although the Arrhenius equation adequately represented the steady state relationships between maximum specific growth rate and temperature, it failed to do so for the transient periods. The experimentally observed magnitude of the change in growth rates during the transient periods was always less than that predicted by the Arrhenius equation. Similar lags have been observed with *E. coli* by Topiwala and Sinclair (1971), and with *Spirosoma* sp. by Wirsen and Jannasch (1970).

While the variation of growth characteristics with temperature has been shown to play a role in competition among micro-organisms under steady-state conditions at different temperatures, such an effect (see p. 189) has not yet been demonstrated for physiological responses to temperature changes. An organism exhibiting a rapid physiological response to a certain shift of temperature will have a competitive advantage over a more slowly responding organism. While seasonal temperature changes in the ocean or other large bodies of water would appear too slow to significantly affect microbial competition, temperature fluctuations in the surface layers of some soils or shallow and poorly mixed waters indeed fall within an order of magnitude where different responses have been observed in continuous culture experiments. Another environment of rapid temperature fluctuations that might result in selective enrichments is the vicinity of conventional and nuclear power plants and their cooling installations.

In steady-state cultures of *Spirosoma* sp. it was found that the substrate affinity increased with decreasing temperature (Wirsen and Jannasch, 1970). When the temperature was lowered from 30 to 15°C, the K_s value (glucose-limiting) decreased from 3·24 to 0·30. This indication of psychrophilic qualities gives the organism a decisive growth advantage over others that do not respond similarly. It may be a characteristic of a typical aquatic micro-organism living in an enrichment where the levels of limiting substrate tend to be low and relatively constant.

III. Mixed Culture Studies

A. Chemostat Enrichments

Novick and Szilard (1950b) first used, and originally designed (1950a), the chemostat for the *selective cultivation* of micro-organisms. Their studies were aimed at the suppression or promotion of mutants and at the determination of mutation rates. Powell (1958), and later Renneboog-Squilbin (1967), treated in theoretical papers the inherently selective effect of the chemostat and the chances of persistence and removal of contaminants or mutants. In general, if two species are competing for the same growth-limiting substrate, the one attaining the higher growth rate, under the given conditions, will compete successfully and ultimately displace the slower-growing competitor.

This principle infers a new type of enrichment technique. While the classical enrichment procedures in closed culture systems depend upon a pronounced substrate specificity of the organism to be enriched, the chemostat may be used to select from a variety of similar species in the presence of *one* growth-limiting substrate by changing either its concentration in the reservoir, the dilution rate, temperature, oxygen tension or other environmental parameters. In other words, the growth rate of several species, limited by the same substrate, may be affected by

TABLE 1. Arrangement of four separate enrichment experiments, in continuous culture, at two different retention times and two different substrate concentrations, in sea-water media (see Fig 7)

Retention time	Glycerol (S_R)	
($1/D$)	10 mg/l	500 mg/l
5 h	I	II
30 h	III	IV

external conditions, thereby favouring growth of one or the other competitor without using substrate specificity as the primary selective factor. The fastest growing species will prevail because of the constant and indiscriminate removal of cells by dilution. Only if interactions, other than competition for the same limiting substrate, occur between species, may they attain a mixed culture steady state or a continuous transient state, e.g. oscillations.

Table 1 outlines an elementary exercise where two different dilution rates and two different concentrations of glycerol as the limiting substrate were combined in four separate enrichment experiments (H. W. Jannasch, unpublished). The chemostat was inoculated with raw seawater, and artificial seawater supplemented with ammonia and phosphate (to assure growth limitation by the carbon source) was the basal medium in the reservoir. In 3-h to 10-h intervals, colony counts were taken on seawater–glycerol agar. The results of the four experiments are shown in Fig. 7. In the experiments I and II, on the one hand, and III and

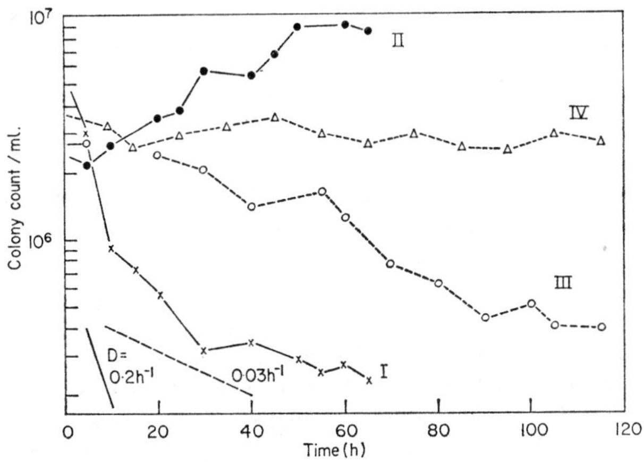

FIG. 7. Colony numbers on streak plates from enrichment cultures carried out in four different chemostat experiments (see Table 1).

IV on the other, the final number of colonies after 12 and 4 retention times, respectively, was strongly affected by the substrate concentration. It was also found that in the experiments II and III the majority of colonies (85–95%) appeared to represent just *one* species. This was not true in the experiments I and IV. The organisms thus enriched were isolated, purified and identified as belonging to the genera *Pseudomonas* sp. (from Experiment II) and *Achromobacter* sp. (from Experiment III).

In a second series of two experiments, a mixture of equal parts of organisms from these two isolates was inoculated in the chemostat run at the same two dilution rates but only *one* substrate concentration in the reservoir (100 mg glycerol/l). Colonies of the two species could be counted separately on agar plates. Figure 8 shows the experimental separation of the two isolates: *Pseudomonas* sp. prevailing at a dilution rate of 0·2 h^{-1}, *Achromobacter* sp. at $D = 0{\cdot}03$ h^{-1}.

More extensive enrichment experiments were done with four concentrations of the limiting carbon source (lactate) and four dilution rates (Table 2). This permitted a "fractionation" of a natural bacterial

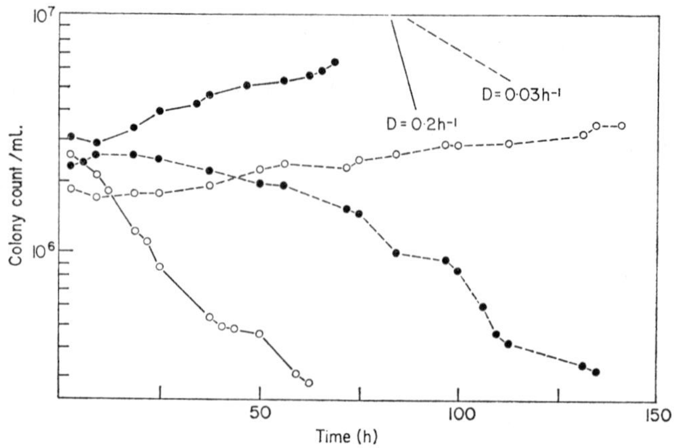

FIG. 8. Colony numbers (on streak plates) of *Pseudomonas* sp. (●) and *Achromobacter* sp. (○) in mixed population grown at two different dilution rates in the chemostat, indicating reciprocal exclusion.

TABLE 2. Genera of marine bacteria enriched from offshore seawater at four different dilution rates and four different concentrations of a carbon source added separately. A, no predominance developed; B, heavy wall growth; C, visible turbidity (from Jannasch, 1967b).

Dilution rate (h^{-1})	Concentration of lactate in the reservoir (mg/l)			
	0·1	1·0	10	100
0·05	B	*Vibrio*	A, B, C	A, B, C
0·01	*Achromobacter*	*Micrococcus*	*Spirillum*	*Spirillum*
0·25	A	*Pseudomonas*	*Pseudomonas*	*Aerobacter*
0·50	A	A	*Aerobacter*	*Aerobacter*

population into species according to their substrate affinity, i.e., their growth constants (K_s and μ_m) with respect to the particular limiting substrate.

B. COMPETITION AND MUTUAL EXCLUSION

The enrichment experiments described above were based on the competition of organisms for a growth-limiting nutrient. A plot of growth rate versus substrate concentration for two isolates (Fig. 9) obtained from the above-mentioned enrichment study (Jannasch, 1967b) shows an intersection indicating that, depending on the dilution rate, either organism may be successful in competing with the other. From the higher substrate affinity of *Spirillum* (101) (lower growth constants) it can be predicted that *Pseudomonas* (201) will be ultimately displaced if the chemostat is run at dilution rates or substrate concentrations below the point of the intersecting curves, and vice versa. This has been verified in many experiments. With regard to microbial ecology, this simple process of competitive exclusion has been interpreted in a manner similar to Winogradski's concept of "autochthonous" and "zymogenous"

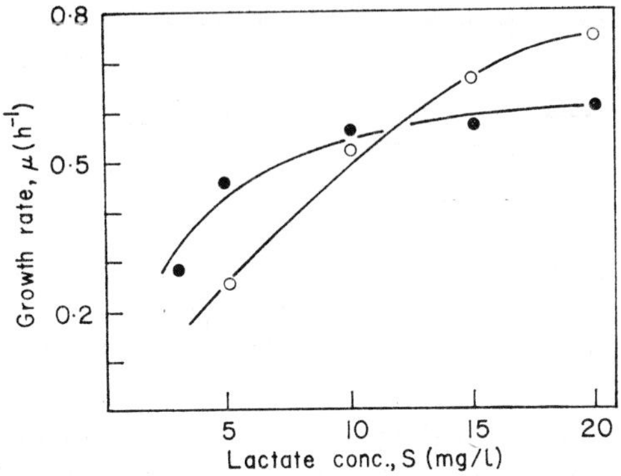

FIG. 9. Specific growth rate versus concentration of limiting substrate for *Pseudomonas* (201) (○) and *Spirillum* (101) (●) (from Jannasch, 1967b).

types of micro-organisms in soil (Jannasch, 1967a). The results also imply that isolating micro-organisms from seawater on agar plates with the commonly used types and concentrations of nutrients will result in a selective enrichment for those organisms that are probably not actively growing under natural conditions, i.e. at extremely low substrate concentrations. Species exhibiting autochthonous growth characteristics may be enriched in open flow systems only where the competitive conditions favour the organism with the lower growth constants. In a more recent study, Meers (1971) reported a case where a change of the dilution rate selected for one of two competitors in a similar fashion.

Competition experiments with a mixed population of psychrophilic *Pseudomonas* sp. and a facultatively psychrophilic *Spirillum* sp. grown at four different temperatures (Harder and Veldkamp, 1971) demonstrated cases where either species outgrew the other for the entire range of dilution rates (Fig. 10). This occurred at 16°C and −2°C respectively. The curves are schematic and based on two measurements each at the growth rates indicated by the arrows. When studied at two substrate concentrations in the reservoir (1·0 and 0·05 mg lactate/l), the results were similar.

The limited survival of *E. coli* in seawater has commonly been studied from the aspect of direct microbial interaction. If, however, a continuous

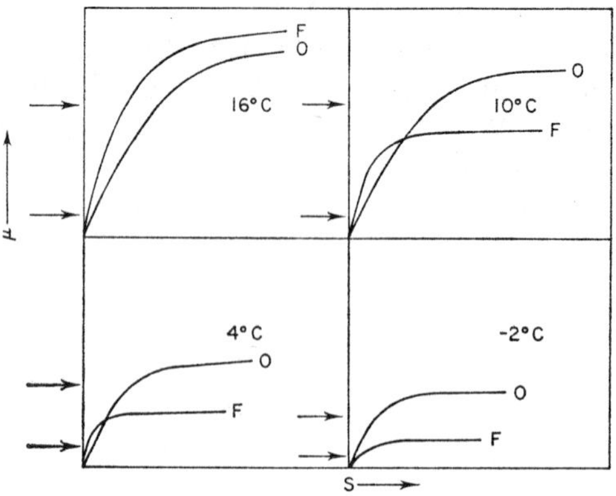

FIG. 10. Schematic representation of specific growth rate (μ) versus concentration of limiting substrate (S) at four different temperatures for *Pseudomonas* sp. (O) and *Spirillum* sp. (F), schematic (from Harder and Veldkamp, 1971).

and indiscriminate removal of cells (e.g. by zooplankton grazing) is assumed, *E. coli* may simply be eliminated by competition for an unknown limiting growth factor. This possibility was experimentally demonstrated in mixed culture with a typical marine isolate (Fig. 11).

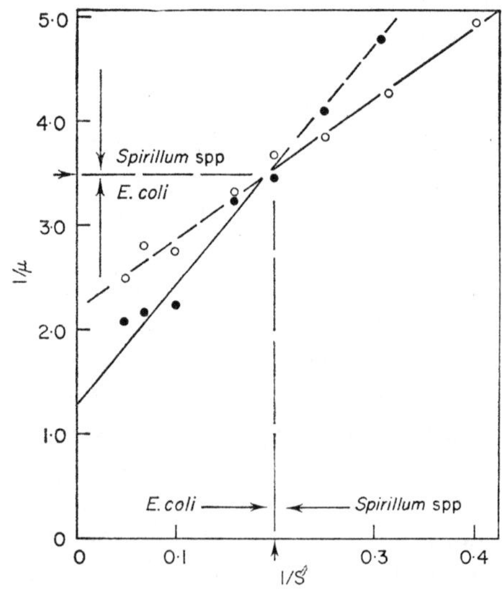

FIG. 11. Reciprocal plot of specific growth rate (μ in h^{-1}) versus concentration of limiting substrate (S in mg lactate/l) demonstrating the range of successful competition between *Escherichia coli* (●) and *Spirillum* sp. (○) (from Jannasch, 1968).

Escherichia coli as well as *Spirillum* sp. could be made to succeed or to be displaced at will by varying both the dilution rate and the concentration of the limiting substrate in the reservoir. Such a competitive elimination (Jannasch, 1968) of a species in a natural environment would be asymptotic. Other observations on the survival of *E. coli* in seawater (Mitchell *et al.*, 1967) indicate that direct inhibitory or bactericidal effects participate in the elimination process.

The effect of sulphide as a substrate, as well as an inhibitor, on growth was found to be the decisive factor in the competition between photosynthetic bacteria in a chemostat (van Gemerden and Jannasch, 1971). In an apparently pure culture of *Chromatium vinosum*, originally grown

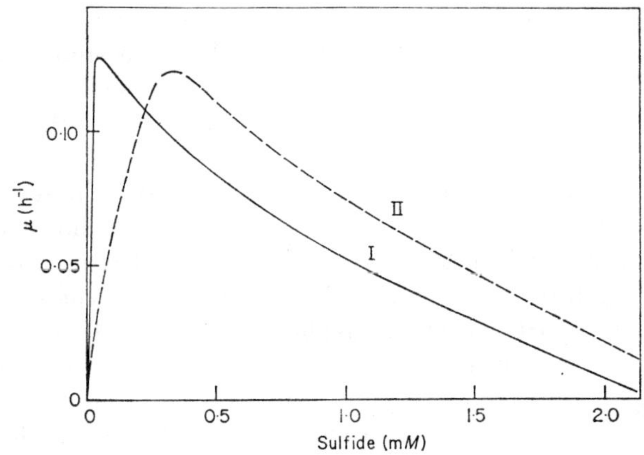

FIG. 12. Specific growth rate (μ) versus concentration of sulphide as the limiting substrate for two competing strains of purple sulphur bacteria; curve II is hypothetical (from Veldkamp and Jannasch, 1972; data from van Gemerden and Jannasch, 1971).

and maintained in bottle cultures, a strain of different pigmentation developed and became predominant in continuous culture at sulphide concentrations of 0·2 m*M*. The ranges at which sulphide as a limiting electron source, and at the same time an inhibitor, determines the outcome of the competition is shown in Fig. 12. Recently, it has been found that algal cultures, limited by carbon dioxide in the chemostat, show similar patterns of competition (Goldman, 1972).

C. OTHER TYPES OF INTERACTION

Next to the direct competition for the limiting substrate, feedback and product-stimulated growth processes (see p. 179) lead to a large variety of microbial interactions determining the predominance and the succession of species in mixed populations. Again, open flow systems offer means of simplifying complex interactions making them amenable to analytical studies. Bungay and Bungay (1968) made an attempt to list microbial interactions systematically according to their nature, such as: commensalism (one member benefiting while the other is unaffected), mutualism (each member benefiting from the other), amensalism (one member adversely changing the environment for the other), parasitism (one member stealing from the other), predation (one member ingesting the other) and neutralism (lack of interaction). This terminology is not extensively used in this discussion because of the general difficulties of attaching meaningful values to the types of often obscure interactions, and because of the relative few thoroughly studied cases in which the nature of interaction is entirely clear and reproducible.

In the preceding section it was shown that the simple and direct competition for the limiting substrate excludes the establishment of steady states in which two or more species remain in the chemostat. The theoretical exception to this rule is a chemostat culture run at the particular dilution rate and substrate concentration indicated by the point of intersection in Figs. 9 and 11. In practice, however, steady states frequently can be established without difficulty, in various mixed populations, indicating a metabolic interaction of some sort. As a principle, two species will remain in the chemostat and approach a joint steady state if the "primary" species, limited by an external factor (e.g. by the substrate concentration), will grow slower than the "secondary" species which is limited by a metabolic product of the former. The first example of such an association was mentioned by Powell (1958) and observed with a mixed culture of *Bacterium cloacae* (primary species) and *Pseudomonas pyocyanea*.

Contois and Yango (1964) studied a number of similar associations with partly predictable steady states. *Dictyostelium discoideum* was feeding on *Aerobacter aerogenes* which, as the primary species, was limited by ammonia. The population of *A. aerogenes* decreased to about one-hundredth of its initial concentration when the dilution rate was increased from 0·05 to 0·15 h^{-1}. In another case, a coli-phage was added to a glucose-limited culture ($S_R = 0{\cdot}6$ g/l) of *E. coli* at a dilution rate of 0·2 h^{-1}. While the primary population maintained itself at a very low level, as compared to the phage-free control, it supported a constant phage titer of about 10^7 p.f.u./ml. Another case, an interaction between *A. aerogenes* and a yeast originally isolated as a contaminant, may stand for many similar experiments (Fig. 13). The yeast was inoculated in a steady state culture of the bacterium run at a dilution rate of 0·075 h^{-1}. At point (1), the flow was stopped in order to allow the population density of the yeast to increase. At point (2), the dilution rate was reset at 0·1 h^{-1} and a new steady state established. At point (3), a vitamin mixture was added to the culture, and to the reservoir, to give a concentration of B-vitamins of 0·02 g/ml. At point (4), the dilution rate was set at 0·27 h^{-1} and a new steady state again attained. In cases of contamination, commensalism is the most common type of interaction. Shindala *et al.* (1965) report a case where *Proteus vulgaris* was found to profit from metabolic products of *Saccharomyces cerevisiae*.

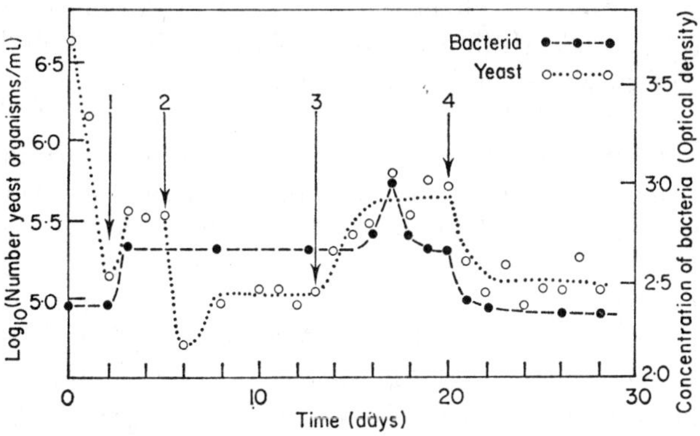

FIG. 13. Populations of *Aerobacter aerogenes* and a yeast in mixed culture in the chemostat; for explanation see text (from Contois and Yango, 1964).

Contois and Yango's studies on bacteria-phage systems were resumed by Noack (1968) in a theoretical treatment of prey–predator or host–parasite interactions. As discussed in more detail by Veldkamp and Jannasch (1972), Horne (1970) became engaged in an interesting controversy with Paynter and Bungay (1969) concerning the stability of bacteria-phage systems in continuous culture with respect to dilution rates and the accumulation of mutants.

While in cases of mere competition for the limiting substrate the relative initial population size of the two competitors is of no importance (except for the time until take-over and displacement is completed), more complex interactions between species may result in threshold values for the initial cell densities. This fact is directly related to the existence of threshold concentrations of limiting nutrients and population densities in pure cultures (see p. 175). Meers and Tempest (1968) studied mixed magnesium-limited cultures of *Bacillus megaterium* and *Torula utilis*, growing in a chemostat. If the initial proportion of yeast cells to the bacterial cells was greater than 25%, *T. utilis* succeeded in displacing *B. megaterium*. The opposite took place when the yeast inoculum amounted to only 5% of the total initial cell number. In the following supplementation of Monod's equation:

$$\mu = \mu_m \frac{1 - kp}{1 - p} \frac{s}{K_s - s} \tag{10}$$

p is the concentration of the assumed growth-stimulating metabolite and k is a constant. Meers and Tempest suggest that only the maximum growth rate is affected in this case of product-stimulated growth. From the observation that the effect did not occur in the presence of excess magnesium (and growth limitation by glucose) they conclude that the mechanism of growth stimulation was related to magnesium uptake.

The following metabolic interaction between anaerobic bacteria rests on the production and utilization of hydrogen. In the continuous culture of anaerobic micro-organisms, a flow of oxygen-free gas has the drawback of removing gaseous metabolic products and may affect the transformation quantitatively and qualitatively by shifting the equilibrium of reactions. Overcoming this problem by using a chemostat (suggested by Kafkewitz *et al.* (1973)), Ianotti *et al.* (1973) grew two isolates from rumen fluid (*Vibrio succinogenes* and *Ruminococcus albus*) in mixed continuous culture. The former organism obtains energy for growth by coupling the oxidation of hydrogen or formate with the reduction of fumarate to succinate. *Ruminococcus albus*, a cellulolytic bacterium, produces ethanol, acetate, formate and hydrogen from cellobiose. In mixed culture, growth and fumarate reduction by *V. succinogenes* was strictly dependent upon the hydrogen production by *R. albus*. At the same time, the ethanol production by *R. albus* was eliminated with a corresponding increase of acetate formation. This was taken as a proof of Hungate's (1966) hypothesis that the concomitant utilization of hydrogen in mixed culture "can cause electrons to be shifted away from the production of a typical fermentative product, e.g. ethanol, to give the more oxidized product, acetate". The normal function of the rumen will involve methane-producing bacteria as the primary hydrogen-utilizing organisms. A similar "mutualistic" interrelationship based on amino acid production and utilization was studied in continuous culture by Hoover and Lipari (1971).

While the formation of a steady state in the chemostat most readily facilitates the understanding of a metabolic interaction, this may also be true in cases where more or less sustained oscillations are observed. Contois and Yango (1964) grew in mixed cultures two mutants of *A. aerogenes* which differed in the capability to form lactic acid and to tolerate low pH values. Oscillations of the pH as well as the total population density resulted. Yeoh *et al.* (1968) described a mixed continuous culture of *Proteus vulgaris* and *Bacillus polymyxa* each presumably limited by a vitamin excreted by the other. The resulting oscillation was explained by the additional production of an inhibitor for *B. polymyxa*. Bungay (1968) included cases of oscillations in a systematic description of observed microbial interactions.

A considerable amount of new information on microbial interactions has been obtained in recent years by Tsuchiya and his co-workers. Megee *et al.* (1972) studied interactions between *Saccharomyces cerevisiae* and *Lactobacillus casei* which fulfilled the definitions of competition for a common substrate, commensalism and mutualism at different conditions and different stages of the culture. As in earlier studies (Jannasch, 1967b), it was possible to predict mixed-culture steady states from growth constants obtained in batch culture experiments. During more detailed experimentation on prey–predator interactions with *Dictyostelium discoideum* and *E. coli* (Tsuchiya *et al.*, 1972), oscillations of the population densities were observed that eventually, however, approached steady states. The Lotka-Volterra relationship (Gause, 1934) was modified to describe the observed oscillations. In Fig. 14, the theoretical curves are superimposed over the measured population densities and the glucose concentration limiting the growth of *E. coli* as the prey organism.

The controversy on "inherent oscillations" of prey–predator systems

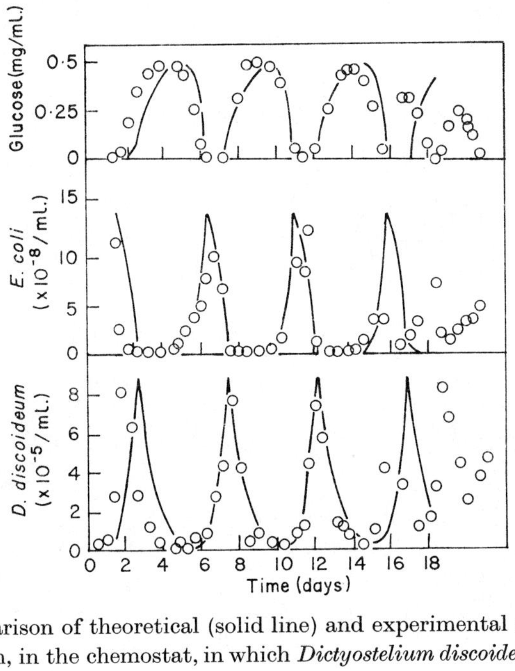

FIG. 14. Comparison of theoretical (solid line) and experimental data in a prey–predator system, in the chemostat, in which *Dictyostelium discoideum* was preying on *Escherichia coli* (from Tsuchiya *et al.*, 1972).

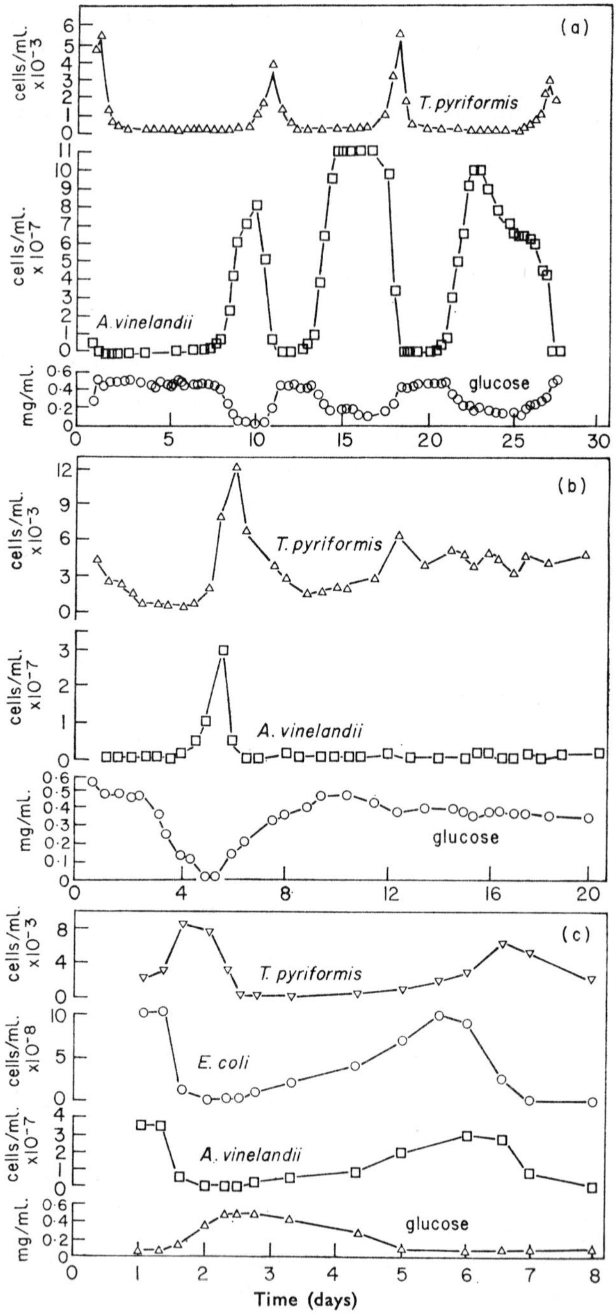
(a)
cells/mL. x10⁻³
T. pyriformis
cells/mL. x10⁻⁷
A. vinelandii
mg/mL.
glucose
(b)
cells/mL. x10⁻³
T. pyriformis
cells/mL. x10⁻⁷
A. vinelandii
mg/mL.
glucose
(c)
cells/mL. x10⁻³
T. pyriformis
cells/mL. x10⁻⁸
E. coli
cells/mL. x10⁻⁷
A. vinelandii
mg/mL.
glucose
Time (days)

of higher organisms (Slobodkin, 1961), their dependence on, or independence of, initial population densities and other parameters, seems to be amenable to experimental proof in microbial systems where decisive factors such as growth rate, dilution rate and substrate concentration are controllable experimental factors. Along this line, Jost *et al.* (1973) proceeded to more complex systems, defining a branched food chain as "food web". Using a differential Coulter Counter technique (Shuler *et al.*, 1972), it was possible to follow the population changes of three organisms separately in a mixed culture: *E. coli* and *Azotobacter vinelandii*, both limited by glucose, and *Tetrahymena pyriformis*, a ciliate feeding on both bacterial species.

Omitting one of the three members of this system led to a competition model or a food chain. In a mixed culture of *E. coli* and *A. vinelandii*, the former displaced the latter independent of dilution rate or glucose concentration in the reservoir. No use of the nitrogen-fixing capability of the less successful competitor was made in these experiments. Mixed populations of *A. vinelandii* and *T. pyriformis* attained stable oscillations at high dilution rates (Fig. 15a) while a low dilution rate resulted in damped oscillations (Fig. 15b). An explanation of this effect is based on the ratio between the saturation constants of prey and predator.

The entire food web of the three organisms appears to result in damped oscillations at a fairly high dilution rate (Fig. 15c). Heavy wall growth prevented running the culture for a reasonable length of time in order to study its behaviour at lower dilution rates. As compared with the competition model between the two bacterial species, the presence of the predator allowed the two competitors to co-exist. This fact can be taken as a simple experimental proof for the general ecological principle that the more stable a system the more complex it is likely to be.

D. Multistage Culture Systems

A high degree of complexity can be attained when mixed populations of micro-organisms are studied in multistage continuous culture. A plausible application of dual-stage chemostat cultures is a prey–predator system where the first stage provides the constant feed for the

---

Fig. 15(a) Sustained oscillations of a glucose–*Azotobacter*–*Tetrahymena* food chain in a chemostat run at a dilution rate of 0·17 h^{-1} (from Jost *et al.*, 1973).

(b) Damped oscillations in a glucose–*Azotobacter*–*Tetrahymena* food chain in a chemostat run at a dilution rate of 0·025 h^{-1} (from Jost *et al.*, 1973).

(c) Oscillations in an *Azotobacter*–*Escherichia coli*–*Tetrahymena* food web in a chemostat run at a dilution rate of 0·135 h^{-1} (from Jost *et al.*, 1973).

predator in the second. Curds and Cockburn (1971) grew *Klebsiella aerogenes*, limited by sucrose, to feed the ciliate *Tetrahymena pyriformis*. Steady states were established in both stages, in contrast to similar prey–predator studies that were conducted in single stage chemostats, and resulted in oscillations of the two populations (see p. 195).

Enrichments in a three-stage chemostat led to some interesting results (H.W. Jannasch, unpublished). Filter-sterilized seawater was supplemented with 1 m*M* glycerol and run through a series of three culture vessels with volumes in the ratio of 1:3:8 (Fig. 16), the retention times

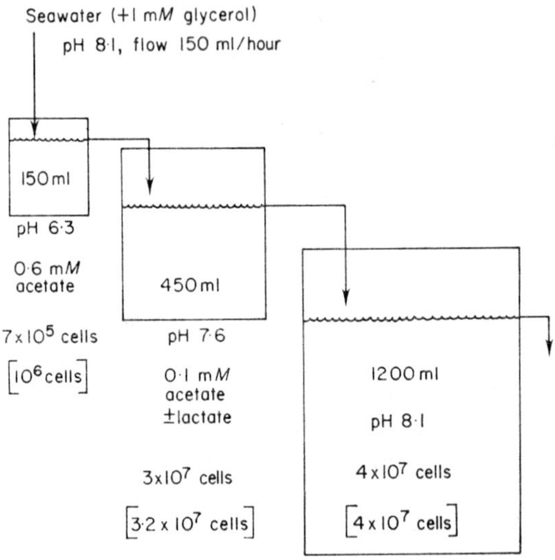

FIG. 16. Three-stage chemostat enrichment run with glycerol-supplemented and filter-sterilized seawater, inoculated with unsterilized seawater, after 5 days of operation; dilution rates: 0·5, 0·17, and 0·062 h^{-1}. Population density (organisms/ml) is shown in brackets and was obtained at a constant pH value of 8·1.

being 2, 6 and 16 h. The first vessel was inoculated with unsterile seawater. The population density and the pH reached approximate constancy after 3–4 days in the first and after 5–6 days in the second vessel. The considerable drop of pH appeared to be due to the production of lactic and acetic acids, and was followed by acid consumption and restoration of the initial pH of seawater in the third vessel. Carry-over of organisms and secondary products from the first two vessels made it difficult to account for the population, and remaining substrate or intermediates, in the third vessel. Its population represented the characteristics of the stationary phase of a batch culture. When the resulting pH change in the culture vessels was eliminated by a pH-stat device (the slight affect on the dilution rate being ignored), an entirely different population was established.

The results of this experiment and the subsequent isolation and characterization of the enriched organisms, gives some information on the mode by which a certain substrate is degraded in seawater and how the process is affected if the pH remains unchanged. The experiment also demonstrates a fundamental drawback of continuous culture studies on natural populations: the imposed dilution rate has no equivalent under natural conditions. The removal of cells in the natural environment (by sedimentation, grazing, lysis and parasitism) is probably highly inconstant and selective. However, multistage chemostat experiments run with low substrate concentrations and at low dilution rates, may quite realistically produce the particular association of micro-organisms that are responsible for the successive breakdown of a complex substrate in natural waters.

A most promising field of applied microbiology, closely related to microbial ecology, concerns the design of multistage systems for the successive breakdown of complex waste materials. Abson and Todhunter (1961) describe a three-stage treatment system for industrial wastewater containing various phenolic compounds at toxic levels. The first stage promotes an exclusive enrichment of phenol-resistant and phenol-decomposing bacteria of the *Pseudomonas* type, aided by nutrient feedback from later stages of the system. This is followed by the oxidation of thiocyanate, cyanide, and thiosulphate (largely by *Thiobacillus*-type bacteria) in the second stage. Vigorous nitrification occurred in the third stage, part of the effluent being fed back into the first stage. The general aim was to increase the rate and efficiency of the microbial attack on waste materials of a certain composition by providing optimal conditions for specific microbial populations engaged in a particular step during the decomposition process. This can be done by (a) adjustments (pH and nutrients, for example) in feedback fashion or by additions from external sources, (b) controlling the relative dilution rates of the different stages by the tank volumes, (c) providing homogeneously mixed or heterogeneous conditions, (d) recirculation of cell mass, and by other means. The direct cross-fertilization of ecological studies, for instance in the field of acid mine waters research, with those of applied microbiology has been beneficial in many cases.

E. Mutants in Continuous Culture

It was discussed earlier (see p. 186) that the establishment of a contaminant as a successful competitor in the chemostat will depend on the particular growth conditions of the culture. Powell (1958) and

Renneboog-Squilbin (1967) presented mathematical models for this process. Both authors contend that the behaviour of contaminants applies to that of mutants as well. Theoretically, a pure culture may turn into a mixed culture if a mutant arises that is favoured by the particular conditions of growth and becomes a competing population. Novick and Horiuchi (1961) and Horiuchi *et al.* (1962) used continuous culture as an effective means of selecting for certain desirable types of mutants.

Concerning the significance of mutation and subsequent selection on a particular fermentation process, the evidence is equivocal. On the one hand, strain degradation during continuous antibiotic production (Reusser, 1961) and the loss of desirable characteristics of a brewing yeast (Thorne, 1968) were observed. On the other hand, the production of butanediol (Pirt and Callow, 1958) and antibiotics (Bartlett and Gerhardt, 1959; Sikyta *et al.*, 1959) was carried out over long periods of time without unfavourable changes in the characteristics of the microbial culture. Given (a) the very strong selective pressure in favour of mutants able to grow slightly faster, owing either to a lower substrate constant or a higher maximum specific growth rate, and (b) the relatively limited knowledge of the relationships between growth rate and the production of a desired metabolite, it must be considered that continuous cultures carried out over extended periods of time may lead to a decrease in the production of a desired metabolite. Each case must be evaluated experimentally on an *ad hoc* basis.

When mixed substrates are fed under substrate-limited conditions, mutants capable of simultaneous utilization of the substrates will be selected. This has been shown by Silver and Mateles (1969) with *E. coli* growing on mixtures of glucose and lactose. The mutants which overgrew the parent population were constitutive for β-galactosidase, but were still subject to catabolite repression. Nevertheless, they were capable of utilizing lactose simultaneously with glucose, which the parent population could do only at very low dilution (growth) rates.

F. Technological Approaches

A considerable amount of experience and data in continuous culture work has been gained in the field of microbial technology. Some of this material is related directly or indirectly to problems that are of ecological concern.

Next in interest to the possible establishment of mutants in open flow systems are certain technical means of running mixed cultures for the most efficient production of valuable substances from waste materials in continuous processes. By choosing a certain combination of microorganisms and culture conditions assuring the dominance and stability of the desired population versus possible contaminants, single cell proteins (SCP) may be produced under non-aseptic conditions. The savings in costs for sterile design and construction of equipment and the sterilization of media constitutes a considerable economic benefit. Furthermore, the need for higher yields and better performance of the mixed populations in this process led to a number of empirical studies.

Professor M. J. Johnson and his group at Wisconsin have been particularly active in exploring the use of mixed cultures as potential sources of SCP. Using a mixed yeast culture, which proved to be composed of *Candida intermedia* and *Candida lipolytica*, they investigated growth on *n*-alkanes and gas oil (Miller and Johnson, 1966). While *C. lipolytica* alone was unable to grow on an unsupplemented mineral salts *n*-alkane medium, the mixed culture grew well. Furthermore, the mixed culture was superior to *C. intermedia* alone in the utilization of long-chain (C_{22}) alkanes. The doubling time of the mixed culture growing on docosane was 3·0 h, whereas that of *C. intermedia* alone was 6·5 h. Cell yields of the mixed culture ranged from 74 to 89%, slightly higher than those of *C. intermedia* alone. The mixed cultures appeared stable in that the ratio of types of cells present throughout the fermentations remained the same.

In another study (Wodzinski and Johnson, 1968), a mixed culture of *Pseudomonas* sp. and *Nocardia* sp. was found to grow well on 3-methylheptane, while neither organism alone grew at all. Cell yields for the mixed culture of about 79%, and doubling times of 4·5 h, were obtained. Similarly, a number of mixed cultures capable of growth on methane were obtained by enrichment culture (Vary and Johnson, 1967). The fastest-growing culture, composed of two Gram-negative rods, had a doubling time of about 3 h and gave cell yields of 65–70%. The isolated organisms would grow separately on mineral agar plates in the presence of methane, but not in liquid culture with methane.

In continuous culture on methane, a mixed population composed of two types of Gram-negative rods grew well at 45°C (Sheehan and Johnson, 1970). The culture proved stable over a period of 3 years. Nitrate was used as the nitrogen source, because ammonia led to significant contamination by nitrifying organisms. Cell yields of about 61%, doubling times of 2·3 h, and cell concentrations up to 12 g/l were obtained.

Mixed cultures of thermophiles, capable of growth on hexadecane up to 65°C were obtained by enrichment techniques (Sukatsch and Johnson, 1972). The culture obtained at 55°C was a mixture of three organisms: (a) a non-motile, Gram-positive rod, (b) a smaller, non-motile, Gram-positive or Gram-variable rod, and (c) a motile, Gram-negative rod. Enrichment at 65°C yielded a binary mixture of a large non-motile, Gram-positive rod, and a smaller Gram-negative to Gram-variable rod of questionable motility. Colonies of the individual organism grew on agar plates only if streaked out in mixtures. Individual colonies could not be transferred to another plate.

Haggström (1969) reported on a mixed culture, composed of Gram-negative rods, that grew well on methanol with a doubling time of about 3·2 h and cell yields of about 40%. The components of the mixed culture could be distinguished by colonial appearance and their ability to form pigments. Their ratio was stable with time. None were isolated as pure cultures.

A mixed culture of *Cellulomonas* sp. and *Alcaligenes* sp. grew on cellulose very much more rapidly than *Cellulomonas* sp. alone (Srinivasan and Han, 1969). Various explanations were proposed, based on the removal by the *Alcaligenes* of factors which limited the growth or activity of the cellulolytic *Cellulomonas* sp.

G. Heterogeneous Systems

With very few exceptions, natural habitats of micro-organisms are highly heterogeneous. Therefore, experimentally useful reproductions of complex environments for ecological studies are desirable. The heterogeneity of such systems may involve the discontinuity of flow, the inconstancy of temperature, chemical composition of the media, liquid–solid phases, and natural microbial populations of unknown composition. The absence of complete mixing in cultures of interacting species may infer certain characteristics of multistage systems and intricate feedback mechanisms. Herbert (1961b) and Malek (1967) tried to bring order in this maze of variabilities.

At first sight, it seems futile to create experimental systems which are as complex and unpredictable as natural environments themselves and do not offer advantages for kinetic analyses. Historically therefore, the main purpose of using such systems has been to grow certain fastidious organisms or specific mixed populations that could not be maintained in more defined cultures. Unknown and complex growth requirements are met either by intricate physico-chemical conditions or by metabolic interactions between species. Certain organisms or mixed populations with a high sensitivity to lag or stationary phase phenomena of closed culture systems may grow in open-flow cultures exclusively.

As an example, field observations on the strong dependence of growth of *Sphaerotilus natans* on temperature and the rate of flow in channels or pipes have been confirmed in heterogeneous continuous culture systems (Ordal and Palmer, 1964). The observed high oxygen requirements of this filamentous organism may be related to its habitual growth in attached cell aggregates. When grown in flow cultures, the organisms utilized a larger variety of substrates than in shake-flask culture (Stokes, 1954). Dias and Huekelekian (1967) and Dias *et al.* (1968) found a strict calcium requirement of *S. natans* in a heterogeneous flow culture, suggesting this to be the basis for successful or unsuccessful growth on "spent sulphite liquor" waste in river water of various chemical compositions (Baalsrud, 1967).

The typical "veils" of *Thiovulum* sp. commonly occur in marine flow-through aquaria on the surface of anaerobic sediments or decaying organic matter. The cultivation of these organisms has only been possible under conditions where hydrogen sulphide as well as dissolved oxygen are available. Since the biological oxidation of sulphide competes with chemical autoxidation, a more or less constant flow of oxygenated seawater over a sulphide-producing layer of sediment is necessary. For culturing a mechanically purified (washed) cell suspension of *Thiovulum* sp., La Riviere (1962) suggests a partially closed system (Fig. 17), where sulphide

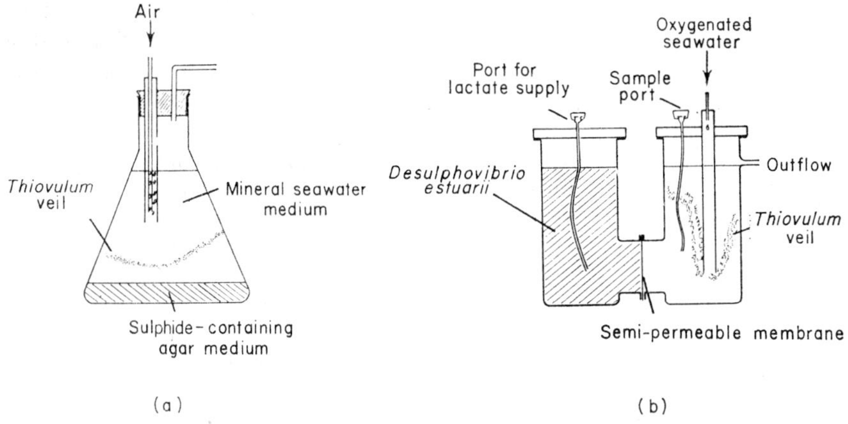

FIG. 17. Devices for growing purified cell suspensions of *Thiovulum* ((a), from La Riviere, 1962; (b), from Jannasch *et al.*, 1970).

diffuses from an agar layer on the bottom of an Erlenmeyer flask into the overlaying liquid mineral medium. A slow stream of air bubbles supplies oxygen from the surface layer. The absence of a flowing liquid medium makes this system susceptible to overgrowth by *Thiobacillus* species. The prior purification of cell suspensions of *Thiovulum* by washing is rendered difficult by slime layers producing the typical veil structure of growing populations.

On the other hand, this rigid structure offers advantages for liquid flow cultures that will remove contaminants by flushing. From a number of systems that were studied (Jannasch *et al.*, 1970), the one depicted in Fig. 17b was the most successful in providing actively growing cells for a period of about a year. Sudden decreases of the population and danger of washout occurred when the veils disintegrated periodically for unknown reasons and the cells became freely suspended. The source of sulphide was a culture of sulphate-reducing bacteria intermittently supplied with lactate and separated from the flow-through chamber by a membrane. This technique appeared to provide a constant and optimal rconcentation of sulphide for growth of *Thiovulum*. There were indications that organic growth factors were also contributing to the maintenance of the culture.

In spite of the heterogeneity of the system, a quantitative approach was possible in the following case. Competition between the natural microbial population of seawater and a known bacterial isolate was studied in a chemostat run with unsterilized seawater and controls of autoclaved seawater. From the difference between the dilution rate and the washout rate of the test strain, the growth rate of the latter was calculated (see p. 177). Table 3 gives the mean generation times of the test strain in the presence of the natural microbial population of seawater and demonstrates the competition for an added carbon source at two different concentrations.

TABLE 3. Mean generation times of *Spirillum* sp. in untreated (unsterilized) and autoclaved inshore seawater, with and without an additional carbon source. Retention time in all experiments was 6 h (from Jannasch, 1969)

Seawater	Glycerol (m*M*)	Mean generation time (h)
Untreated	None	127
Autoclaved	None	98
Untreated	0·1	81
Autoclaved	0·1	17
Untreated	3·0	45
Autoclaved	3·0	3·5

Beginnings have been made in studying pure and mixed bacterial populations in dual-phase systems such as water–oil and water–mineral particles (Button, 1969). Humphrey and Erickson (1972) discussed the kinetics of such dispersed systems. Punčochář (1971) studied growth of bacteria attached to glass surfaces at different flow rates in glucose- and starch-limited media. In mixed populations and low concentrations of dissolved carbon sources, a selection for bacterial species was recorded that grew preferably attached to suspended particles (Jannasch and Pritchard, 1973). Such approaches are of special concern in marine microbiology where the relatively low concentrations of dissolved organic matter in seawater led to a common assumption that heterotrophic bacteria in this environment may be metabolically active only when attached to suspended particles.

Continuous soil percolation columns qualify as open-flow systems and have been used for measuring activities of the indigenous mixed microbial populations. In a series of publications (largely listed by Macura and Kunc, 1965) the results of experiments on qualitative population changes and quantitative assessment of nitrification and the turnover of substrates like glucose and glycine were reported. Limitation by carbohydrates was found to result in a steep increase of the population of nitrogen-fixing bacteria.

An obvious example of a natural continuous flow system is the rumen. Reconstruction *in vitro* has been attempted in various experimental systems inoculated with rumen fluid and fed with ground hay and saliva concentrates (Steward *et al.*, 1961; Rufener *et al.*, 1963; Hungate, 1966; Wright, 1971). Intermittent feeding and periodic removal of a dialysis bag containing cation exchange resin characterized transient stage conditions (Slyter *et al.*, 1964). Yet little changes of the DNA concentration and the numbers of protozoa indicated that semi-steady states of the mixed population had been attained.

Biological waste treatment is the classic example of a technological application of open-flow systems with mixed microbial populations, and recent work in this area has been reviewed by Paynter and Bungay (1971) and Chiu *et al.* (1972). As compared to pure cultures, the heterogeneous populations present in activated sludge show a high variability of responses that complicates the interpretation of data (Gaudy and Ramanathan, 1971; Chian and Mateles, 1968). In general terms, however, a similar type of behaviour is observed. "Steady states" were established when mixed microbial populations of sewage were grown on glucose as the only substrate (Storer and Gaudy, 1969). At a threefold increase of glucose in the reservoir "the transient values of the specific growth rate constant were not consistent with those predicted by the Monod equation at various substrate concentrations observed during the transient state. Experimental evidence of growth rate hysteresis was obtained". In another study, Gaudy and Gaudy (1969) grew natural populations obtained from clarifier effluent under carbon-limited conditions with L-lysine as the carbon source, and the effects of adding glucose and other sugars were observed. In the transient period, degradation of lysine was inhibited, presumably owing to catabolite repression, and this inhibition persisted until glucose was removed from the feed and the system returned to the steady state. Peil and Gaudy (1971) found that Monod's equation appeared to describe the actual data found in mixed population and mixed substrate systems fairly well. The usefulness of such highly simplified models and some mathematical adjustments based on statistical analysis of the data, as well as studies on the stability of heterogeneous systems, have been discussed by Ramanathan and Gaudy (1971).

Using a mixed feed of glucose and butyrate and carbon-limited conditions, with river water inoculum, a population consisting primarily of a pseudomonad and a coliform was obtained (Chian and Mateles, 1968; Mateles and Chian, 1969). At dilution rates below 0·7 h^{-1}, all of the glucose and most of the butyrate was consumed, and the coliform organisms amounted to about 75% of the population. Above this dilution rate, butyrate uptake was reduced substantially, and the residual glucose

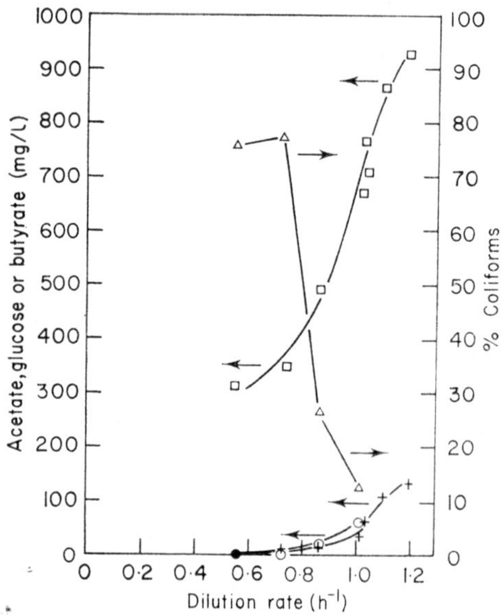

FIG. 18. Substrate utilization curves for mixed continuous cultures of a coliform and a pseudomonad growing on a mixed feed of glucose and butyrate as limiting carbon sources. □, butyrate; ○, glucose; +, acetate produced; △, percent coliforms (from Mateles and Chian, 1969).

rose slightly. This is depicted in Fig. 18. In parallel to the inhibition of butyrate uptake, acetate excretion was noted, supporting the conclusion that inhibition of uptake was due to catabolite repression and/or inhibition.

In this connection, studies on the particular bacterial flora of the mouth (Gritchley, 1970) should be mentioned. When a fresh sample of human dental plaque was used as an inoculum for an anaerobic, pH-controlled chemostat (Ellwood *et al.*, 1972), the original composition of types (streptococci, diplococci, catenabacteria, and a diptheroid-type bacillus) remained qualitatively the same but changed quantitatively dependent upon the dilution rate. The latter also had a strong influence on the fermentative processes and the pH.

Studies in general ecology often reveal a tendency to consider springs, rivers, sewage plants, microcosms and ecosystems as open-flow systems capable of attaining steady state conditions and amenable to relatively simple mathematical analysis. However, the absence of exponential growth and anything but complete mixing, the formation of niches, complex interactions, inconstancy of external conditions, and so on, produce heterogeneity of a degree that renders the continuous culture concept (especially the chemostat concept) irrelevant for a useful description or analysis. In cases of high reproducibility, data fitting curves and equations can be useful, and averaged "constants" may indeed serve a descriptive function as pointed out by Stumm-Zollinger and Harris (1971).

IV. Acknowledgements

It is a pleasure to acknowledge the many helpful and stimulating discussions with our colleagues C. O. Wirsen, J. H. Tuttle, and P. E. Holmes. The most recent literature surveys compiled by the colleagues of the Institute of Microbiology, Czechoslovak Academy of Sciences in Prague (Řičica, 1971, 1972) have been extensively used. One of us (H.W.J.) was supported by the National Science Foundation, grants GA 33405 and GA 29665. Contribution No. 3104 of the Woods Hole Oceanographic Institution.

References

Abson, J. W. and Todhunter, K. H. (1961). *Society of Chemical Industry, Monograph* **12**, 147.

Baalsrud, K. (1967). *Water Pollution Control* **66**, 97.

Bartlett, M. C. and Gerhardt, P. G. (1959). *Biotechnology and Bioengineering* **1**, 359.

Beijerinck, M. W. (1921–1940). Verzamelde Geschriften, Nijhof, Den Haag.

Bergter, F., Knöll, H. and Noack, D. (1968). *Folia Microbiologica* **14**, 308.

Boddy, A., Clarke, P. H., Houldsworth, M. A. and Lilly, M. D. (1967). *Journal of General Microbiology* **48**, 137.

Brooks, R. and Sikyta, B. (1967). *Applied Microbiology* **15**, 224.

Bull, A. T. (1972). *Journal of Applied Chemistry and Biotechnology* **22**, 261.

Bungay, H. R. and Bungay, M. L. (1968). *Advances in Applied Microbiology* **10**, 269.

Bungay, H. R. (1968). *Chemical Engineering Progress Symposium Series* **64**, 19.

Button, D. K. (1969). *Limnology and Oceanography* **14**, 95.

Chian, S. K. and Mateles, R. I. (1968). *Applied Microbiology* **16**, 1337.

Chiu, S. Y., Erickson, J. E., Fan, L. T. and Kao, I. C. (1972). *Biotechnology and Bioengineering* **14**, 207.
Contois, D. E. and Yango, L. D. (1964). *American Chemical Society* (*Abstracts*) 148th Meeting (manuscript).
Curds, C. R. and Cockburn, A. (1971). *Journal of General Microbiology* **66**, 95.
Dalton, H. and Postgate, J. R. (1969a). *Journal of General Microbiology* **54**, 463.
Dalton, H. and Postgate, J. R. (1969b). *Journal of General Microbiology* **56**, 307.
Dawes, I. W. and Mandelstam, J. (1970). *Journal of Bacteriology* **103**, 529.
Dean, A. C. R. and Rogers, P. L. (1967). *Biochimica et Biophysica. Acta* **148**, 267.
Dean, A. C. R., Pirt, S. J. and Tempest, D. W. (1972). "Environmental Control of Cell Synthesis and Function." Academic Press, London and New York.
Demain, A. L. (1972). *Journal of Applied Chemistry and Biotechnology* **22**, 345.
Dias, F. F. and Heukelekian, H. (1967). *Applied Microbiology* **15**, 1083.
Dias, F. F., Okrend, H. and Dondero, N. C. (1968). *Applied Microbiology* **16**, 276.
Dyr, J., Protiva, J. and Praus, R. (1958). *In* "Continued Cultivation of Microorganisms", p. 210. Czechoslovak Academy of Sciences, Prague.
Ellwood, D. C., Longyear, V. M. C. and Hunter, J. R. (1972). *Journal of General Microbiology* **71**, x.
Fencl, Z. (1966). *In* "Theoretical and Methodological Basis of Continuous Culture of Microorganisms", p. 69. Academic Press, London.
Fencl, Z., Řičica, J. and Kodesová, J. (1972). *Journal of Applied Chemistry and Biotechnology* **22**, 405.
Gaudy, C. P. L. and Gaudy, A. F. (1969). *Applied Microbiology* **18**, 790.
Gaudy, A. F. and Ramanathan, M. (1971). *Biotechnology and Bioengineering* **13**, 113.
Gause, G. (1934). "The Struggle for Existence". Williams and Wilkins, Baltimore.
Gilley, J. W. and Bungay, H. R. (1967). *Biotechnology and Bioengineering* **9**, 617.
Goldman, J. C. (1972). Thesis, University of California, Berkeley.
Gorini, L. (1960). *Proceedings of the National Academy of Sciences* (*Washington*) **46**, 682.
Gould, G. W. and Lees, H. (1960). *Canadian Journal of Microbiology* **6**, 299.
Gritchley, P. (1970). *Journal of Dental Research* **49** (Supplement), 1283.
Häggström, L. (1969). *Biotechnology and Bioengineering* **11**, 1043.
Harder, W. and Veldkamp, H. (1971). *Antonie. voan Leeuvenhoek* **37**, 51.
Harrison, D. E. F. (1972). *Journal of Applied Chemistry and Biotechnology* **22**, 417.
Harrison, D. E. F. and Loveless, J. E. (1971). *Journal of General Microbiology* **68**, 35.
Herbert, D. (1958). *7th International Congress of Microbiology, Symposium*, p. 381.
Herbert, D. (1961a). *Symposium of the Society of General Microbiology* **11**, 391.
Herbert, D. (1961b). *Society of Chemical Industry Monograph* **12**, 21.
Herbert, D. (1964). *In* "Continuous Cultivation of Microorganisms", p. 23. Czechoslovak Academy of Sciences, Prague.
Herbert, D., Elsworth, R. and Telling, R. C. (1956). *Journal of General Microbiology* **14**, 601.
Hill, S., Drozd, J. and Postgate, J. R. (1972). *Journal of Applied Chemistry and Biotechnology* **22**, 541.
Hinshelwood, C. N. (1946). "The Chemical Kinetics of the Bacterial Cell". Clarendon Press, Oxford.
Holzberg, I., Finn, R. K. and Steinkraus, K. H. (1967). *Biotechnology and Bioengineering* **9**, 414.
Hoover, W. H. and Lipari, J. J. (1971). *Journal of Dairy Science* **54**, 1662.
Horiuchi, T., Tomizawa, J. and Novick, A. (1962). *Biochimica et Biophysica Acta* **55**, 152.

Horne, M. T. (1970). *Science* **168**, 992.

Humphrey, A. E. and Erickson, L. E. (1972). *Journal of Applied Chemistry and Biotechnology* **22**, 125.

Hungate, R. E. (1966). "The Rumen and its Microbes". Academic Press, New York and London.

Hunter, K. and Rose, A. H. (1972). *Journal of Applied Chemistry and Biotechnology* **22**, 527.

Ianotti, E. L., Kafkewitz, D., Wolin, M. J. and Bryant, M. P. (1973). *Journal of Bacteriology* (in press).

Jannasch, H. W. (1963). *Archiv für Mikrobiologie* **45**, 323.

Jannasch, H. W. (1965a). *Laboratory Practice* **14**, 1162.

Jannasch, H. W. (1965b). *Biotechnology and Bioengineering* **7**, 279.

Jannasch, H. W. (1967a). *Limnology and Oceanography* **12**, 264.

Jannasch, H. W. (1967b). *Archiv für Mikrobiologie* **59**, 165.

Jannasch, H. W. (1968). *Applied Microbiology* **16**, 1616.

Jannasch. H. W. (1969). *Journal of Bacteriology* **99**, 156.

Jannasch, H. W. and Jones, G. E. (1959). *Limnology and Oceanography* **4**, 128.

Jannasch, H. W. and Pritchard, P. H. (1972). *Mem. Inst. Ital. Idrobiol.* **29** Suppl., 289.

Jannasch, H. W. and Wirsen, C. O. and Eimhjellen, K. (1970). *Bacteriological Proceedings* (abstr.) G179, 42.

Jost, J. L., Drake, J. F., Fredrickson, A. G. and Tsuchiya, H. M. (1973). *Journal of Bacteriology* **113**, 834.

Kafkewitz, D., Ianotti, E. L., Wolin, M. J. and Bryant, M. P. (1973). *Applied Microbiology* (in press).

Kubitschek, H. E. (1970). "Introduction to Research with Continuous Culture". Prentice Hall, Englewood Cliffs, N.J. U.S.A.

Knorre, W. A. (1968). *Biochemical and Biophysical Research Communications* **31**, 812.

La Riviere, J. W. M. (1962). *In* "Marine Microbiology", p. 61. C. C. Thomas.

Larsen, D. H. and Dimmick, R. L. (1964). *Journal of Bacteriology* **88**, 1380.

Lees, H. and Postgate, J. R. (1973). *Journal of General Microbiology* **75**, 161.

Kuedeking, R. and Piret, E. L. (1959). *Biotechnology and Bioengineering* **1**, 393.

Luscombe, B. M. and Gray, T. R. G. (1971). *Journal of General Microbiology* **69**, 433.

Maaløe, O. and Kjeldgaard, N. O. (1966). "Control of Molecular Synthesis". Benjamin, Inc., New York.

Magasanik, B. (1957). *Annual Reviews of Microbiology* **11**, 221.

Magasanik, B. (1961). *Cold Spring Harbor Symposium on Quantitative Biology* **26**, 249.

Magasanik, B., Magasanik, A. K. and Neidhardt, F. C. (1959). Ciba Foundation Symposium, London, p. 334.

Macura, J. and Kunc, F. (1965). *Folia Microbiologica* **10**, 125.

Macura, J. and Kotkova, M. (1953). *Československá Biologie* **2**, 41.

Malek, I. (1967). *In* "Microbial Physiology and Continuous Culture", p. 1. H.M.S.O., London.

Malek, I. and Fencl, Z. (1966). "Theoretical and Methodological Basis of Continuous Culture of Microorganisms". Academic Press, New York and London.

Marr, A. G., Nilson, E. H. and Clark, D. J. (1963). *Annals of the New York Academy of Sciences* **102**, 536.

Masurekar, P. (1968). M.A. Dissertation, Massachusetts Institute of Technology, Cambridge, U.S.A.

Mateles, R. I. and Chian, S. K. (1969). *Environmental Science and Technology*, **3**, 569.

Mateles, R. I., Chian, S. K. and Silver, R. S. (1967). *In* "Microbial Physiology and Continuous Culture", p. 232. H.M.S.O., London.
Maxon, W. D. (1955). *Applied Microbiology* **3**, 110.
McGrew, S. B. and Malette, M. F. (1962). *Journal of Bacteriology* **83**, 844.
Meers, J. L. (1971). *Journal of General Microbiology* **67**, 359.
Meers, J. L. (1973). *CRC Critical Reviews in Microbiology* **2** (in press).
Meers, J. L. and Tempest, D. W. (1968). *Journal of General Microbiology* **52**, 309.
Megee, R. D., Drake, J. F., Fredrickson, A. G. and Tsuchiya, H. M. (1972). *Canadian Journal of Microbiology* **18**, 1733.
Mian, F. A., Fencl, Z. and Prokop, A. (1970). *Folia Microbiologica* **15**, 213.
Miller, T. L. and Johnson, M. J. (1966). *Biotechnology and Bioengineering* **8**, 549 and 567.
Mitchell, R., Yanjofsky, S. and Jannasch, H. W. (1967). *Nature* **215**, 891.
Monod, J. (1942). "Recherches sur la Croissance des Cultures Bacteriennes". Masson & Cie, Paris.
Monod, J. (1947). *Growth* **11**, 223.
Monod, J. (1950). *Annals de l'Institut Pasteur* **79**, 390.
Moser, H. (1958). *Carnegie Institute Washington, Publication* 614.
Munson, T. O. and Burris, R. H. (1969). *Journal of Bacteriology* **97**, 1093.
Noack, D. A. (1968). *Journal of Theoretical Biology* **18**, 1.
Novick, A. (1955). *Annual Reviews of Microbiology* **23**, 123.
Novick, A. (1958a). *In* "Continuous Culture of Microorganisms", p. 29. Academia, Prague.
Novick, A. (1958b). *Folia Microbiologica* **4**, 395.
Novick, A. and Horiuchi, T. (1961). *Cold Spring Harbor Symposium of Quantitative Biology* **26**, 239.
Novick, A. and Szilard, L. (1950a). *Science* **112**, 715.
Novick, A. and Szilard, L. (1950b). *Proceedings of the National Academy of Sciences (Washington)* **36**, 708.
Ordal, E. J. and Palmer, F. J. (1964). *In* "Continuous Culture of Micro-organisms", p. 133. Academia, Prague.
Paynter, M. J. B. and Bungay, H. R. (1969). *In* "Fermentation Advances", p. 323. Academic Press, New York and London.
Paynter, M. J. B. and Bungay, H. R. (1971). *In* "Biological Waste Treatment", p. 51. Interscience, New York.
Peil, K. M. and Gaudy, A. F. (1971). *Applied Microbiology* **21**, 253.
Perret, C. J. (1960). *Journal of General Microbiology* **22**, 589.
Pirt, S. J. (1972). *Journal of Applied Chemistry and Biotechnology* **22**, 55.
Pirt, S. J. (1965). *Proceedings of the Royal Society*, B **163**, 224.
Pirt, S. J. and Callow, D. S. (1958). *Journal of Applied Bacteriology* **21**, 188.
Pokorna, M., Kockova-Kratochvilova, A., Stuchlik, V. and Hronska, L. (1967). *Folia Microbiologica* **12**, 447.
Postgate, J. R. (1967). *Advances in Microbial Physiology* **1**, 1.
Postgate, J. R. and Hunter, J. R. (1962). *Journal of General Microbiology* **29**, 233.
Postgate, J. R. and Hunter, J. R. (1963). *Journal of Applied Bacteriology* **26**, 295.
Powell, E. O. (1958). *Journal of General Microbiology* **18**, 259.
Powell, E. O. (1967). *In* "Microbial Physiology and Continuous Culture", p. 34. H.M.S.O., London.
Powell, E. O. (1972). *Journal of Applied Chemistry and Biotechnology*, **22**, 71.
Punčochář, P. (1971). *Folia Microbiologica* **16**, 507.
Rahn, O. (1932). "Physiology of Bacteria". P. Blakiston, Philadelphia.

Ramanathan, M. and Gaudy, A. F. (1971). *Biotechnology and Bioengineering* **13**, 125.

Renneboog-Squilbin, C. (1967). *Journal of Theoretical Biology* **14**, 74.

Regan, D. L. and Roper, G. H. (1971). *Biotechnology and Bioengineering* **13**, 815.

Reusser, F. (1961). *Applied Microbiology* **9**, 366.

Řičica, J. (1971). *Folia Microbiologica* **16**, 389.

Řičica, J. (1972). *Folia Microbiologica* **17**, 398.

Řičica, J., Necinova, S., Stejskalova, E. and Fencl, Z. (1967). *In* "Microbial Physiology and Continuous Culture", p. 196. H.M.S.O., London.

Rufener, W. H., Nelson, W. O. and Wolin, M. J. (1963). *Applied Microbiology* **11**, 196.

Ryu, D. Y. and Mateles, R. I. (1968). *Biotechnology and Bioengineering* **10**, 385.

Schaezler, D. J., McHarg, W. H. and Busch, A. W. (1971). *In* "Biological Waste Treatment", p. 107. Interscience, New York.

Shehata, T. and Marr, A. G. (1971). *Journal of Bacteriology* **107**, 210.

Sheehan, B. T. and Johnson, M. J. (1970). *Applied Microbiology* **21**, 511.

Shindala, A., Bungay, H. R., Krieg, N. R. and Culbert, K. (1965). *Journal of Bacteriology* **89**, 693.

Shuler, M. L., Aris, R. and Tsuchiya, H. M. (1972). *Applied Microbiology* **24**, 384.

Sikyta, B., Doscocil, J. and Kasparova, J. (1959). *Biotechnology and Bioengineering* **1**, 379.

Silver, R. S. and Mateles, R. I. (1969). *Journal of Bacteriology* **97**, 535.

Sinclair, C. G., King, W. R., Ryder, D. H. and Topiwala, H. H. (1971). *Biotechnology and Bioengineering* **13**, 451.

Slobodkin, L. B. (1961). *In* "Growth and Regulation of Animal Populations", Holt, Rinehart and Winston, New York.

Slyter, L. L., Nelson, W. O. and Wolin, M. J. (1964). *Applied Microbiology* **12**, 374.

Srinivasan, V. R. and Han, Y. W. (1969). *In* "Cellulases and their Application", p. 447. American Chemical Society, Washington.

Standing, C. N., Fredrickson, A. G. and Tsuchiya, H. M. (1972). *Applied Microbiology* **23**, 234.

Steward, D. G., Warner, R. G. and Seeley, H. W. (1961). *Applied Microbiology* **9**, 150.

Stokes, J. L. (1954). *Journal of Bacteriology* **67**, 278.

Storer, F. F. and Gaudy, A. F. (1969). *Environmental Science and Technology* **3**, 143.

Stumm-Zollinger, E. and Harris, R. H. (1971). In "Organic Compounds in Aquatic Environments", p. 555. M. Dekker, Inc., New York.

Sukatsch, D. A. and Johnson, M. J. (1972). *Applied Microbiology* **23**, 543.

Teissier, G. (1942). *Revue de Science (Paris)* **80**, 209.

Tempest, D. W. (1970). *In* "Methods in Microbiology", p. 259. Academic Press.

Tempest, D. W., Herbert, D. and Phipps, P. J. (1967). *In* "Microbial Physiology and Continuous Culture", p. 240. H.M.S.O., London.

Thorne, R. S. W. (1968). *Journal of the Institute of Brewing* **74**, 516.

Topiwala, H. H. and Hamer, G. (1971). *Biotechnology and Bioengineering* **13**, 919.

Topiwala, H. H. and Sinclair, C. G. (1971). *Biotechnology and Bioengineering* **13**, 795.

Tsuchiya, H. M., Drake, J. F., Jost, J. L., and Fredrickson, A. G. (1972). *Journal of Bacteriology* **110**, 1147.

van Gemerden, H. and Jannasch, H. W. (1971). *Archiv. für Mikrobiologie* **79**, 345.

van Niel, C. B. (1949). *In* "The Chemistry and Physiology of Growth", p. 97. Princeton University Press.

van Uden, N. (1968). *Archiv für Mikrobiologie* **62**, 34.

Vary, P. S. and Johnson, M. J. (1967). *Applied Microbiology* **15**, 1473.
Veldkamp, H. and Kuenen, J. G. (1973). *In* "Modern Methods in the Study of Microbial Ecology," Bulletin 17, Swedish National Science Research Council, Stockholm (in press).
Veldkamp, H. and Jannasch, H. W. (1972). *Journal of Applied Chemistry and Biotechnology* **22**, 105.
Watson, T. G. (1970). *Journal of General Microbiology* **64**, 91.
Winogradski, S. (1949). "Microbiologie du Sol". Masson & Cie, Paris.
Wirsen, C. O. and Jannasch, H. W. (1970). *Bacteriological Proceedings* (Abstract) G118, p. 32.
Wodzinski, R. S. and Johnson, M. J. (1968). *Applied Microbiology* **16**, 1886.
Wright, P. L. (1971). *Journal of Dairy Science* **54**, 688.
Yeoh, H. T., Bungay, H. R. and Krieg, N. R. (1968). *Canadian Journal of Microbiology* **14**, 491.
Zines, D. O. and Rogers, P. L. (1971). *Biotechnology and Bioengineering* **13**, 293.
Zwaig, N. and Lin, E. C. C. (1966). *Science* **153**, 755.

Membrane Associated Enzymes in Bacteria

MILTON R. J. SALTON

Department of Microbiology
New York University School of Medicine
New York, N.Y. 10016 U.S.A.

I. Introduction

Membranes have been isolated from a great variety of organisms and cellular organelles, and chemically they are essentially lipid–protein structures. Many sophisticated techniques have been employed in studying their molecular architecture and the internal and surface environments determining their structure and functions. There is general agreement as to the basic construction of cellular membranes of various origins. Much of the membrane lipid exists in bilayer form, thus confirming the early model for biomembrane structure proposed by Danielli and Davson (1935). Lipid constituents of cell membranes have been intensively investigated, and the variety of lipid-soluble components, the structures of phospholipids and glycolipids, and the nature of their fatty acids are well documented for both eukaryotic and prokaryotic membranes (Kates, 1964; Macfarlane, 1964; van Deenen, 1965; O'Leary, 1967; Lennarz, 1970; Law and Snyder, 1972). Although much attention has been focused on membrane lipids and the mode of their organization in cell membranes, most isolated cellular membranes contain a higher ratio of protein to lipid. Until comparatively recently very little precise information had been available on the numbers and variety of proteins in cell membranes, apart from studies of the detection and localization of specific enzymes in membranes. With the extensive characterization of membrane lipids, much of the current biochemical interest in cell membranes has shifted to membrane proteins, their "solubilization", isolation and identification (Wallach, 1969; Steck and Fox, 1972). One of the techniques which has stimulated much interest in identifying and characterizing cell-membrane proteins has been the polyacrylamide gel-electrophoresis of sodium dodecyl sulphate-dissociated proteins and separation of the protein subunits in the presence of the dissociating detergents (Shapiro *et al.*, 1967; Weber and Osborn, 1969; Fairbanks *et al.*, 1971). Aspects of the application of these methods to membrane enzymes will be discussed further. By means of this technique, it is assumed that most of the membrane proteins are dissociated into their individual polypeptide subunit chains, and that these are separated on the basis of their molecular weights to yield a characteristic pattern for the variety of protein subunits in a given membrane preparation. Such analytical procedures based on the methods developed by a number of investigators (Shapiro *et al.*, 1967; Fairbanks *et al.*, 1971; Weber and Osborn, 1969; Neville, 1971) have all pointed to the general complexity of the protein constituents in cell membranes. Although the sodium dodecyl sulphate polyacrylamide-gel electrophoretic analysis of membrane proteins has been invaluable in membrane studies, this method reveals little about the specific functions of the enzymes in these structures. However, it has served to stimulate new interest in characterizing cell-membrane proteins and glycoproteins, and identifying the origins of the proteins associated with isolated cell-membrane preparations. As more membrane enzymes are purified to homogeneity and their subunit structures established, the complex patterns of bands seen in the sodium dodecyl sulphate-polyacrylamide electrophoretic analyses of whole membranes will become interpretable in terms of specific membrane-enzyme subunits. Moreover, this methodology has provided a most valuable way of monitoring cell-membrane preparations during their isolation, and facilitates a more critical approach to the problems of studying the association of enzymes with cell membranes in general and more specifically with bacterial membranes, which is the topic of this review.

The problems of localizing enzymes in cellular membrane structures are of course not unique to bacteria, and the methods applicable to "higher" eukaryotic cells can also be used with the "simpler" prokaryotic membrane systems. Thus, cytochemical staining for the localization of membrane enzymes *in situ* can be performed on all types of cells. Adenosine triphosphatases have thus been localized on the inner surface of erythrocyte membranes (Marchesi and Palade, 1967) and in the envelope surface of the bacterial cell (Voelz and Ortigoza, 1968) by cytochemical staining with lead salts and examination of thin sections in the electron microscope. Similar cytochemical studies at the electron microscope level have also been performed in the peripheral localization of cell-surface or periplasmic nucleases and phosphatases (Nisonson *et al.*, 1969; Cheng *et al.*, 1971). Localization of amino-acid transport proteins in the envelope of *E. coli* has also been achieved by using fluorescent-labelled antibody and cytochemical staining with peroxidase-antibody conjugates (Nakane *et al.*, 1968). It has, however, been difficult to deduce from these studies whether such enzymes or binding proteins are confined to the inner and/or outer membranes of the Gram-negative cell envelope or to the periplasmic space. On the other hand, the localization of electron-transport components, such as succinate dehydrogenase in bacteria, by cytochemical techniques has been less successful. Thus van Iterson and Leene (1964) concluded that the respiratory components were localized in the mesosomes on the basis of tetrazolium and tellurite staining, whereas Sedar and Burde (1965) found that succinate-dependent tetrazolium staining occurred uniformly in both plasma and mesosome membrane structures. It will be recalled that early attempts to apply mitochondrial staining techniques to bacteria led to anomalous results due to intracellular coalescence of formazan and invalid conclusions, as pointed out by Weibull (1953a). Cytochemical staining for enzyme localization in the bacterial cell has thus been less definitive, especially where it has been dependent on light- or phase-contrast microscope techniques. This is not surprising in view of the smaller dimensions of the bacterial cell compared to eukaryotic cells, and the general absence from bacteria of membranous organelles possessing compartmentalized functions, such as mitochondria, endoplasmic reticulum, lysosomes, and Golgi bodies. However, as indicated above, localization techniques such as those involving ferritin, peroxidase- or fluorescent-labelling of antibodies specific for membrane enzymes or transport proteins can be used in conjunction with electron microscopy or fluorescence microscopy for bacteria and other cells alike.

Most of our knowledge of enzyme localization in cells has come from direct determination of the distribution of enzymes following cell disruption and separation of "particulate", membrane and/or organelle fractions (Benedetti and Emmelot, 1968; de-Thé, 1968; Racker, 1970). From these studies with isolated membranes or membranous organelles, such as mitochondria in the case of eukaryotic cells, a body of information on membrane-localized enzymes has been built up. Thus certain enzymes have become regarded as typical markers for plasma, microsomal or mitochondrial membranes. Implicit in many of the investigations is the assumption that the membrane-associated state of a given enzyme reflects the natural localization of the enzyme in the cell and does not arise from a redistribution of the enzyme during the course of cell disruption and fractionation procedures. The isolation of membranes (or membranous organelles in eukaryotic cells) involves procedures which take the cells out of their natural environments and which place them in what is usually a totally foreign *milieu* (e.g. physiologically and chemically artificial buffer systems) for cell disruption and membrane separation. The need to do this for isolation of the membrane structures poses difficult problems for all studies with isolated membranes. As a consequence of this need for selecting an artificial environment for isolation, the loss of enzymes normally associated with membranes *in vivo* could occur, or the formation of unnatural associations due to the binding of certain enzymes may result. These problems are by no means confined to higher cells, and they are equally applicable to studies of membrane isolation and enzyme distribution in bacteria. The use of agents which "stabilize" membranes, such as divalent cations, heavy metal ions, glutaraldehyde or other cross-linking reagents, could result in membrane associations which do not exist in the intact cell. Since the cohesive forces in all types of membranes probably span a wide spectrum from weak to very strong associations of protein–protein, protein–lipid and lipid–lipid interactions in the membrane (Wallach, 1969, 1971), the loss of the most weakly associated enzymes could conceivably be fairly selective. The latter phenomenon would be more difficult to control and monitor during membrane isolation than a general weakening of the membrane structure which would result in the loss of more than one enzyme as well as lipid markers from the membrane. Moreover, attempts to lessen the contamination of membrane fractions with cytoplasmic enzymes by washing procedures could also lead to a progressive modification and weakening of the membrane. To minimize these various effects on membrane integrity and molecular associations, careful monitoring of the fractions and studies of enzyme distributions are needed for a variety of cells. The various factors considered above illustrate some but not all of the limitations within which one works during the isolation and enzymic characterization of membranes. Some of these considerations have been applied to the bacterial membrane system of the Gram-positive organism *Micrococcus lysodeikticus*, and the distribution of "cytoplasmic" and "membrane" enzymes has been established and discussed by Nachbar and Salton (1970a) and by Salton and Nachbar (1970).

With the recognition of these general difficulties inherent in membrane isolation and the limitations in extrapolating the state of the isolated membrane to its state *in situ* in the intact cell, we can simply document our knowledge of what has been discovered about the experimental distribution of enzymes between membrane (i.e. membrane-associated) and cytoplasm in bacteria. Therefore, in this review, enzymes which appear to be "firmly associated" with the membrane will be discussed. It is realized of course that, at the present stage of our knowledge of membrane structure–function relationships, it is difficult to assign a quantitative parameter to the term "firmly associated". It is in essence an operational term indicating those membrane associations which survive a standardized membrane isolation and washing procedure applied to a particular species.

Two membrane regions have been recognized in the majority of bacteria, namely the limiting plasma or "protoplast" membrane and the mesosomes (Fitz-James, 1960; Salton, 1971a,b). There are of course other specialized intracellular membranes in bacteria related to their special physiology (e.g. photosynthetic, nitrogen-fixing and nitrifying bacteria). Most of the studies with bacterial membrane enzymes pertain to the plasma membranes or envelopes, and it has been recognized for some time that these are multifunctional membrane structures performing a variety of cellular functions including transport, biosynthetic and energy-transduction processes (Salton, 1967a, 1971a,b). The distribution of such functions and enzymes between the plasma membrane and mesosome fractions is less well documented, since techniques for the isolation of mesosome vesicles free from plasma membranes have only been developed relatively recently (Reaveley, 1968; Reaveley and Rogers, 1969; Ellar and Freer, 1969; Popkin *et al.*, 1971; Owen and Freer, 1972). Thus the majority of membrane fractions from Gram-positive bacteria which have been studied in the past are probably of the plasma membrane type, since the methods of preparation by differential centrifugation at medium gravitational forces (about 20,000–30,000 $\times g$) of cell lysates usually result in the removal of mesosome structures in the supernatant fractions (which is discussed further by Salton, 1971a,b). Electron micrographs of negatively stained membrane fractions prepared in this manner from a variety of Gram-positive bacteria substantiate the impression that they are relatively free of mesosome vesicles (Salton, 1971a,b).

With Gram-negative bacteria, on the other hand, methods for the separation of the plasma membranes (inner) from the outer membranes have only been available quite recently (Schnaitman, 1970, 1971; Osborn *et al.*, 1972) so that much of the earlier work on enzyme localization in these organisms was usually performed on envelope fractions (i.e. inner and outer membrane structures, with or without the peptidoglycan components). As will be discussed in greater detail subsequently, the presence of a number of enzymes in the plasma membranes and not in the outer membrane structures of several Gram-negative bacteria has now been confirmed (Bell *et al.*, 1971; White *et al.*, 1971). Mesosomes in Gram-negative bacteria are generally less conspicuous than in Gram-positive organisms, and their purification and separation from envelopes or the individual inner and outer membranes do not appear to have been achieved. Although tellurite- and tetrazolium staining of mesosomes has been observed in Gram-negative bacteria (van Iterson and Leene, 1964; van Iterson, 1965; Ryter and Jacob, 1966), the direct localization of enzymes in these mesosomes must await further attempts to separate them in organisms possessing these membranous structures.

In addition to the inner and outer envelope membranes and mesosome membranes of Gram-negative bacteria, the specialized intracellular membranes, especially those of photosynthetic and nitrifying bacteria, have attracted considerable attention. The photosynthetic membrane vesicles (chromatophores) can be separated from the bacterial envelopes, and much work has been done on the characterization and functions of these structures (Oelze and Drews, 1972). The differential characterization of envelope membranes and photosynthetic membranes has been greatly facilitated in those species in which formation of the photosynthetic apparatus can be induced by a shift from aerobic growth in the dark to anaerobic growth in the light (Gorchein *et al.*, 1968; Oelze and Drews, 1972). Although the presence of bacteriochlorophyll and associated photosynthetic components has been clearly established in the chromatophore membrane systems of bacteria, very little attention has been paid to the presence of other enzymes. In the past one of the problems in the full characterization of these fractions has been the degree of contamination of the membrane preparations with cell-wall components (Gorchein *et al.*, 1968) and the need for specific markers in outer membrane, inner plasma membrane and chromatophore vesicle fractions. The isolation of homogeneous preparations of all three types of membranes in the photosynthetic bacteria would provide valuable information on the distribution of enzymes in these organisms. Similarly, special methods for the isolation of other intracellular membranes such as those of the nitrifying bacteria (Remsen *et al.*, 1967; Watson and Remsen, 1970) and the extra intracellular membrane "whorls" of strains of *Escherichia coli* (Greenawalt and Weigand, 1972; Weigand and Greenawalt, 1972; Weigand *et al.*, 1973) will ultimately lead to a clearer understanding of the enzymes localized in these membranes. These specialized bacterial membranes and chromatophores will not be discussed further in this review except with reference to the presence of specific enzymes.

As mentioned earlier in this introduction, the bacterial membrane is a multifunctional structure, performing a variety of cellular functions including transport of metabolites across the membrane, electron transport, biosynthesis of phospholipids and other membrane lipid components, and synthesis of macromolecules of walls including lipopolysaccharides, peptidoglycans, teichoic acids, and capsular polysaccharides. Because of their "associations" with ribosomes it has been suggested that membranes play an important role in protein synthesis in bacteria (Hendler, 1968). It should be emphasized that the full significance of this association for synthesis of membrane protein, export of proteins from the cells, and formation of cellular protein in general has yet to be critically explored and evaluated in prokaryotic cells. Bacterial membranes also serve as the anchoring point and probably also the replicating site for DNA synthesis. Many of these important membrane functions in bacteria, such as transport and DNA replication, have been the subjects of extensive studies and reviews in the past few years (Kaback, 1972; Harold, 1972; Simoni, 1972; Lark, 1969). This review will therefore be confined largely to a discussion of specific enzymes associated with bacterial membranes, and no attempt will be made to elaborate upon the broader aspects of the mechanisms of important functions such as transport, DNA replication and segregation and the association of ribosomes with membranes. More emphasis will accordingly be placed on the properties of the membrane-bound and released or solubilized enzymes of bacterial membranes.

II. Bacterial Membrane Adenosine Triphosphatases

Of all the bacterial membrane-associated enzymes, the ATPases have proved to be the most amenable to release from the membranes and the easiest to purify to a state of homogeneity. Thus, from a variety of bacterial membranes, it has become apparent that the ATPases can be released from their membrane-associated state by relatively "mild" perturbations of the membranes. In this respect the bacterial membrane ATPases resemble the mitochondrial enzymes, and these two classes of ATPase contrast markedly with the more tightly integrated [$Na^+ + K^+$]-stimulated Mg^{2+}-ATPases of erythrocyte and microsomal membranes. The latter types of ATPase require much more severe disruption of the

TABLE 1. Occurrence of Membrane ATPases in Various Gram-Positive and Gram-Negative Bacteria and the Effects of Cations on the Enzymes

	Type of preparation	Divalent cation requirement and activation	Effects of Na^+ and K^+	References
GRAM-POSITIVE				
Bacillus megaterium	Membrane-bound and soluble	$Ca^{2+} > Mg^{2+}$	None or little	Ishida and Mizushima (1969a)
Bacillus stearothermophilus	Membrane-bound and soluble	$Mg^{2+} > Ca^{2+}$	Not significant	Hachimori *et al.* (1970)
Bacillus subtilis	Membrane-bound	$Ca^{2+} > Mg^{2+}$	Stimulation at high concentration	Rosenthal and Matheson (1973)
Lactobacillus arabinosus	Cell-wall membrane	Mg^{2+}	Unknown	Cole and Hughes (1965)
Lactobacillus fermenti	Membrane-bound	$Mg^{2+} > Ca^{2+}$	Unknown	Neujahr (1970)
Micrococcus lysodeikticus	Membrane-bound and soluble	$Ca^{2+} > Mg^{2+}$	None; inhibition at high concentration	Muñoz *et al.* (1969)
Staphylococcus aureus	Membrane-bound	$Mg^{2+} > Ca^{2+}$	Stimulation at high concentration	Cross and Coles (1968)
Streptococcus faecalis	Membrane-bound and soluble	Mg^{2+}	Stimulation at high concentration	Abrams *et al.* (1960); Abrams (1965)
GRAM-NEGATIVE				
Escherichia coli	Membrane-bound and soluble	$Mg^{2+} > Ca^{2+}$	Slight effects; inhibitory together	Evans (1969, 1970); Kobayashi and Anraku (1972)
Marine pseudomonad	Envelope	Mg^{2+}	No activation	Drapeau and MacLeod (1963)
Vibrio parahaemolyticus	Envelope	Mg^{2+}	Stimulation at high concentration	Hayashi and Uchida (1965)
Vitreoscilla sp.	Envelope and soluble	Mg^{2+}	Stimulation at high concentration	Burham and Hageage (1967)

membrane structures for their release and "solubilization" (Hokin, 1969). Investigations of bacterial membrane ATPases soon followed the development of procedures for the isolation of membranes following the dissolution of the cell wall and protoplast formation (Weibull, 1953b). Isolated "membrane ghosts", "particle" or "granule" fractions from lysozyme lysates of several Gram-positive bacteria have exhibited ATPase activity (Georgi *et al.*, 1955; Weibull *et al.*, 1962; Abrams *et al.*, 1960; Ishikawa and Lehninger, 1962). Since these early studies, ATPases have been detected in a variety of membranes and envelope preparations from Gram-positive and Gram-negative bacteria as illustrated in Table 1. Virtually all of the ATPases from bacterial membranes have been found to be dependent on divalent cations (Mg^{2+} and/or Ca^{2+}) though a few claims have also been made for stimulations by [$Na^+ + K^+$] (Hafkenscheid and Bonting, 1968; Hayashi and Uchida, 1965). Most of the reports of the occurrence of these ATPases have been based on the detection and properties of the membrane-bound forms of the enzyme in the various bacteria (Table 1). However, as discussed below, the ATPases of several bacteria have been obtained in soluble form and they appear to be relatively homogeneous. By analogy with the mitochondrial ATPases, it is clear that information on the properties of both the membrane-associated state and the purified, soluble enzymes or enzyme complexes will be needed to understand fully the functions and characteristics of this important group of membrane enzymes. Moreover, studies of the differential response of the membrane-bound and soluble forms of the enzyme to stimulators such as trypsin (Muñoz *et al.*, 1969; Salton and Schor, 1972) and polyanions (Ishikawa *et al.*, 1965) and inhibitors of the carbodiimide class (Harold *et al.*, 1969; Abrams and Baron, 1970) will provide an insight into the micro-environment and molecular associations of the enzyme, the mechanisms of regulation of its activities, the property of "latency" and its functions on the bacterial membrane.

A. Release, Solubilization and Purification of ATPases

Ishikawa and Lehninger (1962), in their studies of oxidative phosphorylation by membranes of *Micrococcus lysodeikticus*, found ATPase activity in the sonicated fragments. These membrane fragments when "shocked" by exposure to distilled water released a soluble protein fraction which contained the ATPase as well as other factors. Density-gradient centrifugation of fractions containing ATPase activity gave both ATPase and coupling factor in a 13 *S* peak (Ishikawa, 1966). Further purification and the use of other criteria for establishing the homogeneity of this bacterial enzyme awaited later studies by Muñoz *et al.*

(1968a, 1969). Following solubilization of the ATPase by the distilled water "shock" (Ishikawa and Lehninger, 1962), Abrams (1965) developed a selective release procedure by repeated washing of the membranes of *Streptococcus faecalis* in the absence of Mg^{2+}. The ATPase was recovered as a soluble enzyme which, upon sedimentation analysis in a sucrose gradient, gave an estimated sedimentation coefficient of 12·9 *S*. Electrophoresis of the solubilized ATPase in starch gel showed one zone of enzyme activity and other protein bands (Abrams, 1965).

The cation depletion of membranes by washing in the absence of Mg^{2+} (Abrams, 1965) and the distilled water "shock" (Ishikawa and Lehninger, 1962) have provided the basis for the selective release of ATPases from several other bacterial membrane systems. With some modifications of such procedures, Muñoz *et al.* (1968a) demonstrated a dramatic release of ATPase activity from membranes of *M. lysodeikticus* which had been washed in 0·03 *M* tris-HCl buffer at pH 7·5 in the absence of Mg^{2+} prior to "shocking" in buffer of low ionic strength (3 m*M*-tris). In this way, the bulk of the ATPase activity was released from the membranes. Moreover, the structural changes co-incident with this release of ATPase suggested that the activity was associated with the uniform particles released from the membrane into the supernatant "shock" wash fluid (Muñoz *et al.*, 1968b; Nachbar and Salton, 1970a, b). Evidence for the identity of the membrane particles and ATPase will be discussed in Section II C (p. 245). In addition to the selective release of ATPase from membranes of *Strep. faecalis* and *M. lysodeikticus*, release and solubilization of the ATPases have been achieved with two strains of *Bacillus megaterium* (Ishida and Mizushima, 1969a, b; Mirsky and Barlow, 1971) and with *E. coli* K 12 by Kobayashi and Anraku (1972). The method of Abrams (1965) also resulted in the release of ATPase in the fifth wash of sphaeroplast membranes of another strain of *E. coli* (NRC 482) as reported recently by Davies and Bragg (1972). Essentially the same principle of release of ATPase from membrane ghosts of *Bacillus stearothermophilus* was used by Hachimori *et al.* (1970) except that divalent-cation depletion and solubilization were accomplished by dialysis against 0·1 *M*-EDTA in buffer and then 0·1 *M*-EDTA in distilled water. Release and solubilization of the membrane ATPases by variations of the selective methods of enzyme release, thereby avoiding the use of surface-active agents, has been a major achievement in characterizing what is, albeit, a more "peripheral" class (Singer, 1971) of bacterial membrane enzyme. Unlike most of the other membrane enzymes, the ATPases can be handled in much the same way as globular proteins and, from evidence thus far available, they do not appear to be associated with firmly bound lipids. Although it is not necessary to use reagents and conditions which are ordinarily needed to dissociate strong lipid–protein interactions for the release of ATPase, this does not mean that active enzyme cannot be obtained by these more drastic procedures. Indeed, it should be recalled that Weinbaum and Markman (1966) dissociated envelopes of *E. coli* with sodium dodecyl sulphate (SDS) and were able to detect ATPase entering polyacrylamide gels by substrate-dependent staining with lead salts. Specific detection of the enzyme, upon polyacrylamide gel electrophoresis of the soluble ATPases released by the selective procedures avoiding detergents, has proved most valuable in purification of these enzymes (Muñoz *et al.*, 1968a, 1969). Sodium dodecyl sulphate can be used to release ATPase but, because of its ability to denature proteins, it is necessary to control appropriately the ratios of protein and SDS and time of exposure during membrane dissociation. It is thus a far less suitable agent for ATPase release. Moreover, in high concentration SDS will yield inactive enzyme subunits (Whiteside and Salton, 1970). Nevertheless, Evans (1970) was able successfully to solubilize *E. coli* K 12 ATPase with low concentrations of sodium lauryl sulphate (SLS). This was achieved by treating sphaeroplast membrane ghosts (obtained using lysozyme and EDTA) in 0·01 *M*-tris-HCl buffer containing 0.04% SLS at pH 9·0 in the presence of 0·2 m*M*-$MgCl_2$ for 10 minutes at 37°C followed by centrifugation at 25,000 × *g* for 20 minutes at 24°C. The ATPase activities were quantitatively recovered in the supernatant fractions, and subsequent purification steps were performed at 24°C because of the cold lability of the SLS-solubilized enzyme. An alternative method of releasing ATPase from *E. coli* K 12, which avoids the use of SDS, has recently been published by Kobayashi and Anraku (1972). The cells are disrupted by sonication in 2 m*M*-tris-HCl at pH 7·2, the membrane fraction resuspended in the dilute tris buffer, and the ATPase recovered in the wash fluid. This latter procedure thus simulates the Mg^{2+}-depletion and low ionic-strength shock wash procedure which is so successful with the release of ATPases from membranes of several Gram-positive bacteria.

On the other hand, dissociation of membrane lipid and protein by extraction with *n*-butanol in a two-phase system as originally introduced by Morton (1950), and applied to erythrocyte membranes by Maddy (1964), has been used for many years in the author's laboratory for release of ATPase from membranes of *M. lysodeikticus* into an aqueous phase. This procedure has been used as an alternative method for releasing ATPase, and has led to the observation that the ATPase released by *n*-butanol is chemically more homogeneous than the shock-wash ATPase complex which usually possesses one or more firmly associated minor protein components (Salton and Schor, 1972). Although other aliphatic alcohols can be used for the dissociation of membrane proteins and lipids in two-phase systems, *n*-butanol was superior to *t*-amyl alcohol and 3-pentanol, so far as release and stability of the ATPase in the solvent-saturated aqueous phases during the application of this procedure (Nachbar *et al.*, 1972).

Bacterial ATPases can thus be released from isolated membranes by three principal procedures: (1) selective release at low ionic strengths following cation depletion; (2) sodium dodecyl sulphate (or sodium lauryl sulphate)-mediated dissociation of membranes; and (3) organic-solvent extractions of lipids and release of ATPase into the aqueous phase in two-phase systems. Sonication of isolated membranes was shown to release a major component which enters polyacrylamide gels (Salton, 1967b), and this was subsequently found to be the membrane ATPase complex. Thus, a variety of release procedures can be used as the starting point for ATPase purification. The selective, low-ionic strength, shock-wash procedure has obvious advantages over the other two procedures in that it avoids the presence of potentially deleterious agents, such as sodium dodecyl sulphate and organic solvents. Following release from the membrane and separation of residual membrane structures or derived particles, fairly conventional methods of purification can be applied to this enzyme. In their earlier studies of purification of ATPase from *Strep. faecalis*, Abrams and Baron (1967) concentrated the enzyme by ammonium sulphate precipitation, and separated the ATPase by sucrose density-gradient centrifugation. In later work, the enzyme was obtained in higher yield by procedures including a heat-treatment step, chromatography on DEAE cellulose, followed by ammonium sulphate precipitation, gel filtration on Agarose A 0·5 and a second cycle of chromatography on DEAE cellulose. Under these conditions, the purification was 85-fold. The ATPase of *M. lysodeikticus* was purified about 50-fold by two cycles of gel filtration on Sephadex G-200 (Muñoz *et al.*, 1969). A combination of DEAE-cellulose chromatography, hydroxyapatite chromatography and gel filtration was used in the purification of ATPase from *B. stearothermophilus* (Hachimori *et al.*, 1970). Mirsky and Barlow (1971) used a batch procedure for adsorption of the ATPase from *B. megaterium* KM to DEAE cellulose and elution with 0·25 *M*-ammonium sulphate. Based on the specific activities of the initial lysate and final fraction from DEAE cellulose, a purification of about 500-fold was achieved, but the final yield of protein was quite small. An alternative method of purification on glycerol gradients was subsequently used by these authors, but the degree of purification was not reported and the specific activities of the final preparations were lower than those previously found (Mirsky and Barlow, 1972). Ishida and Mizushima (1969b) had used DEAE chromatography and protamine sulphate precipitation in their earlier purification of *B. megaterium*

ATPase. A protamine sulphate precipitation step was also found to be useful in purification of the ATPase from *E. coli* K 12 released from the membrane by SLS (Evans, 1970). Agarose gel filtration was also employed in purification of the *E. coli* enzyme, and the final specific activities indicated a 25- to 40-fold purification. Kobayashi and Anraku (1972) purified the ATPase from *E. coli* K 12 which was released into the wash fluid from membranes prepared by sonic disruption of the cells, by chromatography on DEAE cellulose. With this method, the purification factor was 62-fold. With both methods of purification of the *E. coli* ATPase, many manipulations had to be performed at room temperature (about 24°C) because of its cold lability. In more recent studies from the author's laboratory, purifications of about 100-fold of the ATPase from *M. lysodeikticus* have been achieved (Oppenheim and Salton, 1973; M. T. Schor and M. R. J. Salton, unpublished data).

From these studies it would appear that purifications of about 50- to 100-fold can be generally expected, although in one instance it was as high as 500-fold (Mirsky and Barlow, 1971). The reasons for such a large difference in purification are obscure, but may be a reflection of the extreme variability in the initial specific activities of the lysates or crude membrane fractions due to the "latency" exhibited by the membrane-bound form of the enzyme. Degrees of purification can thus represent true purification as well as unmasking effects resulting from dissociation of the enzyme from membranes and/or inhibitors and regulatory proteins. This is a problem which is by no means unique to ATPases, and assays of enzymic activities in whole membranes usually underestimate the true initial total activity of the enzyme, e.g. $NADH_2$ dehydrogenase (Nachbar and Salton, 1970b), succinate dehydrogenase (Pollock *et al.*, 1972) and cardiolipin synthetase (De Siervo and Salton, 1971), unless an assay procedure giving complete dissociation and unmasking can be used.

1. Homogeneity, Subunit Structure and Amino-Acid Composition of Purified ATPases

Apart from the enzymic specificity which will be discussed in Section II B (p. 234), ATPases purified from bacterial membranes by procedures described above have been subjected to one or more of the various criteria generally used in judging the homogeneity of proteins and enzyme preparations. Two criteria have been used with most of the purified membrane ATPase preparations, namely the detection of a single sedimenting peak in the ultracentrifugal analysis of the enzyme fractions and a single or major band of protein and/or specific ATPase activity on electrophoresis in the standard polyacrylamide gel systems. By one or both of these criteria, membrane ATPase preparations of *Strep. faecalis* (Abrams and Baron, 1967; Schnebli and Abrams, 1970), *B. megaterium* (Ishida and Mizushima, 1969a,b; Mirsky and Barlow, 1971, 1972), *M. lysodeikticus* (Muñoz *et al.*, 1968a, 1969), *B. stearothermophilus* (Hachimori *et al.*, 1970) and *E. coli* (Kobayashi and Anraku, 1972) have been judged as being homogeneous. Although some of these preparations have given single peaks of protein or activity in the ultracentrifuge and by chromatography and gel filtration, two preparations, those from *B. megaterium* (Ishida and Mizushima, 1969a,b) and *M. lysodeikticus* (Whiteside and Salton, 1970), have given two closely migrating bands of specific ATPase activity in the standard polyacrylamide gel electrophoresis systems. Moreover, with *M. lysodeikticus* ATPase, such preparations have appeared to be homogeneous by two other criteria, namely homogeneity on reaction with specific ATPase antibody by double diffusion, and immunoelectrophoresis tests (Whiteside and Salton, 1970; Fukui *et al.*, 1971; Oppenheim and Salton, 1973) and the uniformity of the particles seen in the electron microscope (Muñoz *et al.*, 1968b; Whiteside and Salton, 1970; Oppenheim and Salton, 1973). The detection of a single line of precipitate in double-diffusion tests of the purified ATPases reacted against specific antiserum, and the production of a single arc of precipitate in immunoelectrophoresis tests of the purified fractions (Whiteside and Salton, 1970; Fukui *et al.*, 1971; Oppenheim and Salton, 1973), provide an additional immunological criterion for homogeneity.

Perhaps one of the most sensitive methods for judging protein heterogeneity and, at the same time, subunit structure has been the sodium dodecyl sulphate-polyacrylamide gel electrophoresis technique which resolves the individual polypeptide chains on the basis of their molecular weights (Shapiro *et al.*, 1967; Weber and Osborn, 1969; Fairbanks *et al.*, 1971). Unfortunately, identical techniques for protein dissociation and polyacrylamide gel electrophoresis have not been used in all studies, so that comparisons of subunit structure and protein heterogeneity of the various ATPase preparations are difficult to evaluate. Abrams and Baron (1967) found five protein bands in membrane ATPase from *Strep. faecalis* in gels containing 8 *M*- or 10 *M*-urea. In the presence of mercaptoethanol or dithiothreitol, only three bands were found in ratios of approximately 2 : 2 : 1. In later studies of the subunit structure of ATPase from *Strep. faecalis*, two bands were obtained upon electrophoresis of the enzyme dissolved in 8 *M*-urea+0·1 *M*-dithiothreitol in polyacrylamide gel containing 8 *M*-urea, and the subunit molecular weight determined by sedimentation equilibrium in 6 *M*-guanidine hydrochloride was 33,000 daltons (Schnebli *et al.*, 1970). The subunit molecular weight of 33,000 daltons was in good agreement with the minimum molecular weight of

32,800 daltons calculated from the amino-acid composition. It was concluded that the enzyme consists of 12 subunits (Schnebli *et al.*, 1970). Although two kinds of subunits were detectable and clearly separable by polyacrylamide gel analysis in 8 *M*-urea, the fact that only one molecular-weight class of subunit was obtained by sedimentation equilibrium in 6 *M*-guanidine hydrochloride can only be interpreted as meaning that the α and β subunits (Abrams and Baron, 1967; Schnebli *et al.*, 1970) have very similar molecular weights but may possess charge differences. Sodium dodecyl sulphate–mercaptoethanol polyacrylamide gel electrophoresis studies by the technique of Weber and Osborn (1969) have not been reported for this enzyme.

The homogeneity and subunit structures of two other ATPases from bacterial membranes have been investigated by sodium dodecyl sulphate–polyacrylamide gel electrophoresis. Adenosine triphosphatases from both *B. megaterium* (Mirsky and Barlow, 1971) and *M. lysodeikticus* (Fukui and Salton, 1972) gave single bands in polyacrylamide gels containing 0·1% sodium dodecyl sulphate. However, applying the Weber and Osborn (1969) method of dissociating ATPase from *M. lysodeikticus* in 1% sodium dodecyl sulphate–1% mercaptoethanol and electrophoresis in gels containing 0·1% sodium dodecyl sulphate and 0·1% mercaptoethanol, two distinct bands were obtained (Salton and Schor, 1972). Moreover, two subunit bands were also obtained from purified ATPase preparations which had been either reduced and alkylated or subjected to performic acid oxidation (Salton and Schor, 1972). Two subunit bands were also subsequently found in ATPase from *B. megaterium* by dissociation with sodium dodecyl sulphate–mercaptoethanol as reported recently by Mirsky and Barlow (1973), Thus, both of these ATPases have shown the presence of two subunits in approximately 1 : 1 ratios (based on scans of the Coomassie blue staining) and, in both instances, the molecular weights of the subunits differ by less than 10%. No additional bands were found in the ATPase from *B. megaterium* dissociated and examined under these conditions (Mirsky and Barlow, 1973). On the other hand, the ATPase complex from *M. lysodeikticus* released by the shock-wash treatment showed the presence of one or more additional minor bands by the sodium dodecyl sulphate–polyacrylamide gel electrophoresis technique (Salton and Schor, 1972) and thus resembled the ATPase from *Strep. faecalis* (Abrams and Baron, 1967). The ATPase released by the *n*-butanol extraction procedure possessed only the two subunit bands and no associated polypeptide chains and, therefore, appears to be chemically more homogeneous than the shock-wash ATPase from *M. lysodeikticus* membranes. The principal minor band found in the shock wash of *M. lysodeikticus* ATPase corresponds to the fast-moving component identified in previous studies

(Fukui *et al.*, 1971; Fukui and Salton, 1972). The presence of the additional protein(s) associated with the shock-wash ATPase probably confers ability to rebind to the depleted membranes and the trypsin stimulation property, both of which are not exhibited by ATPase preparations released by *n*-butanol (Salton and Schor, 1972). It is likely that the fast-moving component protein of membrane and ATPase complex from *M. lysodeikticus* corresponds to the nectin or binding protein found in the ATPase system from *Strep. faecalis* by Baron and Abrams (1971), and the latter presumably is responsible for the γ protein band found in the subunit analysis by Abrams and Baron (1967). As suggested by Salton and Schor (1972) it is likely that the shock-wash ATPase complex is a mixture of ATPase molecules, with and without the additional junction protein or nectin, due to the "spontaneous" dissociation of this protein upon storage. The fact that the ATPase activity which remained unadsorbed to the depleted membranes did not exhibit trypsin stimulation would certainly be in accord with the presence of the two types of ATPase (Salton and Schor, 1972). Such a dissociation of the ATPase system from *M. lysodeikticus* would account for the apparent rise in specific activities during the first few days of storage at 4°C and the decline in the magnitude of trypsin stimulation and the double bands of enzymic activity encountered in the standard polyacrylamide gel electrophoresis (i.e. in the absence of sodium dodecyl sulphate–mercaptoethanol dissociation), and a similar explanation could also account for the double-banding observed by Ishida and Mizushima (1969b).

From these studies it would appear that bacterial ATPases possess two subunit polypeptide chains of very similar or slightly different molecular weights, and that the native ATPase complex may possess one or more associated proteins giving rise to additional minor bands on electrophoresis in sodium dodecyl sulphate–mercaptoethanol polyacrylamide gels. The complexity of the ATPase would thus be dependent on its mode of dissociation from the membrane and the extent to which further dissociation of the proteins of the complex had occurred during storage and purification. Although the membrane ATPases may behave as homogeneous proteins when judged by a number of criteria (ultracentrifugation, immunoelectrophoresis, single bands in standard polyacrylamide gel electrophoresis, single peaks by gel filtration and chromatography) their analysis for subunit structure by sodium dodecyl sulphate-association may reveal the presence of additional associated proteins, some of which may be of considerable functional significance for the attachment and functioning of the enzyme on the membrane. In this respect, the more complex shock-wash ATPases of the bacterial membrane resemble the mitochondrial ATPases which, in addition to their major subunit polypeptide chains, possess a number of other

TABLE 2. Comparison of Molecular Weights and Subunit Structures of Bacterial Membrane ATPases

Organism	Molecular weight (daltons)	S_w	Subunit structure: Number[a]	Subunit structure: Molecular weight(s)[b]	Associated protein(s)	References
Bacillus megaterium KM	379,000	13·6	2	68,000 65,000	None	Mirsky and Barlow (1973)
Bacillus stearothermophilus	280,000	11·9	—	—	—	Hachimori *et al.* (1970)
Escherichia coli K 12	100,000[c]	—	—	—	—	Evans (1970)
	>400,000[c]	—	—	—	—	Kobayashi and Anraku (1972)
Escherichia coli NRC 482	365,000 390,000	—	—	—	—	Davies and Bragg (1972)
Micrococcus lysodeikticus	—	14–15	2	62,000 60,000	Present	Muñoz *et al.* (1969) Salton and Schor (1972)
Streptococcus faecalis	385,000	13·4	2	33,000	Present	Abrams and Baron (1967) Schnebli *et al.* (1970)

[a] This refers to the number of different subunits detectable by polyacrylamide gel electrophoresis of dissociated ATPases; the value does not indicate the total number in each ATPase particle.

[b] Molecular weights were determined by sodium dodecyl sulphate–polyacrylamide gel electrophoresis or by ultracentrifugation of dissociated enzyme.

[c] Estimated molecular weights.

— Data not reported.

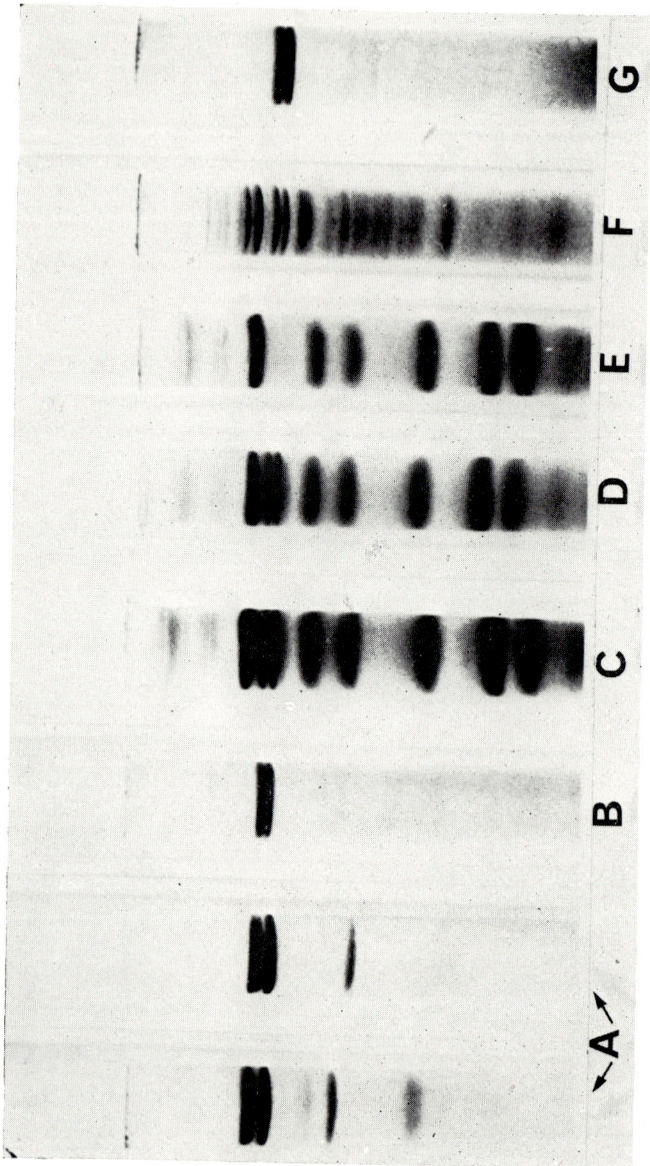
A
B
C
D
E
F
G

identifiable components (Tzagoloff, 1971; Knowles and Penefsky, 1972). Moreover, the small differences in molecular weight between the individual subunits of each of the bacterial ATPases and those observed with yeast mitochondrial ATPase (Tzagoloff, 1971) illustrate yet another striking similarity in the properties of these enzymes.

Both sodium dodecyl sulphate–polyacrylamide gel electrophoresis and ultracentrifugation studies have yielded valuable criteria on which to judge the homogeneity of ATPase preparations as well as giving the molecular weights of the component subunits and the undissociated enzymes. Data on the molecular weights and subunit structures of the ATPases are presented in Table 2. The separation of subunit bands and detection of additional proteins in the sodium dodecyl sulphate–polyacrylamide gel electrophoresis studies of ATPase from *M. lysodeikticus* is illustrated in Figure 1 (Salton and Schor, 1972). Moreover, identification of the subunit bands of the ATPase has enabled us to identify two of the bands in the complex pattern of proteins in whole membranes as originating from this membrane enzyme (see gel F in Fig. 1). Determination of the subunit structures of other membrane enzymes will greatly facilitate a meaningful interpretation of the protein bands obtained by this technique.

The interpretation of the subunit structure of the bacterial ATPases has been greatly assisted by electron micrographs of negatively stained enzyme preparations. The first purified bacterial ATPase studied in this way was that of *M. lysodeikticus* (Muñoz *et al.*, 1968b), and the ATPase

---

FIG. 1. Subunits of the ATPase from *Micrococcus lysodeikticus* membrane and additional bands due to associated proteins of the shock-wash ATPase complex have been identified by sodium dodecyl sulphate–polyacrylamide gel electrophoresis. The components in the Coomassie blue-stained gels obtained by the Weber and Osborn (1969) method are as follows: A. Shock-wash ATPase preparations showing two major subunits and additional minor protein bands. The principal minor band has a molecular weight close to that of pepsin (i.e. about 35,000 daltons) and is probably similar to the nectin (Baron and Abrams, 1971) and fast-moving component (Fukui and Salton, 1972), the likely functions of which are as junction proteins for the ATPase complex. B. *n*-Butanol-type ATPase showing two subunits identical to those found in shock-wash ATPase. C. Shock-wash ATPase together with protein standards; bands from the top but one to the bottom of the gel are, in order, bovine serum albumin, ATPase subunits, ovalbumin, pepsin, myoglobin and haemoglobin; the principal minor band overlaps with pepsin. D. Adenosine triphosphatase reduced with dithiothreitol and alkylated with iodoacetamide, together with standard proteins as in gel C. The identity of bands is as for gel C. Similar results were obtained with both types of ATPase and with performic acid-oxidized preparations. E. Protein standards alone, as in gels C and D. F. Subunit pattern for whole membranes of *Micrococcus lysodeikticus*. G. *n*-Butanol-type ATPase showing two major subunits and corresponding bands in whole membranes (gel F). From Salton and Schor (1972).

particles were found to possess six peripheral units surrounding a central structure. The striking similarity of the appearance of the bacterial ATPase to the mitochondrial particles (Stiles and Crane, 1966; Racker, 1967) was pointed out by Muñoz *et al.* (1968b). Similar results were later found for ATPases from *B. megaterium* (Ishida and Mizushima, 1969b) and *Strep. faecalis* (Schnebli *et al.*, 1970). The molecular weights, subunit structure, and appearance of the enzyme particles in the electron microscope are compatible with a 12-subunit structure as proposed by Schnebli *et al.* (1970). The additional "central" unit seen in the shock-wash ATPase from *M. lysodeikticus* (Muñoz *et al.*, 1968b) could be interpreted as involving the attachment protein(s), but further investigations are needed to resolve this and the apparent absence of such a structure in the *Strep. faecalis* ATPase (Schnebli *et al.*, 1970).

The relationships between molecular weights, subunits visible by electron microscopy of the ATPase particles, subunits detectable in polyacrylamide gel electrophoresis, and the presence of associated proteins which may be involved in the binding and regulation of the enzyme need fuller exploration. The data establishing essentially one molecular weight subunit class of about 33,000 daltons for the *Strep. faecalis* ATPase agree well with the amino-acid analysis, molecular weights of the protein in the ultracentrifuge, and the six visible subunits as revealed by electron microscopy (Schnebli *et al.*, 1970). Thus, each of the six subunit structures seen in negatively stained preparations of the ATPase would be composed of two subunits each of molecular weight 33,000 daltons and give a good fit to the model proposed by Schnebli *et al.* (1970). As seen in Table 2 (p. 229), however, two subunits are found in ATPases of *M. lysodeikticus* and *B. megaterium*, and the values for their molecular weight classes are almost double those of the *Strep. faecalis* ATPase. In both of these ATPases, the molecular weights of the subunits are of the order of 60,000 daltons, as determined by the Weber and Osborn (1969) sodium dodecyl sulphate–polyacrylamide gel technique. If the six spherical subunits visible in the negatively stained preparations of the ATPases represent the true molecular anatomy of the enzyme, then such a structure would accommodate $3 + 3$ of the 62,000 and 60,000 molecular weight subunits for the ATPase from *M. lysodeikticus* (Salton and Schor, 1972), $3 + 3$ for the 68,000 and 65,000 molecular weight subunits of the ATPase from *B. megaterium* (Mirsky and Barlow, 1973), and $6 + 6$ of the 33,000 molecular weight subunits of the ATPase from *Strep. faecalis* (Schnebli *et al.*, 1970). It is clear that more sophisticated information is needed on the molecular anatomy (e.g. high resolution electron micrographs) and precise subunit structure and arrangement of subunits in the ATPase particle. Moreover, the ATPase preparations will have to be defined more carefully in terms of associated proteins, and more quantitative information on the ratios of the component subunits will be required despite the fact that present data from gel scans (Mirsky and Barlow, 1973; Salton and Schor, 1972) suggest about a 1 : 1 ratio for the two separable subunits.

The amino-acid compositions of the ATPases of *Strep. faecalis*, *B. megaterium*, *B. stearothermophilus* and *M. lysodeikticus* have been determined and the collected data are summarized in Table 3. There do not appear to be any unusual features about their amino-acid compositions and, as seen in Table 3, the percentages of hydrophobic amino-acid residues are not excessively high. This of course does not mean that hydrophobic amino acid-rich areas of the polypeptide do not exist in the

TABLE 3. Amino-Acid Composition of Bacterial Membrane ATPases

Residue	Mole per cent: *Streptococcus faecalis*	*Bacillus megaterium*	*Bacillus stearo-thermophilus*	*Micrococcus lysodeikticus*
Alanine	8·4	8·8	6·2	7·0
Valine	6·8	8·9	7·3	7·8
Leucine	9·3	9·3	8·2	9·7
Isoleucine	6·2	6·3	8·0	5·4
Proline	3·9	4·3	4·2	4·5
Phenylalanine	3·1	3·3	4·9	2·7
Methionine	2·3	2·0	—	1·8
Glycine	8·7	9·0	4·5	5·5
Serine	6·3	5·6	3·8	5·0
Threonine	6·7	5·9	4·9	6·3
Half-cystine	0·33	0·1	—	0·23
Tyrosine	3·3	2·7	4·0	3·40
Aspartate	10·0	8·5	9·5	10·2
Glutamate	13·0	13·4	14·3	15·0
Arginine	4·5	5·6	7·9	8·5
Lysine	6·1	5·1	6·3	4·1
Histidine	1·7	1·6	2·9	2·0
Tryptophan	—	0·4	—	1·0
% Hydrophobic amino acids	31·0	32·9	32·4	31·8

Data for the enzyme from *Strep. faecalis* are from Schnebli *et al.* (1970); those for *B. megaterium* from Mirsky and Barlow (1973); those for *B. stearothermophilus* from Hachimori *et al.* (1970). Those for *M. lysodeikticus* are unpublished data of M. T. Schor, M. Heincz and M. R. J. Salton. A molecular weight of 350,000 has been assumed for these calculations.

The percentages of hydrophobic amino-acid residues are calculated from the values for Val, Met, Ile, Leu, Tyr, Phe and Trp.

protein subunits, for these may be essential for the assembly of a complex enzyme particle such as that seen in the electron microscope (Muñoz *et al.*, 1968b; Ishida and Mizushima, 1969b; Schnebli *et al.*, 1970). Tryptic digestion of purified ATPase of *M. lysodeikticus* has been performed; a number of specific peptide spots has been observed in finger prints, and about 15 of the peptides appear to be common to other membrane fractions (Fukui *et al.*, 1971). These common peptides obtained by tryptic digestion account for as much as 50% of the peptides detected in the trypsin digests (Fukui and Salton, 1972). Very similar results were obtained with three yeast mitochondrial proteins subjected to tryptic digestion and finger printing as reported by Yang and Criddle (1970). In both instances, the number of common peptides in the digests showed good agreement with the expected number of sensitive bonds as determined by amino-acid analysis (Fukui *et al.*, 1971; Yang and Criddle, 1970). Further characterization of specific and common peptides from tryptic digests is needed especially in relation to the individual subunits and associated proteins. Such data will be important in evaluating the significance of the apparently common peptides.

As a class of bacterial membrane-associated enzymes, the ATPases present an excellent opportunity for examining the functional and structural relationships of a protein which can be fairly readily released, purified, and characterized in terms of its amino-acid composition, subunit structure, peptide mapping, and immunochemical properties and cross-reactivities (Whiteside *et al.*, 1971). The large-scale isolation and crystallization of such an enzyme would offer new opportunities for investigating the primary structure of subunits and the X-ray crystallography of enzyme–substrate interactions of an important class of membrane enzyme.

B. Enzymic Characterization of Bacterial ATPases

1. *Substrate Specificity*

Membrane ATPases, purified by the variety of methods reviewed above and subjected to the various criteria for establishing their homogeneity, have been characterized for their substrate specificities, metal-ion requirements, pH value optima and response to activating and inhibitory agents. In a few instances the purified, soluble and membrane-bound ATPases have been examined for their cold lability and for their response to stimulatory effects of heat or trypsin.

By definition, the nucleoside triphosphate of adenine is the substrate for these enzymes, but many of the bacterial ATPases have also exhibited appreciable hydrolysis of GTP. Other nucleoside triphosphates including

TABLE 4. Substrate Specificity Profiles of a Selection of ATPases from Bacterial Membranes

		Relative Activity								References
		ATP	GTP	UTP	CTP	ITP	ADP	AMP	PP_i	
Bacillus megaterium KM	(Ca^{2+})	100	72	3·5	9	100	0	—	0	Mirsky and Barlow (1971)
Bacillus megaterium B939	(Ca^{2+})	100	32	18	5	—	0	0	0	Ishida and Mizushima (1969a)
Bacillis stearothermophilus	(Mg^{2+})	100	90	30	16	—	0	0	0	Hachimori *et al.* (1970)
Bacillis subtilis	(Ca^{2+})	100	235	180	0	—	0	—	—	Rosenthal and Matheson (1973)
Escherichia coli K 12	(Mg^{2+})	100	79	2·9	<1	—	<1	<1	2·0	Kobayashi and Anraku (1972)
Escherichia coli NRC 482	(Mg^{2+})	100	62	27	19	41	19	0	—	Davies and Bragg (1972)
	(Ca^{2+})	100	22	9	1	26	0	0	—	Davies and Bragg (1972)
Micrococcus lysodeikticus	(Ca^{2+})	100	—	—	1·5	0	0	0	0	Muñoz *et al.* (1969)
Streptococcus faecalis	(Mg^{2+})	100	60	6	0	—	0	—	—	Abrams (1965)

Data are for soluble ATPases with the exception of those for *B. subtilis*; — indicates that data not given.

CTP, UTP, TTP and ITP are hydrolysed to a lesser extent. In general, ADP is not attacked and indeed it is usually inhibitory for the hydrolysis of ATP. Accumulation of ADP during the course of hydrolysis of ATP is responsible for the non-linear kinetics exhibited by the ATPase from *E. coli*, studied by Roisin and Kepes (1972). However, the reaction remained linear for a longer time when ADP was rephosphorylated by pyruvate kinase by adding this ATP-generating system to the ATPase assay mixture (Roisin and Kepes, 1972). Neither AMP nor inorganic pyrophosphate is hydrolysed by the ATPases. The latter substrate is however attacked by inorganic pyrophosphatases which also appear to be membrane associated (Klemme *et al.*, 1971). The specificity profiles of a number of the bacterial ATPases are seen from the data summarized in Table 4. From this summary, it is evident that the purine nucleoside triphosphates, ATP and GTP, are the best substrates in those instances where the latter nucleoside has also been tested.

The stoicheiometry of the reaction has been investigated for several of the bacterial ATPases, using labelled ATP (either [^{14}C-] and/or γ-[^{32}P]ATP), and identification of the products (Schnebli and Abrams, 1970; Kobayashi and Anraku, 1972) clearly established that the enzymes hydrolyse the terminal pyrophosphate bond of ATP according to the following reaction:

$$\text{ATP} \rightarrow \text{ADP} + \text{P}_i$$

Exchange reactions between $^{32}P_i$ and ATP have been examined, and it has been concluded that no exchange occurs either in the P_i-ATP or ADP–ATP reaction performed with highly purified ATPases (Schnebli and Abrams, 1970; Kobayashi and Anraku, 1972). On the other hand, Schnebli and Abrams (1970) found that ADP–ATP exchange reactions do occur with washed membrane preparations but at only 5% of the ATPase reaction rate. Such an exchange catalysed by the membrane preparations could have been due to the presence of other enzymes, although the possibility that the membrane-bound ATPase could be responsible for this and that the soluble enzyme was unable to perform this exchange reaction was not ruled out (Schnebli and Abrams, 1970). The fractions from *E. coli* used by Roisin and Kepes (1972) did catalyse an ADP-activated ATP–ADP exchange, and the shift from hydrolysis was stimulated by high concentrations of Mg^{2+} and orthophosphate. The parallelism between decreased hydrolysis and exchange was invoked in favour of the ATPase carrying out both reactions. However, the preparations were closer in structure to the membrane-particle bound form, and this conclusion would have to be re-evaluated with purer ATPase preparations. The reverse reaction, i.e. the formation of ATP from

ADP and P_i, was investigated for the ATPase from *E. coli* K 12 by Kobayashi and Anraku (1972) but no detectable ^{32}P-ATP was formed.

A search for an acid-stable phosphorylated enzyme intermediate of the purified ATPase from *Strep. faecalis* failed to reveal such a reaction intermediate (Schnebli and Abrams, 1970). However, in a subsequent study, a phosphorylated form of the enzyme was detected (Abrams and Nolan, 1972).

Product-inhibition studies have shown that the ATPases are competitively inhibited by orthophosphate and by ADP. With the *Strep. faecalis* enzyme, the K_i value for orthophosphate was 10 mM and the K_i value for ADP was 0·7 mM (Schnebli and Abrams, 1970). The corresponding K_i value for ADP with the ATPase from *E. coli* was 0·3 mM (Kobayashi and Anraku, 1972). The K_m values for ATP were 2·5 mM for the *Strep. faecalis* enzyme (Schnebli and Abrams, 1970) and 0·6 mM and 0·5 mM for the ATPase from *E. coli*, respectively (Kobayashi and Anraku, 1972; Roisin and Kepes, 1972).

2. *Effects of Cations*

Unlike the ouabain-sensitive, $[Na^+ + K^+]$-stimulated, Mg^{2+}-ATPases (Skou, 1965), the bacterial membrane ATPases are generally Mg^{2+}- and/or Ca^{2+}-activated enzymes. They thus resemble the mitochondrial ATPases rather than the ion-transport ATPases which respond to $[Na^+ + K^+]$ stimulation. It should be noted, however, that some degree of alkali-cation stimulation of membrane ATPase activities of several bacteria has been reported. Hafkenscheid and Bonting (1968) observed a $[Na^+ + K^+]$-activated ATPase in *E. coli* subjected to freeze-drying and homogenization in 1·5 M-urea. However, the Mg^{2+}-ATPase activity was more prominent in such preparations. Varying degrees of stimulation by $[Na^+ + K^+]$ have also been reported for an enzyme from a *Vibrio* species (Hayashi and Uchida, 1965), *Staph. aureus* (Gross and Coles, 1968), and *B. subtilis* (Rosenthal and Matheson, 1973) in addition to the *E. coli* preparation mentioned above. These monovalent cation effects have generally been found for ATPase activities associated with the membranes. So far as one can tell at the present time, such stimulatory effects have not been demonstrable with the highly purified, soluble bacterial ATPases. Indeed, in a number of instances, Na^+ and/or K^+ appear to be inhibitory rather than stimulatory, and several of the ATPases are relatively insensitive to these monovalent cations (Evans, 1969, 1970; Kobayashi and Anraku, 1972; Davies and Bragg, 1972; Muñoz *et al.*, 1969).

Both membrane-bound and soluble forms of bacterial ATPases show maximal activities with the divalent cations Mg^{2+} or Ca^{2+}. Each ATPase system appears to exhibit its own unique properties and responses to divalent cations, depending on the bacterial species. Thus the ATPases of *Strep. faecalis*, *B. megaterium* and *E. coli* show dependency upon Mg^{2+} and maximal enzymic activities with this cation (Abrams, 1965; Ishida and Mizushima, 1969a,b; Evans, 1970; Kobayashi and Anraku, 1972; Roisin and Kepes, 1972). The ATPase of *M. lysodeikticus*, on the other hand, was maximally active with Ca^{2+} although it responded to a lesser extent to Mg^{2+} (Muñoz *et al.*, 1969). There appears to be a great deal of individuality in the responses of the ATPase activities of membranes or soluble forms of the enzyme to the presence of Mg^{2+} and/or Ca^{2+} and other cations, stimulating agents such as trypsin, or inhibitory agents (Abrams, 1965; Muñoz *et al.*, 1969; Evans, 1969, 1970; Lastras and Muñoz, 1971; Kobayashi and Anraku, 1972; Roisin and Kepes, 1972). Moreover, the complexity of the responses can be compounded even further by effects of pH value (Muñoz *et al.*, 1969; Roisin and Kepes, 1972). The interpretation of the influence of monovalent or divalent cations, or combinations thereof, upon the activity of the ATPase as it exists on the membrane is clearly very difficult and of necessity largely speculative. Various explanations of the complexities of the responses have been invoked, including conformational changes, and regulatory and co-operative effects in the membrane environments. All of these explanations are eminently appropriate but much more experimental sophistication and evidence will be needed to give real substance to the complexities of these responses. At the present time it is evident that subtle changes in the enzyme and its interaction with its substrate and inhibitors are occurring both with the membrane-associated and soluble forms of the ATPase in their various ionic environments. There is little information on the nature and quantity of ions associated with membranes and their associated responses, so the complexity of adding further ions to those already bound will inevitably lead to complex responses. The difficult task now remains to document the responses in measurable parameters of protein structure instead of placing them under a self-explanatory verbal umbrella of "conformational changes".

Divalent cations clearly play an important role in the attachment of the enzyme to its natural "receptor" in the native membrane. Moreover, Mg^{2+} is an essential ionic requirement for the re-association of some forms of the ATPase with the enzyme-depleted membrane (Abrams, 1965; Ishida and Mizushima, 1969b) and it has been observed that Mg^{2+} is more effective than Ca^{2+} in the recoupling of the ATPase to the depleted membranes. The ATPase of *Strep. faecalis* is dependent on the presence of the nectin protein (Baron and Abrams, 1971), the equivalent of the junction proteins in mitochondria (Tzagoloff, 1971), and divalent cation for re-association with the membrane. Preparations of ATPase devoid of the nectin or junction protein appear to have completely lost the ability to rebind to the membranes even in the presence of divalent cations (Baron and Abrams, 1971; Salton and Schor, 1972). Polyamines can substitute for divalent cations in re-association studies (Abrams, 1965) but the requirements for the nectin or coupling factors appear to be highly specific (Baron and Abrams, 1971).

3. *Optimum pH Values*

The pH value optima of a number of the ATPases have been studied with membrane-associated and purified enzymes, and the results are summarized in Table 5. As shown in this summary the bacterial ATPases are optimally active in a broad range from pH 7 to 9.

TABLE 5. Optimum pH Values for Activity of Bacterial ATPases

	Optimum pH values	
Bacillus megaterium	8·0–10·0	(Ca^{2+} : ATP, 1 : 2·5)
	7·2	(Ca^{2+} : ATP, 1 : 1)
Bacillus stearothermophilus	8·0	
Bacillus subtilis	8·5–9·0	
Escherichia coli K 12	9·5	
Micrococcus lysodeikticus	7·5–9·0	
Staphylococcus aureus	6·0	
Streptococcus faecalis	8·0	

4. *Inhibitors*

The bacterial ATPases differ from the [$Na^+ + K^+$]-stimulated Mg^{2+}-ATPases and the mitochondrial Mg^{2+}-ATPases in that they show little response to the inhibitory concentrations of ouabain or oligomycin which are highly effective against the other types of ATPase. On the other hand, both mitochondrial and membrane-associated bacterial ATPases have been found to be sensitive to the inhibitory effects of carbodiimides such as *N*,*N′*-dicyclohexyl carbodiimide (DCCD) (Harold *et al.*, 1969; Abrams and Baron, 1970). Differences in response of the membrane-bound and soluble ATPases from *Strep. faecalis* were observed, the purified enzyme exhibiting insensitivity to the carbodiimide (Harold *et al.*, 1969). Later investigations with this ATPase demonstrated the sensitivity of the soluble enzyme from *Strep. faecalis* to other carbodiimides, especially those of a more hydrophobic character than DCCD (Abrams and Baron, 1970). Both membrane-bound and

TABLE 6. Responses of Bacterial ATPases to Various Inhibitors

Organism	Inhibitory compounds	Insensitive to	References
Bacillus megaterium	Azide, pentachlorophenol, dinitrophenol (partial)	Oligomycin, *p*-chloromercuribenzoate	Ishida and Mizushima (1969a)
Escherichia coli	Azide, *N*,*N*′-dicyclohexylcarbodiimide, *p*-chloromercuribenzoate, guanidine (partial)	Oligomycin, ouabain, dinitrophenol	Evans (1970); Kobayashi and Anraku (1972)
Micrococcus lysodeikticus	Azide, 1-anilinonaphthalene 8-sulphonate, phloretin	Oligomycin, ouabain, *p*-chloromercuribenzoate	Muñoz *et al.* (1969); Salton and Schor (1972)
Streptococcus faecalis	*N*,*N*′-dicyclohexylcarbodiimide, Dio 9, guanidines	Oligomycin, ouabain	Harold (1972)

Dio 9 is an antibiotic of unknown structure.

partially purified ATPases from *E. coli* were inhibited, the solubilized enzyme showing greater sensitivity to DCCD (Evans, 1970).

Other compounds exhibiting hydrophobic properties and interactions have been examined for their inhibitory effects on bacterial ATPases. Phloretin, the aglucone of phlorizin, inhibits photophosphorylation reactions and activity of Ca^{2+} and Mg^{2+}-ATPase of the chloroplast (Uribe, 1970). It was also found to be an effective inhibitor of the purified "shock-wash" ATPase complex and of the ATPase released by the *n*-butanol procedure from membranes of *M. lysodeikticus* (Salton and Schor, 1972). These bacterial ATPases were also sensitive to inhibition with the fluorescent membrane probe, 1-anilino-8-naphthalene sulphonate (ANS) (Waggoner and Stryer, 1970; Metcalfe *et al.*, 1971), at concentrations of about

TABLE 7. Relationship between Inhibition of Membrane ATPases by Anti-ATPase from *Micrococcus lysodeikticus* and G + C Contents of Micrococci and Other Bacteria

Micro-organism	Inhibition by excess antibody (%)	G + C content (mol %)
Micrococcus lysodeikticus	100	73·3
Micrococcus tetragenus	89	73·4
Sarcina flava	87	72·4
Micrococcus roseus	84	73·5
Micrococcus roseus R 27	79	—
Sarcina lutea	75	73·0
Micrococcus conglomeratus	36	69.4
Micrococcus varians	35	72·4
Corynebacterium coelicolor	30	65.3
Micrococcus rhodochrous	0	70·4
Micrococcus caseolyticus	0	44·4
Sporosarcina ureae	0	42·9
Bacillus subtilis	0	43·0

Data from Whiteside *et al.* (1971).

0·4 m*M* (Salton and Schor, 1972). Although ANS is less effective than DCCD, such a fluorescent-probe inhibitor may prove to be very useful in studies of conformational changes in the ATPase molecules.

Many of the commoner inhibitors, such as azide, cyanide, dinitrophenol and sulphydryl-blocking reagents, have been examined and a selection of results of inhibitory effects on bacterial ATPases is presented in Table 6.

In addition to these chemical inhibitors it should be noted that antibody specific for the ATPase of *M. lysodeikticus* caused complete inhibition of enzyme activity in the presence of excess antibody (Whiteside and Salton, 1970). The inhibitory effects of the anti-ATPase were non-competitive with respect to the enzyme substrate (Whiteside and Salton, 1970). Incomplete inhibitions by excess antibody with ATPases of closely related micrococci indicated cross-reactivities of the ATPase proteins (Whiteside *et al.*, 1971). The relationships of *M. lysodeikticus* anti-ATPase inhibitions with the G + C contents of these organisms are illustrated in Table 7. Thus inhibitor studies with specific antisera to ATPases may yield valuable immunological correlations with structural homologies of these enzymes when more is known about their primary amino-acid sequences.

5. *Stimulation of ATPase Activities*

Masked or latent Mg^{2+}- and Ca^{2+}-activated ATPase activities in mitochondrial and chloroplast preparations have been known from the earlier work of Racker (1963) and Vambutas and Racker (1965). A marked stimulation of activity occurred upon treatment of organelle subparticles with trypsin. Without giving any details, Ishikawa (1966) mentioned that the Ca^{2+}-ATPase from *M. lysodeikticus* "was activated several fold by brief trypsin digestion". The trypsin-stimulation phenomenon was later confirmed for membrane-bound ATPase from *M. lysodeikticus* by Muñoz *et al.* (1968a, 1969) and bound and soluble ATPase of differing coupling factor activities by Ishikawa (1970). The response to trypsin stimulation of solubilized ATPase or ATPase complexes is probably dependent upon the degree of association with other protein components. Thus the shock-wash ATPase from *M. lysodeikticus*, which had protein components additional to its two subunits as detected by sodium dodecyl sulphate–polyacrylamide gel electrophoresis, was stimulated by trypsin and still possessed rebinding capabilities (Salton and Schor, 1972). In contrast the ATPase devoid of the associated protein(s) (coupling factors or nectin?) was sensitive to trypsin and unable to rebind to depleted membranes (Salton and Schor, 1972).

The trypsin stimulation phenomenon can be visualized as involving the attack on an associated protein of the ATPase complex in such a way that more active sites of the enzyme become available, either by direct removal of the protein from the vicinity or by conformational changes of the enzyme consequent to the proteolytic attack on the protein. The presence of associated regulatory protein(s) which could also govern the interaction and attachment of the enzyme with the membrane would be functionally very important. The fact that it is sensitive to a proteolytic enzyme may be quite incidental and bear little relationship to the mechanisms of regulation of the ATPase *in situ* in the membrane. While direct stimulatory effects of trypsin on the enzyme molecule cannot be excluded, the evidence suggests that it is the result of the state of association of the enzyme with other protein component(s) of the complex. The decline in the magnitude of trypsin stimulation and dissociation of the complex on storage in the cold would be in accord with such an explanation. The final proof will undoubtedly come if reconstitution of ATPase with the coupling protein restores the trypsin stimulation phenomenon in a membrane system such as that of *M. lysodeikticus*.

It should not be inferred from these studies that the stimulatory effects of trypsin have any bond-specificity implications, since pronase, a broad-spectrum proteolytic enzyme, will serve equally well (M. R. J. Salton, unpublished results). Nor should it be inferred that the ATPase protein in its associated states is insensitive to proteolytic enzymes. Indeed, time-course experiments indicate the sensitivity of the enzyme to trypsin especially in the absence of substrate (Muñoz *et al.*, 1969; Ishikawa, 1970). Very similar results were obtained by Neujahr (1970) who used combined proteolysis and lysozyme treatments of *Lactobacillus fermenti*. The protease treatment rendered the cells sensitive to lysozyme and also resulted in the release of about 60–80% of the ATPase activity into a soluble fraction. However, the protcolytic enzymes had a destructive effect on the ATPase when they were used for membrane isolation. Substrate protection of the enzyme was also observed (Neujahr, 1970).

Trypsin stimulation of other solubilized bacterial ATPases does not appear to have been investigated to any extent, so it is difficult to evaluate the generality of this phenomenon. On the basis of the mechanisms of attachment, the need to regulate the hydrolytic activity of the enzyme as it occurs in the native membrane, and the rather low levels of activity of the enzyme in the initial isolated membrane preparations, it would not be surprising if other ATPase systems also exhibited this phenomenon. It may be that factors favouring dissociation of the associated proteins could result in loss of trypsin stimulation and also confer cold sensitivity. Further investigations are clearly needed to clarify the molecular basis of the trypsin stimulatory effects, cold lability, and rebinding abilities of bacterial ATPases, especially with the enzymes which have been purified to homogeneity as indicated by subunit structure.

Other stimulatory effects have been reported including the DNA-dependent type of activity reported by Ishikawa (1966). Such effects have not been obtained under the different conditions of ATPase purification used in our laboratory (Muñoz *et al.*, 1969), and it appears unlikely that the stimulation is highly specific for the DNA molecule but that other polyanions may be able to replace DNA.

In general, there has been little evidence of specific phospholipid or lipid requirements for bacterial ATPases. However, Klemme *et al.* (1971) reported inactivation of chromatophore ATPase as well as inorganic pyrophosphatase and photophosphorylation when chromatophores from *Rhodospirillum rubrum* were treated with phospholipase A. Adenosine triphosphatase activity could be restored to the lipid-deficient particles by addition of egg-yolk lecithin, phosphatidylethanolamine or phosphatidylserine. Other effects of lipid-soluble substances have been reported, including the stabilizing effect of oleic acid on solubilized ATPase from *E. coli* (Evans, 1970) and the role of unsaturated fatty-acid composition of membrane lipids on ATPase activities and its response to Na^+ inhibition in *E. coli* auxotrophs (Farias *et al.*, 1972).

Most studies of bacterial membrane ATPases give no indication of lipid requirements for their maximal activities, and direct stimulatory effects with purified ATPases have not been reported. This does not mean that the enzyme is incapable of interacting with lipid regions of membranes. It is of interest that attempts have been made to analyse the possible role of the ATPase from *Strep. faecalis* in ion transport by interacting soluble ATPase with phospholipid bilayers (Redwood *et al.*, 1969). Increased conductance of the bilayer resulted from the alteration, and the similarity of the dependence upon Mg^{2+}-ATP and $[Na^+ + K^+]$ concentrations suggested that the bilayer-ATPase complex may be similar to the complex as it exists in the membrane of the intact organism (Redwood *et al.*, 1969).

6. *Effects of Temperature on ATPases*

The response of ATPases to effects of temperature are of considerable interest, and one of the most striking properties of the mitochondrial enzyme was its cold lability (Penefsky and Warner, 1965). These investigations clearly defined the conditions which resulted in dissociation of the enzyme from the 11·9 *S* native ATPase to an equilibrium mixture of 11·9 *S*, 9·1 *S* and 3·5 *S* components at low temperatures with an accompanying loss of enzymic activity. Some of these changes were favoured by lowering the protein concentration of the preparations and by the presence of salts. Rewarming the cold-labile dissociated preparations restored about 90% of the activity with almost complete restoration of the 11·9 *S* form (Penefsky and Warner, 1965). Although several bacterial ATPases have exhibited marked cold lability (Abrams, 1965; Evans, 1969, 1970; Mirsky and Barlow, 1971; Kobayashi and Anraku, 1972; Davies and Bragg, 1972) they have not been subjected to a similarly detailed analysis to that described above for mitochondrial ATPase.

The importance of protein concentration and salts on cold lability of the bacterial ATPases has not been examined, and it is possible that some of the cold lability may be due to low protein concentrations.

It is of interest to note that the ATPases from *E. coli* (Evans, 1969) and *M. lysodeikticus* (M. R. J. Salton, unpublished observations) exhibit enhanced activities after pretreatment with heat. The increases in activity, expressed as percentages of the unheated control activities, were of the same order (approximately 120–150%) for the ATPases (heated in the range of 40–60°C) from these two organisms. The ATPase from *Strep. faecalis,* on the other hand, showed no activation (Evans, 1969) although it should be recalled that Schnebli and Abrams (1970) used a heat treatment step of 10 minutes exposure at 55°C without loss of enzyme activity. The results reported for the activity of ATPase from *E. coli* were obtained with the membrane-bound form of the enzyme (Evans, 1969) whereas those from *M. lysodeikticus* were purified, homogeneous enzyme fractions.

It is intriguing to speculate as to whether the apparent heat activation represents regeneration of a fraction of cold-dissociated enzyme rather than a stimulatory effect on fully active "native" enzyme. Surprisingly, the ATPase from *B. stearothermophilus* showed no evidence of activation by preheating (Hachimori *et al.*, 1970). The effects of temperature on the enzyme can thus have important consequences on the activity of ATPases, either by dissociation and loss of activity in the cold, or by restoration of native enzyme and apparent activation by preheating prior to assaying the enzyme fractions. Thus, specific activities of the enzyme preparations could be markedly altered depending on the temperature of storage, the protein concentration, the presence of salts, and the duration of prewarming to room temperature or incubation at 37°C prior to addition of substrate to reaction mixtures. It is evident that closer attention will have to be paid to the effects of temperature on the enzymes prior to assay. At temperature above 60°C, the ATPases are rapidly inactivated by heat and share this property with most enzymes.

C. Localization of ATPases and Membrane Architecture

Attention to the presence of stalk-like structures associated with bacterial protoplast membranes was drawn in the studies of negatively stained preparations of *B. stearothermophilus* by Abram (1965) and of *M. lysodeikticus* by Biryuzova *et al.* (1964). These particulate structures were strikingly similar to those observed on the inner membranes of mitochondria (Racker, 1967). Such particles were not invariably present on washed, isolated bacterial membranes, and the reasons for this were not apparent until the fractions obtained by the selective wash for

ATPase release (Abrams, 1965) were examined in the electron microscope (Muñoz *et al.*, 1968b). An association of ATPase activity with the uniform particles, observed upon purification of the membrane enzyme, suggested that this structure was the site of the enzyme (Muñoz *et al.*, 1968b) thereby resembling the mitochondrial ATPase enzyme (Kagawa and Racker, 1966; Stiles and Crane, 1966). Concomitant loss of the particles from the membranes with ATPase release (Nachbar and Salton, 1970a,b), and the particulate nature of the purified ATPase preparations from several bacteria (Muñoz *et al.*, 1968b; Ishida and Mizushima, 1969b; Schnebli *et al.*, 1970), provided further evidence for the identity of the particles.

Essentially two problems were evident in attempts to elucidate the relationship of ATPases to the architecture of the membrane. The first problem was to obtain evidence as to the distribution of the enzyme on outer and inner faces of the membrane, and the second problem was to identify the membrane-associated particles as the ATPase. The cytochemical localization of erythrocyte membrane ATPase on the inner face of the membrane was established by means of electron microscopy utilizing ATP-dependent deposition of lead phosphate (Marchesi and Palade, 1967). This clearly demonstrated the asymmetric distribution of the enzyme and its occurrence on one face of the membrane. Voelz (1964) and Voelz and Ortigoza (1968) also showed localization of bacterial ATPases on the cytoplasmic membrane by similar cytochemical methods and electron microscopy. The resolution by this technique did not permit firm conclusions about the asymmetric distribution of this enzyme on the plasma membranes of the bacterial cell. Moreover, in both the erythrocyte and the bacterial cell, the cytochemical localization of an enzyme such as the ATPase gives no indication of the number of individual enzymic sites on the membranes. For these reasons, the more sensitive method, using a ferritin-labelled conjugate of antibody specific for the membrane ATPase of *M. lysodeikticus*, was used (Oppenheim and Salton, 1973).

The purified γ-globulin fraction of antiserum to homogeneous preparations of ATPase from *M. lysodeikticus* was conjugated to ferritin essentially by the method of Singer and Schick (1961). The conjugate was separated from unreacted ferritin and γ-globulin, and was shown to possess ability to react with ATPase in immunochemical tests as well as reacting specifically with ATPase on the membrane. The specific labelling of the ATPase with the ferritin–antibody conjugate is illustrated in negatively stained preparations in Fig. 2, and thin sections of control and ferritin–antibody labelled membrane fractions in Fig. 3. The asymmetric labelling of the ATPase in the latter preparations is evident, and the labelling of inside-out vesicles is similar to that observed with

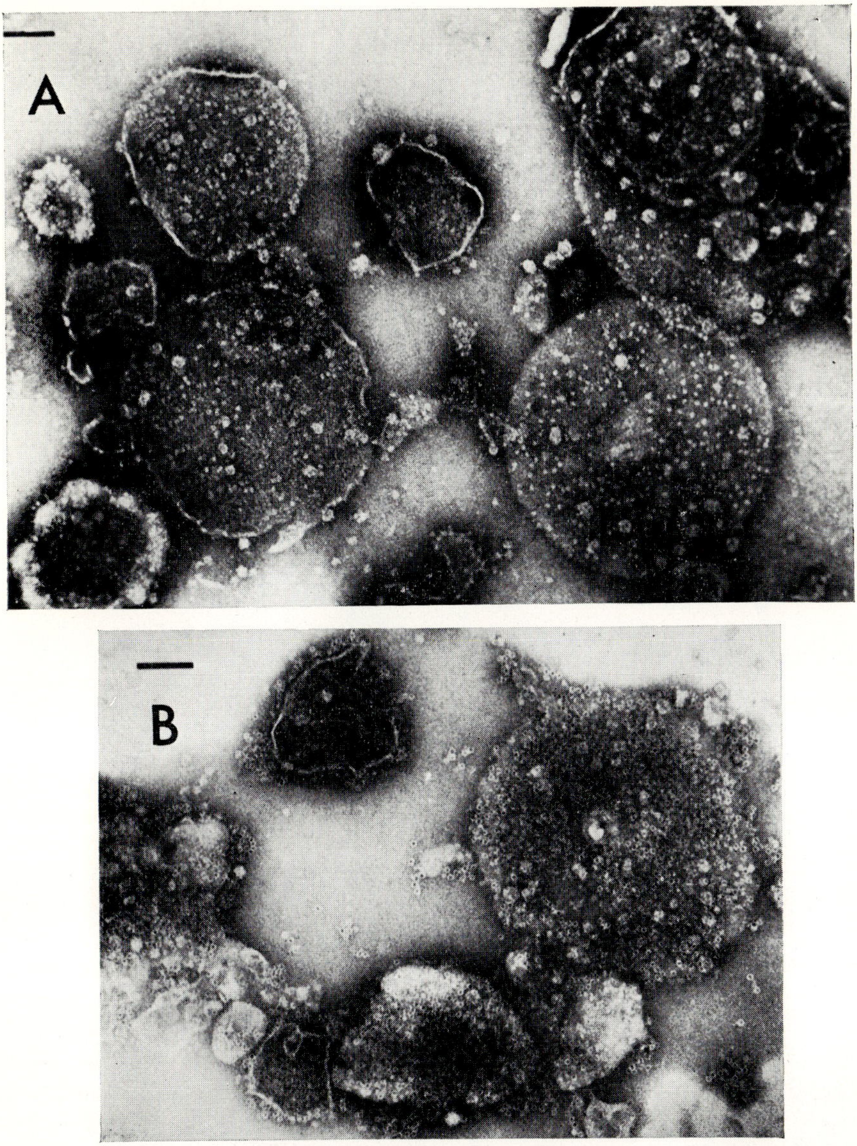

FIG. 2. Localization of ATPase particles on membranes from *Micrococcus lysodeikticus* by use of ferritin conjugated to antibody specific for the ATPase is illustrated in preparations which have been negatively stained with ammonium molybdate and examined in the electron microscope. Untreated, washed membranes in electron micrograph A show the presence of the uniform 10 nm-diameter particles believed to be the ATPase. Labelling of the ATPase with the specific ferritin–antibody conjugate is illustrated in B. The bars in each preparation represent 0·1 μm. From Oppenheim and Salton (1973).

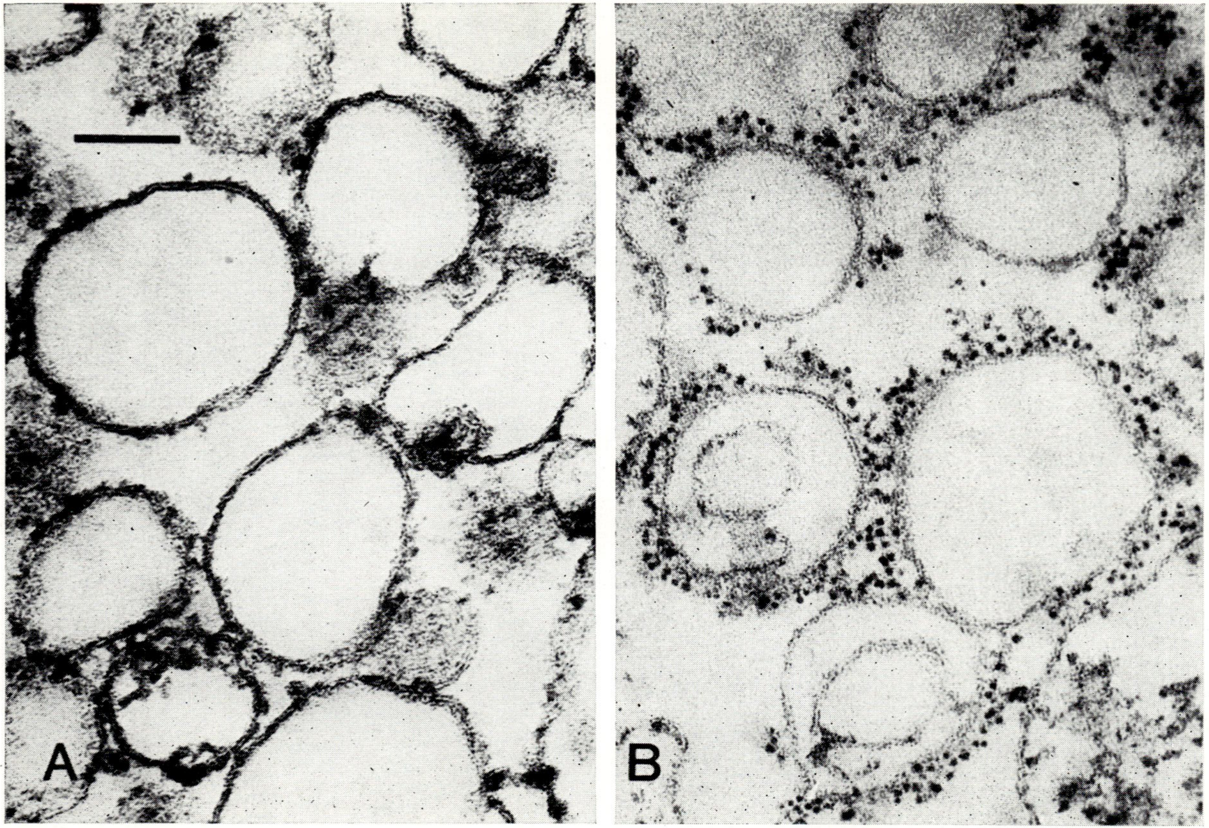

FIG. 3. The asymmetric distribution of ATPase particles on the membranes of *Micrococcus lysodeikticus* is shown in the thin sections of (A) unlabelled preparation and membranes (B) labelled with the ferritin–anti-ATPase conjugate. Many of the vesicles show the inside-out orientation, with labelling on one face of the membrane, a result similar to that obtained with the cytochemical localization of erythrocyte membrane ATPase by Marchesi and Palade (1967). [illegible]

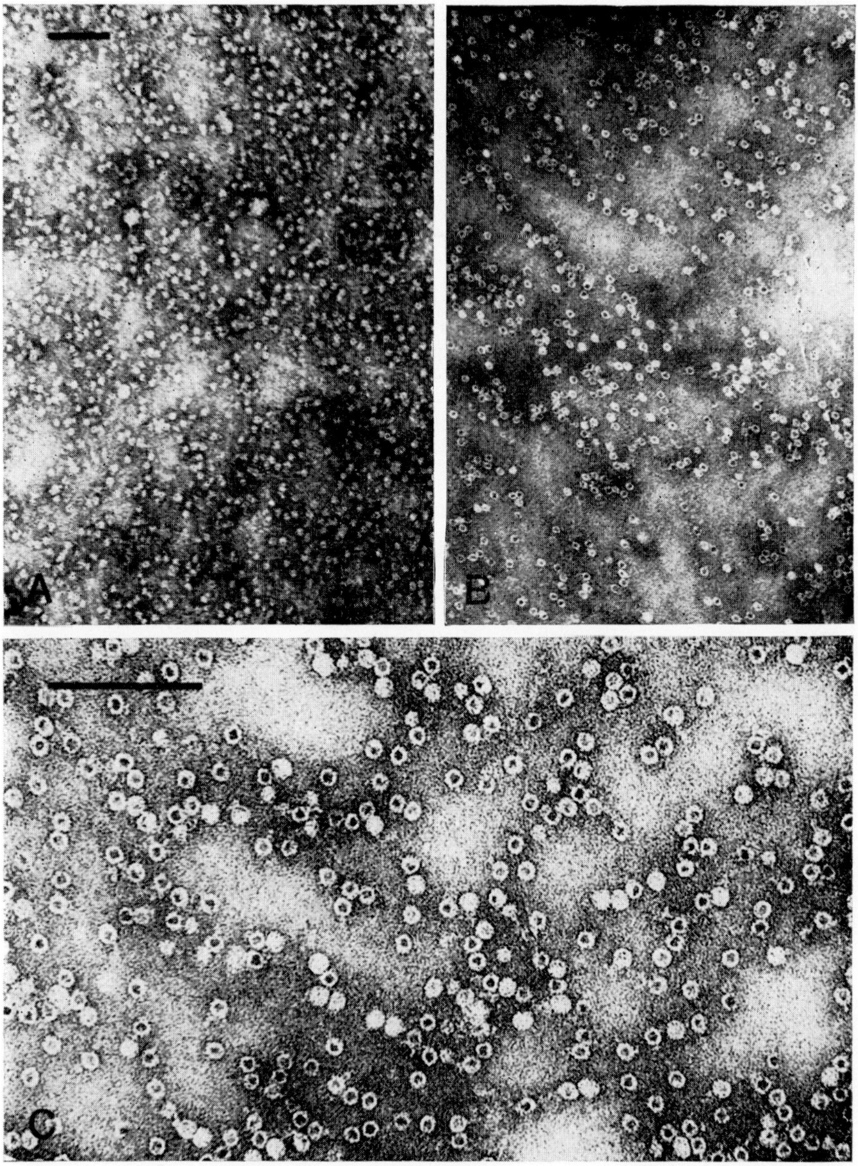

FIG. 4. Adenosine triphosphatase-containing particles released from the membranes of *Micrococcus lysodeikticus* by the selective shock-wash procedure are illustrated in the negatively stained preparation (A). Membranes which had been labelled with the ferritin–anti-ATPase conjugate, washed to remove unreacted label, and then subjected to the shock-wash procedure released the ferritin–antibody-ATPase complex as shown in B and C. The bars represent 0·1 μm. From Oppenheim and Salton (1973).

the cytochemical staining of erythrocyte membrane ATPase studied by Marchesi and Palade (1967). That the ATPase occurs only on one face of the plasma membrane, the inner face, was confirmed by reacting the conjugate with intact protoplasts. Labelling of the outer surface of the membrane of the protoplast could not be observed (Oppenheim and Salton, 1973). Membranes which had been subjected to the shock-wash procedure for the release of ATPase showed no significant labelling by use of the ferritin conjugate. Moreover, membranes which had been specifically labelled with the ferritin–antibody conjugate and then subjected to the selective release procedure yielded ferritin–antibody–ATPase complexes as indicated in Fig. 4. The selective release of enzyme–antibody–antibody–ferritin complexes from membranes would appear to offer new opportunities for specific identification of membrane enzymes and establishing the precise molecular architecture of membranes (Oppenheim and Salton, 1973).

Localization of the ATPase on the inner face of the plasma membrane has also been confirmed by an independent technique by labelling with ^{125}I either with the lactoperoxidase method (Phillips and Morrison, 1970, 1971) or by reaction with ^{125}ICl (Salton *et al.*, 1972). The ATPase was inaccessible to labelling when intact protoplasts were reacted with ^{125}I, but was readily labelled when isolated membranes were used (Salton *et al.*, 1972). However, specific release of the membrane particles, co-incident with the specific inhibition of the ATPase with the ferritin-labelled antibody and identification of the complex by electron microscopy and immunoelectrophoresis, provide definitive evidence of the identification of these membrane-associated particles as the site of ATPase (Oppenheim and Salton, 1973). In addition, the ferritin-labelled antibody interaction with the ATPase molecules on the membrane enables one to estimate the approximate number of sites of the enzyme. Assuming that each electron-dense ferritin particle corresponds to one ATPase molecule as seen in the unstained labelled membrane fragments (Fig. 10b in Oppenheim and Salton, 1973), then about 50 ATPase particles are present in a membrane area of $0{\cdot}1 \times 0{\cdot}1\ \mu m$. Thus, the ferritin labelling technique is not only valuable in identification of specific enzyme protein sites on the membrane but it can also give an approximation of the number of such sites. The method therefore has great potential in identification and quantitation of enzymes located on membranes as well as specific antigenic sites as determined by Nicholson and Singer (1971). The ferritin-labelling technique has also been useful in the *M. lysodeikticus* system in confirming the absence of ATPase from mesosome structures. Labelled ferritin–antibody specific for ATPase failed to react with the mesosome vesicles; moreover, enzymic activity could not be detected under a variety of conditicns (e.g. sonication, trypsin treatment in the presence of substrate) and ATPase antigen could not be detected by double-diffusion tests (Oppenheim and Salton, 1973).

D. Functions of Bacterial Membrane ATPases

Adenosine triphosphatases from bacterial membranes exhibit a number of striking similarities to mitochondrial and chloroplast ATPases. They are generally stable when associated with the membrane structures but, when released and purified in soluble form, many of them exhibit cold lability. To what extent this latter property is due largely to low concentrations of protein in the purified bacterial enzyme preparations is at the present time difficult to evaluate. Certain of the bacterial ATPases show latency and are stimulated by trypsin (Muñoz *et al.*, 1969; Salton and Schor, 1972), a property they share with mitochondrial and chloroplast ATPases (see Harold, 1972). Moreover, their cation requirements (Mg^{2+} and/or Ca^{2+}) and general lack of response to Na^+ and K^+ are similar to those of the mitochondrial ATPases and contrast with the [$Na^+ + K^+$] transport ATPases of mammalian plasma membranes. Unlike the mitochondrial enzyme, the bacterial ATPases are insensitive to oligomycin, but enzymes from both sources are inhibited by azide and carbodiimides and are little affected by ouabain (Harold, 1972). Other features which make the bacterial ATPases generally similar to the mitochondrial enzymes include the appearance of the particles as seen in the electron microscope, molecular weights, some features of their subunit structure, and the many reports of their presence in membrane particulate fractions possessing "coupling factor" activities (Harold, 1972). All of these properties have, therefore, suggested a role in oxidative phosphorylation for the bacterial membrane ATPases analogous to the corresponding enzymes in mitochondria.

No attempt will be made to review details of the variety of bacterial membrane fractions and soluble coupling factors required for oxidation and/or phosphorylation. Aspects of this have been discussed recently by Harold (1972) and it has been noted that latent ATPase activity is a factor common to all of the phosphorylating preparations. There is thus abundant evidence for participation of membrane fractions containing ATPase activity in energy-coupling reactions of the respiratory chain in bacteria (Gel'man *et al.*, 1967; Harold, 1972), but its precise function is still nonetheless not clearly defined and rather poorly understood.

As Harold (1972) has pointed out, many bacterial membrane preparations possess considerable ATPase activity without apparent need for "unmasking". Such hydrolytic activity for ATP would be potentially harmful for the organism and it is evident that this must be carefully regulated *in vivo*. The ATPase could thus serve "to couple the membrane-bound catalysts of oxidative phosphorylation to the synthesis of ATP" on the one hand, and, "on the other, to enable cells to utilize ATP as an energy source for membrane functions (Harold, 1972)". The coupling of ATP metabolism to cation transport through a membrane-associated ATPase has been strengthened by circumstantial evidence of the participation of the ATPase in *Strep. faecalis* in proton expulsion reported by Harold and Papineau (1972). The primary function of this ATPase would thus be the reverse of the mitochondrial enzyme in the generation of a proton gradient by the hydrolysis of ATP rather than the synthesis of ATP by the agency of a proton gradient (Harold and Papineau, 1972). It is clear that further work will be needed to clarify the functions of bacterial membrane ATPases in relation to their role in oxidative phosphorylation and proton extrusion and their bearing on the chemi-osmotic hypothesis of Mitchell (1970). The use of mutants with defective ATPases, similar to the *E. coli* mutant recently described by Butlin *et al.* (1971), will greatly facilitate a clearer understanding of the functions of this important class of membrane enzyme. A key role in oxidative phosphorylation is already suggested for the coupling factor activity (Mg^{2+}-ATPase) from studies with the *unc*- (uncoupled) mutants.

III. Membrane Enzymes Involved in Phospholipid Metabolism

A. Biosynthesis of Membrane Phospholipids

It has been known for many years that virtually all of the cellular phospholipid in bacteria is localized in the plasma membrane and internal membrane systems such as mesosomes, chromatophores and membrane whorls (Vorbeck and Marinetti, 1965; Salton, 1971a). Indeed, this is also the case for other "lipid soluble" components of the cell, including glycolipids, carotenoids, hydrocarbons and polyisoprenoid compounds. Since phospholipids constitute the major class of lipid in bacterial membranes, it is therefore not surprising that the membranes have been recognized as the site of their biosynthesis (Salton, 1971a,b; Cronan and Vagelos, 1972). Whether the enzymes involved in phospholipid biosynthesis are distributed randomly throughout the entire membrane system (including plasma and mesosome membranes) is not known with any certainty at present. There have been suggestions that mesosomes may be the preferential site of phospholipid biosynthesis (Fitz-James, 1967) but these claims do not appear to have been verified. On the contrary, pulse-labelling of phospholipids results in similar activities in the phospholipids in both plasma membrane and mesosomes

(G. Laneelle and M. R. J. Salton, unpublished results; Thomas and Ellar, 1973). However, the possibility still exists that phospholipids are synthesized preferentially in one of the two membrane structures, and rapidly equilibrate by migration throughout the membrane system. On the other hand, multiple sites for phospholipid synthesis throughout the bacterial membrane systems would be in good accord with the evidence for dispersive growth of the membrane (Green and Schaechter, 1972; Mindich and Dales, 1972). With suitable methods for the complete separation of plasma membranes and mesosome vesicles (Ellar and Freer, 1969; Owen and Freer, 1972; Popkin *et al.*, 1971; Reaveley and Rogers, 1969), the distribution of the enzymes between these two membranous structures could now be determined. The ability to separate inner and outer membranes of the envelopes of Gram-negative bacteria has also been of value in determining the distribution of the enzymes between the two membrane structures (Bell *et al.*, 1971; White *et al.*, 1971). Thus, in Gram-negative bacteria, the plasma (inner) membrane and not the outer membrane is the site of a number of enzymes involved in phospholipid synthesis (Bell *et al.*, 1971; White *et al.*, 1971).

Much information is now available on phospholipid biosynthesis, turnover, and degradation in studies with intact cells of a variety of bacteria and especially with investigations of glycerol- and fatty acid-auxotrophs of *E. coli*, *B. subtilis* and other bacteria. The mutant studies have been especially valuable in confirming and establishing the precise biosynthetic pathways, as well as yielding valuable information on regulation of phospholipid synthesis. No attempt will be made to review this aspect of the extensive literature, much of which has been covered in recent contributions (Cronan and Vagelos, 1972; Osborn, 1971). The principal reactions involved in the biosynthesis of bacterial phospholipids are given in the metabolic map in (Fig. 5), which has been modified from Cronan and Vagelos (1972) to take into account the new reaction involved in cardiolipin synthesis. This section of the review will be restricted to a discussion of the membrane-associated enzymes involved in phospholipid biosynthesis for, as Cronan and Vagelos (1972) have pointed out, most of them are found in particulate or cell-envelope fractions from disrupted bacteria. Although acyl carrier protein and enzymes participating in fatty-acid biosynthesis are usually found in the soluble "cytoplasmic" fraction of disrupted bacteria, they do not appear to be firmly associated with the membrane. However, while they may not be an integral part of the membrane structure, their proximity to the membrane would be of functional importance *in vivo* and their location close to the inner face of the membrane has been suggested (van den Bosch *et al.*, 1970).

Phosphatidic acid appears to be the common lipid precursor of the major phospholipids found in bacteria. It is, however, present in only trace amounts in the majority of bacteria, and the results of studies with radioactive labelling suggest a rapid loss of label from phosphatidic acid into other phospholipids. The first enzyme involved in phospholipid synthesis is the *sn*-glycero-3-phosphate acyltransferase involving transfer of two fatty-acid residues from acyl-CoA or acyl-acyl carrier protein to form phosphatidic acid (Lennarz, 1970; Cronan and Vagelos,

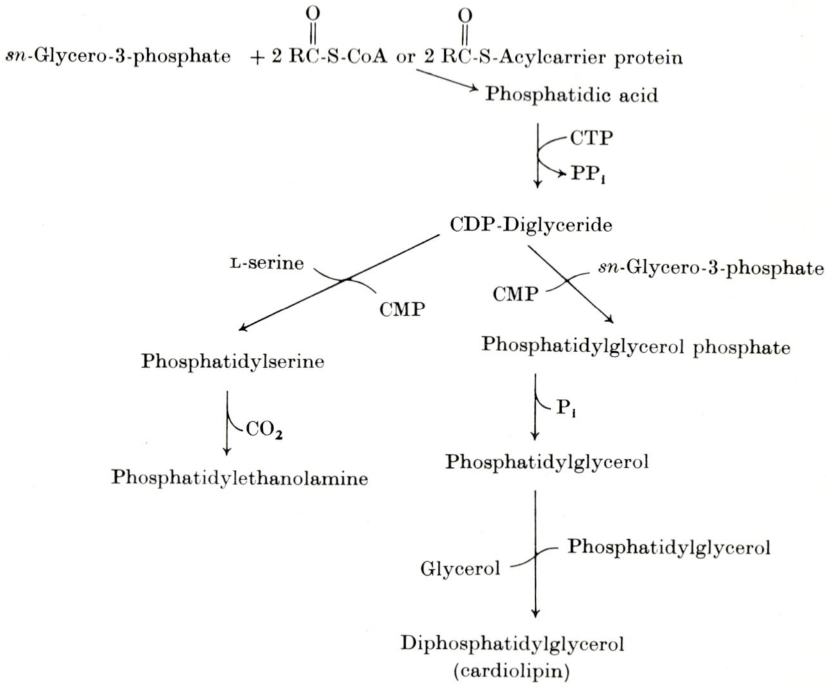

FIG. 5. Pathways of phospholipid synthesis in bacteria.

1972). Particulate fractions from *E. coli* perform this enzymic reaction, and thermosensitive mutants defective in the acyltransferase (Cronan *et al.*, 1970) have been isolated and characterized. Isolation of a mutant of *E. coli* by Hechemy and Goldfine (1971) giving lysophosphatidic acid indicated that at least two different enzymes are required for the conversion of *sn*-glycero-3-phosphate to phosphatidic acid, the first enzyme forming the monoacyl-*sn*-glycerol-3-phosphate (Cronan and Vagelos, 1972). The particulate acyltransferase preparations exhibit a high degree of specificity in their acylation reactions, and it is suggested that these enzymes may exert a strict control to minimize variations in the physical properties of the membrane phospholipids (Esfahani *et al.*, 1969, 1971; Silbert, 1970). There is little information on the purification of this enzyme, which undoubtedly exists in association with endogenous membrane lipid in the usual particulate form used in enzymic studies. This situation is by no means confined to the acyltransferase activities as very few membrane enzymes have been purified to the state of lipid-free protein homogeneity (i.e. as apo-enzymes).

A novel membrane-bound pyrophosphatase in *E. coli*, which hydrolyses CDP-diglyceride to yield phosphatidic acid and CMP, has been recently reported by Raetz *et al.* (1972), but its function is not known at present. Some features of this enzyme will be discussed below in the section on hydrolytic enzymes (p. 261).

Although CDP-diglyceride occupies a key position in the phospholipid biosynthetic pathways (Hill and Lands, 1970; Cronan and Vagelos, 1972) it has not been isolated from bacteria or other tissues. However, the enzymic synthesis of CDP-diglyceride occurs by the following reaction:

$$\text{phosphatidic acid} + \text{CTP} \rightleftharpoons \text{CDP-diglyceride} + \text{PP}_i$$

This reaction was first described in animal tissues and later found in extracts of *E. coli* (Carter, 1968). It has since been detected in membrane preparations of the Gram-positive bacteria *M. cerificans* (McCaman and Finnerty, 1968), *M. lysodeikticus* (De Siervo and Salton, 1971), and a *Bacillus* sp. (Patterson and Lennarz, 1971) and in the isolated cytoplasmic membranes of *E. coli* (White *et al.*, 1971) and *S. typhimurium* (Bell *et al.*, 1971). Thus, the enzyme CTP-phosphatidic acid cytidyltransferase occurs widely in bacteria and other organisms. The activity in *M. lysodeikticus* is exclusively localized in the membrane (De Siervo and Salton; 1970; Nachbar and Salton, 1970a) and is clearly present in the more tightly integrated regions of the membrane than is cardiolipin synthetase (De Siervo and Salton, 1970, 1971). The particulate nature of the bacterial enzyme was first reported by McCaman and Finnerty (1968) and, in *M. cerificans*, the partially purified activity still resided in membranous particles and remained active indefinitely upon freezing (Finnerty, 1971). McCaman and Finnerty (1968) showed that the enzyme activity was dependent on the presence of surface-active agents, and maximal activity was found with a non-ionic detergent, Cutscum. A similar dependency on detergent was reported for the *Bacillus* sp. (Patterson and Lennarz, 1969) and the *M. lysodeikticus* enzyme (De Siervo and Salton, 1971). Triton X-100 could be used instead of Cutscum in the latter organism. Complete purification of this enzyme from bacterial membranes does not appear to have been achieved as yet although Finnerty (1971) obtained a 175-fold purification of the enzyme from *M. cerificans*.

Phosphatidylglycerol accounts for about 5–15% of the cellular phospholipid in *E. coli* (De Siervo, 1969; Cronan and Vagelos, 1972) but it may account for a much higher fraction in other bacteria. In stationary-phase cells of *M. lysodeikticus*, phosphatidylglycerol accounts for about 36% of the total phospholipid (De Siervo and Salton, 1971). However, it should be emphasized that marked changes occur in phospholipid composition throughout the growth sequence of bacteria (Frerman and White, 1967; Cronan, 1968; De Siervo, 1969; White and Tucker, 1969; De Siervo and Salton, 1973) reflecting the turnover, interconversions, and degradation of these membrane constituents. The enzymes involved in the synthesis of phosphatidylglycerol have been partially purified from particulate fractions of *E. coli* by Chang and Kennedy (1967a, b) and the following two reactions have been demonstrated:

CDP-diglyceride + *sn*-glycero-3-phosphate → phosphatidylglycerol phosphate + CMP
phosphatidylglycerol phosphate→phosphatidylglycerol + P_i

Both the glycero-3-phosphate: CMP phosphatidyltransferase and the phosphatidylglycerol phosphate phosphatase are stimulated by Triton X-100 and require Mg^{2+} for activity. The former enzyme does not exhibit a very high level of activity and is thus in marked contrast to the majority of the other enzymes involved in phospholipid biosynthesis.

Cytidine diphosphate diglyceride forms a branch point for the synthesis of phosphatidylglycerol and diphosphatidylglycerol (cardiolipin) on the one hand, and phosphatidylserine and phosphatidylethanolamine on the other. Synthesis of phosphatidylserine which is usually present only in certain bacteria (Kates, 1965; Law, 1967) in rather small amounts is catalysed by the following reaction:

CDP-diglyceride + L-serine → phosphatidylserine + CMP

Reports on the specific localization of this enzyme in the bacterial membrane are at variance. Early studies by Kanfer and Kennedy (1964) suggested that it occurred in a soluble fraction, whereas White *et al.* (1971) reported higher activity in the plasma membrane than in the outer cell wall (the unfractionated envelopes possessed even higher specific activities). Patterson and Lennarz (1971) found this enzyme in the membrane of a *Bacillus* sp. used in their investigations, but were unable to demonstrate the reaction in the cytoplasmic fraction.

The conversion of phosphatidylserine to phosphatidylethanolamine is catalysed by membrane enzymes in both *E. coli* (Kanfer and Kennedy, 1964) and *Bacillus* species (Patterson and Lennarz, 1969, 1971) according to the following reaction:

phosphatidylserine → phosphatidylethanolamine + CO_2

Phosphatidylethanolamine is a major membrane phospholipid in *E. coli* and a number of Gram-positive and Gram-negative bacteria

(Kates, 1964; O'Leary, 1967). The enzyme responsible for synthesis of phosphatidylethanolamine, phosphatidylserine decarboxylase, is one of the few enzymes extracted from the membrane and purified to homogeneity (Wickner and Kennedy, 1971). In common with most enzymes involved in phospholipid biosynthesis, the activity is dependent upon or stimulated by surface-active agents. The studies of White *et al.* (1971) showed that the membrane rather than the wall of *E. coli* had a much higher level of the phosphatidylserine decarboxylase activity.

The final major reaction in the pathway arising from CDP-diglyceride left to discuss is the synthesis of cardiolipin (diphosphatidylglycerol) from phosphatidylglycerol. Evidence by Stanacev *et al.* (1967) suggested that cardiolipin was synthesized from phosphatidylglycerol and CDP-diglyceride by the following reaction:

$$\text{phosphatidylglycerol} + \text{CDP-diglyceride} \rightarrow \text{diphosphatidylglycerol} + \text{CMP}$$

In these studies with particulate preparations of *E. coli* only moderate stimulation of cardiolipin synthesis by CDP-diglyceride was observed (Stanacev *et al.*, 1967). Attempts to demonstrate a CDP-diglyceride-dependent synthesis of cardiolipin with isolated membranes of *M. lysodeikticus* failed, but led to the discovery of an enzyme which catalysed formation of cardiolipin from the phosphatidylglycerol as the sole substrate added to the membrane fractions (De Siervo and Salton, 1970, 1971). Moreover, the enzyme fractions of the membrane carrying out this reaction contained no detectable CDP-diglyceride despite the fact that the membranes of this organism are able to synthesize it from CTP and phosphatidic acid (De Siervo and Salton, 1971). Cardiolipin synthetase activity was released by treating the membranes of *M. lysodeikticus* with 5 m*M* EDTA in 5 m*M* tris-HCl buffer at pH 7·5 (De Siervo and Salton, 1970, 1971). This yielded a low-density particle fraction containing the synthetase activity which catalysed synthesis of one molecule of cardiolipin from two molecules of phosphatidlglycerol (De Siervo and Salton, 1971). Such a reaction had been postulated by Rampini *et al.* (1970) and it is of interest to note that Stanacev and Stuhne-Sekalec (1970) found that cabbage phospholipase D catalysed formation of cardiolipin from two molecules of phosphatidylglycerol. However, in contrast to the cardiolipin synthetase from *M. lysodeikticus* which converted over 90% of the phosphatidylglycerol to cardiolipin, the reaction with the cabbage phospholipase D yielded only 2% cardiolipin and 95% phosphatidic acid.

Subsequent studies by Short and White (1972) with *Staphylococcus aureus* and by Hirschberg and Kennedy (1972) with *E. coli* have established the presence of the cardiolipin synthetase which catalyses phosphatidylation of phosphatidylglycerol in these organisms. There is thus good agreement that at least one pathway for cardiolipin synthesis is according to the reaction first detected in *M. lysodeikticus* membranes (De Siervo and Salton, 1970, 1971). The reaction for cardiolipin synthesis involves almost quantitative conversion of phosphatidylglycerol to cardiolipin with elimination of free glycerol according to the following reaction:

$$2 \text{ phosphatidylglycerol} \rightarrow \text{diphosphatidylglycerol (cardiolipin)} + \text{glycerol}$$

Hostetler *et al.* (1972) concluded that a similar reaction was catalysed by an enzyme from rat liver mitochondria. Although the evidence with the bacterial systems indicates that CDP-diglyceride is not an essential component in the biosynthesis of cardiolipin from phosphatidylglycerol, Stanacev *et al.* (1973) have re-investigated the pathway in rat liver mitochondria, and have concluded that synthesis of cardiolipin occurs according to the reaction originally suggested above for *E. coli* (Stanacev *et al.*, 1967). Stanacev *et al.* (1973) therefore defined the cardiolipin synthetase as a CDP-diglyceride:phosphatidylglycerol phosphatidyltransferase in mitochondria, although they suggest that it need not be the exclusive pathway for biosynthesis in this membranous organelle. Further work with the bacterial membrane enzymes will be needed to determine whether this pathway also exists in these organisms. So far the bacterial-membrane cardiolipin synthetase has not been purified but, in *M. lysodeikticus* membranes, it is readily separable from the CTP-phosphatidic acid cytidyltransferase.

Other derivatives of phosphatidylglycerol found in bacteria are the aminoacylphosphatidylglycerols (Macfarlane, 1962). Lennarz *et al.* (1966) have investigated the biosynthesis of lysylphosphatidylglycerol in *Staph. aureus* from lysyl-tRNA and phosphatidylglycerol. Synthesis of lysyl-tRNA was catalysed by a soluble enzyme but transfer of the lysyl group to phosphatidylglycerol was catalysed by a particulate enzyme. A novel glucosaminyl derivative of phosphatidylglycerol was found in *B. megaterium* membranes (Op den Kamp *et al.*, 1969) but its biosynthetic enzymes do not appear to have been investigated. The functions of these derivatives of phosphatidylglycerol are still largely unknown, but some of the possibilities have been discussed by Lennarz (1970) and Houtsmuller and van Deenen (1964). Formation of the positively charged lysylphosphatidylglycerol could conceivably permit growth at low pH values by minimizing proton entry through the membrane and thereby function in the control of membrane permeability to protons. Although the evidence is consistent with this possible function for an aminoacylphosphatidylglycerol, further confirmation of this hypothesis is needed (Houtsmuller and van Deenen, 1964; Haest *et al.*, 1972).

Phosphatidylinositol has been reported to be present in only few bacteria (Macfarlane, 1962, 1964; Kates, 1964) and, although this phospholipid from certain bacterial membranes behaves in an identical manner to the authentic compound (e.g. De Siervo and Salton, 1971), it has not been isolated and chemically characterized from bacterial membranes. There is no information as to whether it is synthesized in bacteria by the pathway proposed for mammalian phosphatidylinositol (Paulus and Kennedy, 1960). Both phosphatidylinositol and its mannosyl derivatives occur in *Mycobacterium tuberculosis* (Pangborn and McKinney, 1966) and *Mycobacterium phlei* (Brennan and Ballou, 1967). Particulate enzymes have also been implicated in the biosynthesis of mannosylphosphoinositides (Brennan and Ballou, 1967).

In addition to the phospholipids already discussed, there are several others which occur much less frequently in bacteria, e.g. phosphatidylcholine, and methylated analogues of this lipid, phosphatidylethanolamine, plasmalogens in anaerobes (Goldfine, 1968), the diphytanyl ethers of phosphatidylglycerol in the extremely halophilic bacteria (Kates *et al.*, 1966; Joo and Kates, 1969; Kates and Wassef, 1970) and sphingolipid in *Bacteroides melaninogenicus* (LaBach and White, 1969). There is at present very little information on the biosynthesis of many of these less common phospholipids in bacteria. The diphytanyl ether analogues of phosphatidylglycerol are particularly interesting since the phytanyl ether residues replace the acyl fatty-acid residues. Apparently the extremely halophilic organisms, such as *H. cutirubrum*, do indeed possess a malonyl-CoA : acyl carrier protein transacylase which is nearly as active as the corresponding *E. coli* enzyme only in the absence of sodium chloride (Pugh *et al.*, 1971). Unlike the *E. coli* enzyme, the *H. cutirubrum* transacylase is almost completely inhibited at high salt concentrations. Thus the repression of the fatty-acid synthetase system by the high internal concentration of salt in these organisms would account for trace amounts of fatty acids normally found in these bacteria. Chemical synthesis of the key diphytanyl derivatives of phosphatidic acid and CDP-diglyceride will provide the necessary reference compounds for establishing the biosynthesis of this unusual class of bacterial phospholipid (Kates *et al.*, 1971). The biosynthetic pathways for other uncommon bacterial phospholipids have yet to be elucidated. An early step in the synthesis of sphingolipid has been studied by Lev and Milford (1973) and the formation of 3-ketodihydrosphingosine from palmitoyl-CoA and L-serine is catalysed by a "soluble" enzyme released by sonication of *Bacteroides melaninogenicus*.

B. Enzymic Degradation of Phospholipids

Very little is known about the *in vivo* degradation of bacterial membrane phospholipids and the role that membrane enzymes may play in the turnover of the lipids. There have of course been numerous studies on phospholipid and fatty-acid turnover in a number of bacteria (see Cronan and Vagelos, 1972) but the enzymic mechanisms involved are poorly understood. There is substantial evidence for stability of the phosphate groups in some lipids (e.g. phosphatidylethanolamine) while a moderate turnover is found in others (cardiolipin and phosphatidylglycerol). Fatty-acid residues of membrane phospholipids also exhibit great stability with little turnover detectable in growing cells (Cronan and Vagelos, 1972). Conversion of phosphatidylglycerol to cardiolipin could in part account for some of the apparent metabolism of these phospholipids.

The formation and secretion of extracellular phospholipases, such as phospholipase C of *Clostridium perfringens*, has been the subject of many studies, but the origins and functions of these enzymes in the bacterial cell are not at all clear. However, in the past few years, several phospholipases tightly bound to the cell membranes have been detected and this has been of special interest in view of the observations that phage infection appears to activate or elicit the action of phospholipases. Proulx and Fung (1969) found both phospholipase A1 and A2 activities in crude extracts of *E. coli*. This enzyme was also found by Okuyama and Nojima (1969). Scandella and Kornberg (1971) purified the membrane-bound phospholipase A1 from *E. coli* by solubilization with sodium dodecyl sulphate, isoelectric precipitation, acetone fractionation and sodium dodecyl sulphate polyacrylamide gel electrophoresis. The enzyme was purified about 5000-fold to near homogeneity and it was stable in 3% sodium dodecyl sulphate. The enzyme hydrolysed the 1-acyl chain of phosphatidylcholine, phosphatidylethanolamine, phosphatidylglycerol and cardiolipin. Neither cofractionated lipid nor detergent were essential for enzymic activity, but the enzyme aggregated in the absence of the latter. Scandella and Kornberg (1971) suggest that the enzyme may be responsible for phospholipid breakdown and changes in membrane integrity following perturbations due to agents such as infecting phages, colicins and antibody-complement action. A role in acyl-chain turnover is a possibility, although little turnover in logarithmically growing cells is observed, unless it is masked by rapid re-acylation. It is of interest to note that *E. coli* fractions can directly acylate lysophosphoglycerides (Proulx and van Deenen, 1966).

An osmotically-fragile mutant of *B. subtilis* possessed an active membrane-bound phospholipase A1, which together with a cytoplasmic lysophospholipase was responsible for degradation of phospholipids in this organism (Kent and Lennarz, 1972). A potent inhibitor of the enzyme was found in the wild-type *B. subtilis* and, curiously, phospholipase A1 was not detectable in this organism (Kent and Lennarz, 1972).

Osborn *et al.* (1972) developed methods for isolating and characterizing the outer membranes and the cytoplasmic membranes of *Salmonella typhimurium* and, in so doing, they were able to compare the distribution of various enzymic activities between the two membranes. In contrast to many of the enzymes which were localized in the inner (cytoplasmic) membrane, a phospholipase activity appeared to be localized primarily in the outer membrane. This phospholipase has been tentatively identified as a mixture of phospholipase A and lysophospholipase. The phospholipase A and lysophospholipase together catalysed the complete de-acylation of phosphatidylethanolamine, phosphatidylglycerol and cardiolipin to the water-soluble products identified as glycerolphosphorylethanolamine, glycerylphosphorylglycerol and bis(glycerolphosphoryl) glycerol respectively (Osborn *et al.*, 1972). Although the physiological role of the outer-membrane phospholipases is at present unknown, their presence would indicate a potential for active phospholipid metabolism in this structure.

A phospholipase D, which specifically cleaves cardiolipin to phosphatidic acid and phosphatidylglycerol, has been detected in *Haemophilus parainfluenzae* membranes and the activity was enhanced somewhat by surface-active agents (Ono and White, 1970). Astrachan (1973) studied the bond specificity of this phospholipase D, and showed an absolute specificity for the bond between phosphate and the C-3′ of the central glycerol residue in cardiolipin. Ono and White (1970) suggested that the enzyme also acts *in vivo* since the addition of inhibitors of the enzyme to the growth medium for *H. parainfluenzae* resulted in accumulation of cardiolipin and disappearance of phosphatidylglycerol in the membrane. However, in light of our understanding of the action of cardiolipin synthetase, the effects of the inhibitors would have to be re-interpreted.

Another interesting membrane-bound phosphatase involving the lipid-soluble nucleotide, CDP-diglyceride, has recently been reported in *E. coli* (Raetz *et al.*, 1972). This pyrophosphatase carried out the following hydrolysis:

$$\text{CDP-diglyceride} + H_2O \rightarrow \text{CMP} + \text{phosphatidic acid}$$

The enzyme has been released from the membrane fractions with Triton X-100 and has been purified 115-fold. It is devoid of phospholipase D activity and is strongly inhibited by AMP. The function of this pyrophosphatase is not known, but its presence in the membrane could influence the rates and extents of other reactions involving CDP-diglyceride (Raetz *et al.*, 1972).

It is evident that a number of enzymes hydrolysing phospholipids or liponucleotide exist in bacterial membranes, but that their roles in phospholipid metabolism have yet to be clearly defined. As with many other membrane enzymes their activities appear to be masked or latent, and become manifest after perturbation of the membrane structures. The generation of lysophospholipids by phospholipase A is a potentially dangerous event for the cell but it is conceivable that transient weakening of the membrane structure may be required in release or uptake of macromolecules by the cell and may even be important in stages of macromolecular syntheses during growth and division.

IV. Biosynthesis of Glycolipids

Glycolipids were first detected in bacteria (*M. lysodeikticus*) by Macfarlane (1961) and have subsequently been found in many mycoplasmas (Shaw and Baddiley, 1968) and less frequently in Gram-negative organisms (Lennarz, 1970). Biosynthesis of the glycolipids has been investigated primarily in three systems, namely that of dimannosyl diglyceride synthesis in *M. lysodeikticus* (Lennarz and Talamo, 1966), galactosyl-glucosyl diglyceride in *Pneumococcus* (Kaufman *et al.*, 1965), and mono- and diglycosyl diglycerides in *Strep. faecalis* (Pieringer, 1968; Ambron and Pieringer, 1972).

In their studies of the dimannosyl diglyceride from *M. lysodeikticus*, Lennarz and Talamo (1966) found that relatively crude enzyme catalysed the following reaction sequence:

diglyceride + GDP-mannose → α-D-mannosyl-(1 → 3)-diglyceride + GDP mannose →
α-D-mannosyl-(1 → 3)-α-D-mannosyl-(1 → 3)-diglyceride

The enzyme fractions required Mg^{2+} and anionic detergent for activity, and one interesting feature of the *M. lysodeikticus* system was the preference exhibited for diglycerides containing branched-chain fatty acids. Since the fatty acids of this organism are largely of the C_{15}-branched-chain type (Cho and Salton, 1966), this preference for diglycerides containing them is readily understandable.

Biosynthesis of the glycosyl diglycerides studied so far appears to occur through the stepwise transfer of a glycosyl residue from a sugar nucleotide to the hydroxyl group of the diglyceride with subsequent transfer of a second residue to the first glycosyl substituent on the diglyceride. It is of interest to record that, in the course of studies on the biosynthesis of *M. lysodeikticus* mannosyldiglyceride, a mannolipid characterized as mannosyl-1-phosphorylpolyisoprenol (Scher *et al.*, 1968) which functions in the biosynthesis of "membrane-bound" mannan was found. Particulate enzyme preparations of *Strep. faecalis* are also involved in the transfer of glucose residues from UDP-glucose to diglyceride with the resultant formation of monoglucosyl and diglucosyl diglycerides (Pieringer, 1968). A third glucolipid appeared only after diglucosyl diglyceride had been formed and this was characterized as a new type of phosphoglycolipid tentatively identified as a glycerylphosphoryldiglucosyl glycerol lipid (Ambron and Pieringer, 1971). A similar glycerylphosphoryldiglucosyl glycerol-lipid had also been detected in *Mycoplasma laidlawii* by Shaw *et al.* (1970), and in *Strep. faecalis* by Fischer *et al.* (1973).

The functions of these glycolipids in membranes are still largely a matter for speculation. The recent investigations of membrane lipoteichoic acids, some of which appear to be membrane anchored through attachment to glycosyldiglyceride residues (Knox and Wicken, 1973), may provide one of a number of possible functions. The discovery of phosphoglycolipids would tend to support such a function.

V. Membrane-Associated Enzymes Involved in Biosynthesis of Cell-Wall and Capsular Components

The assembly of cell-wall structures as outer shells external to the plasma membranes involves a number of enzymes in the membrane so that the biosynthetic intermediates can be translocated across the membrane to an externally located acceptor site. The discovery of the important class of "lipid carriers", "lipid intermediates", "glycosyl lipid carriers" of the polyisoprenoid type of compound, and their role in biosynthesis of peptidoglycan, teichoic acid, lipopolysaccharide, capsular polysaccharide and mannan has been reviewed extensively in recent years (Lennarz, 1970; Osborn, 1971; Lennarz and Scher, 1972) and no attempt will be made to duplicate this information. Complete pathways for the biosynthesis of cell-wall peptidoglycans, teichoic acids and lipopolysaccharides have been admirably reviewed by Osborn (1971), and the role of the polyisoprenols by Lennarz and Scher (1972). Emphasis will therefore be placed on membrane enzymes not previously discussed and those enzymes recently purified.

A. Peptidoglycan Biosynthesis

Precursors of the cell-wall peptidoglycan polymers are assembled sequentially on UDP-*N*-acetylmuramic acid by cytoplasmic enzymes to form UDP-*N*-acetylmuramylpentapeptide (Strominger, 1969). Membrane preparations of *Staph. aureus* were shown to catalyse a rapid transfer of radioactively-labelled precursor to an acceptor associated with the membrane. Neuhaus (1971) has proposed the name "translocase" for the enzyme transferring the precursor molecule from uridylic acid to the undecaprenyl phosphate carrier in the membrane. Particulate enzymes are involved in all of the steps in wall peptidoglycan synthesis subsequent to the formation of *N*-acetylmuramylpentapeptide-C_{55} isoprenol pyrophosphate from UDP-muramyl pentapeptide and poly-isoprenol phosphate with concomitant release of UMP (Osborn, 1971). Many of the enzymes found in membrane-bound form have not been obtained in soluble form so that the full elucidation of the mechanisms of enzyme action must await purification. Partial solubilization of the phospho-*N*-acetylmuramyl-pentapeptide translocase from *Staph. aureus* has been achieved by Heydanek and Neuhaus (1969) and the lipid dependence and phospholipid and polyisoprenol requirements investigated (Pless *et al.*, 1972). Triton-X-100 has been used in extraction of the enzyme from *Staph. aureus* membranes, and the supernatant fraction after centrifugation at 150,000 × *g* has been filtered on Agarose A-50 m. The translocase activity is excluded, and required a lipid extract of *Staph. aureus* for transfer activity (Pless *et al.*, 1972). The enzyme also catalyses an exchange reaction of UMP into UDP-*N*-acetylmuramyl pentapeptide (Heydanek *et al.*, 1969) and the staphylococcal lipid extract was also required with the partially purified enzyme for this reaction (Pless *et al.*, 1972). However, to date, there is little evidence indicating that this and many other enzymes involved in peptidoglycan synthesis have been obtained as lipid-free, soluble, homogeneous enzyme preparations. Nonetheless, much progress has been made with the difficult task of resolving the activity of the translocase and defining the requirements of lipids for the transfer and exchange reactions. The phospho-mur-*N*Ac-pentapeptide translocase in the membrane of *Micrococcus luteus* has also been investigated by Umbreit and Strominger (1972a). In agreement with the work done with the staphylococcal translocase, the detergent-solubilized enzyme from *M. luteus* required the addition of three lipids, the lipid substrate (C_{55}-isoprenyl-phosphate), neutral lipid and a polar lipid. The reaction catalysed by these solubilized translocases from *Staph. aureus* and *M. luteus* is as follows:

$$\text{UDP-mur-}N\text{Ac-pentapeptide} + C_{55}\text{-isoprenyl-phosphate} \rightleftharpoons C_{55}\text{-isoprenyl-pyrophosphate-mur-}N\text{Ac-pentapeptide} + \text{UMP}$$

Enzymes involved in the action of the isoprenoid lipid in wall biosynthesis have been the subject of recent studies by Higashi and Strominger (1970), Sanderman and Strominger (1972) and Goldman and Strominger (1972). One of the enzymes, the C_{55}-isoprenoid alcohol phosphokinase from *Staph. aureus*, has been purified and ADP identified as the product from ATP. Crude membranes were subjected to extractions with *n*-butanol yielding the multiplicity of phases characteristic of this type of solvent extraction (Nachbar *et al.*, 1972). The enzyme concentrated in the organic solvent precipitated on storage at −20°C, and the enzyme eluted with methanol–butanol mixtures. Chromatography on DEAE-cellulose and elution of the column with butanol–methanol gradients containing ammonium acetate was followed by chromatography on hydroxypropylated Sephadex G-50. It was concluded that the C_{55}-isoprenoid alcohol phosphokinase activity was associated with a single polypeptide apoprotein of molecular weight 17,000 daltons (Sanderman and Strominger, 1972). The amino-acid composition of this unusual membrane enzyme showed a high proportion of hydrophobic amino-acid residues (Sanderman and Strominger, 1971) which probably accounts for its extractability and stability in organic solvents. As with so many membrane enzymes, the surface-active agent Triton X-100 (0·3%) was added to the assay mixture and Span-20, a non-ionic detergent, added to substitute for phospholipid cofactor requirement of the apoprotein.

Another membrane enzyme which catalyses dephosphorylation of the C_{55}-isoprenylpyrophosphate is essential for re-initiation of another cycle of peptidoglycan synthesis. This phosphatase has been purified from membrane particles from *M. lysodeikticus* following solubilization with Triton X-100. The solubilized material, eluted from DEAE cellulose, was rather unstable and yielded four bands in sodium dodecyl sulphate–polyacrylamide gels. Exhaustive hydrolysis of the polyisoprenyl pyrophosphate substrate by membrane particles or solubilized membrane released 55% of the radioactivity in the form of orthophosphate. The phosphatase was not inhibited by inorganic phosphate, inorganic pyrophosphate, ATP, ADP or AMP at final concentration of 5 m*M* (Goldman and Strominger, 1972).

Other steps in the biosynthesis of wall peptidoglycan prior to the enzymic dephosphorylation of the C_{55}-isoprenyl pyrophosphate, which enables the recycling of the lipid intermediate (glycosyl carrier lipid), are also catalysed by particulate membrane enzymes but their purification to homogeneity does not appear to have been achieved. One of the terminal reactions in wall-peptidoglycan synthesis involves a transpeptidase responsible for the cross-linking of adjacent peptides, either by a direct cross-link or through one of the many specialized cross-bridges (Ghuysen, 1968). The presence of a transpeptidase in a particulate preparation from *Bacillus megaterium* has been reported briefly by Wickus and Strominger (1970) and this enzyme catalysed the incorporation of *meso*- or DD-diaminopimelic acid into acceptor peptide.

The status of the transpeptidases has been re-evaluated in the past few years following the discovery of the sensitivity to penicillin of the DD-carboxypeptidases (Leyh-Bouille *et al.*, 1971; Strominger, 1969). The DD-carboxypeptidases from *Streptomyces* have been of particular interest since these have been purified to homogeneity and their substrate specificity profiles established (Leyh-Bouille *et al.*, 1970, 1971).

These enzymes exist as extracellular soluble enzymes, and it has been suggested that they may in fact be the solubilized forms of the transpeptidases which, when integrated in the plasma membrane, would catalyse the last step in peptidoglycan synthesis (Leyh-Bouille *et al.*, 1970, 1971). The hypothesis that these enzymes could function as carboxypeptidases or transpeptidases, depending upon the availability of a nucleophilic acceptor (H_2O or R-NH_2), was tested by Pollock *et al.* (1972), and the purified D-alanyl-D-carboxypeptidases of *Streptomyces* were indeed found to be able to carry out a true transpeptidation with suitable donor peptides. Moreover, the hydrolytic action of the DD-carboxypeptidases and their abilities to carry out transpeptidation exhibited similar sensitivities to penicillin G (Pollock *et al.*, 1972). The reasons for the soluble nature of the *Streptomyces* enzymes are not clear at the present time. However, a membrane-bound penicillin-sensitive D-alanyl-D-alanine carboxypeptidase has been solubilized with Triton X-100 and purified from membranes of *B. subtilis* (Umbreit and Strominger, 1972b). The pure enzyme was depleted of detergent, and showed a requirement for detergent or lipid. At the present time there is no information on the ability of this DD-carboxypeptidase to catalyse transpeptidation reactions. Enzyme preparations released from envelope membranes of *E. coli* K 12 by BRIJ 36T and partially purified by ampicillin-Sepharose affinity chromatography exhibit both DD-carboxypeptidase and transpeptidase activities (Pollock *et al.*, 1973). These results open the way for an evaluation of the dual functions of such enzymes in the membranes of *E. coli*, and should lead to a clarification of whether one or more enzymes are involved in the DD-carboxypeptidase-transpeptidase activities.

B. Biosynthesis of Lipopolysaccharides and Polysaccharides

1. *Lipopolysaccharides*

Bacterial lipopolysaccharides have been investigated intensively at the chemical and biochemical levels, and isolation of mutants blocked in the biosynthesis of core and *O*-antigen regions of the macromolecules have contributed much to our understanding of their structures and their biochemical genetics. Extensive reviews on all aspects of the lipopolysaccharides have appeared in recent years (Weinbaum *et al.*, 1971; Lennarz, 1970; Osborn, 1971; Rothfield and Romeo, 1971). Pathways for biosynthesis of core and *O*-antigen polysaccharides have been reviewed in detail by Osborn (1971) and Rothfield and Romeo (1971). Membrane-bound enzymes, lipids, and glycosyl carrier lipids (undecaprenols) play important roles in the biosynthetic processes (Wright *et al.*,

1967; Osborn, 1971; Rothfield and Romeo, 1971; Bell *et al.*, 1971; Lennarz and Scher, 1972).

Several of the glycosyl transferases involved in synthesis of the core region of *Salmonella typhimurium* lipopolysaccharide have been obtained in pure form and, under appropriate conditions with lipopolysaccharide acceptor, the following two reactions are catalysed:

UDP-glucose + lipopolysaccharide → glucosyl-lipopolysaccharide + UDP
UDP-galactose + glucosyl-lipopolysaccharide → galactosylglucosyl-lipopolysaccharide + UDP

For this *in vitro* system, both the acceptor lipopolysaccharide and a phospholipid (phosphatidylethanolamine is specifically required) must first interact to form a multimolecular complex. The enzyme proteins can then be incorporated into the lipopolysaccharide–phosphatidylethanolamine complex to form ternary complexes of the transferase system (Rothfield, 1971). The enzymes appear to remain in the reconstituted structure following the completion of their transferase reactions, and it is suggested that migration of the components in the membrane is essential for the further biosynthetic reactions utilizing the lipopolysaccharide as substrate.

The assembly of oligosaccharide chains on the glycosyl carrier lipids, and their extension and transfer to the core region, have been investigated extensively in the laboratories of Osborn and Robbins, and the details of the enzymic steps involved have been reviewed by Osborn (1971).

2. *Polysaccharides*

Several other cell-surface polysaccharides have attracted attention at the biosynthetic level. One of these is the mannan of *M. lysodeikticus* which is believed to be associated with the membrane. Investigations by Scher *et al.* (1968) established a role for undecaprenyl phosphate as an intermediate in the glycosylation reactions. The enzyme from *M. lysodeikticus* membrane catalyses the transfer of hexose to the undecaprenol phosphate with production of nucleotide diphosphate rather than nucleotide monophosphate. Thus, the mannose is linked through a phosphodiester bond rather than a pyrophosphoryl linkage as in peptidoglycan and *O*-antigen synthesis (Lennarz and Scher, 1972). The mannose units appear to be transferred to the non-reducing sites of the endogenous mannan acceptor, but the enzymic steps have not yet been elucidated (Scher and Lennarz, 1969).

Biosynthesis of the capsular polysaccharide of *Klebsiella aerogenes* has been investigated by Troy *et al.* (1971) and a lipid intermediate identified as undecaprenyl phosphate. Galactosyl units are added through the sugar nucleotide, and UDP-galactose and mannosyl and glucuronyl units added to the galactosyl pyrophosphoryl-undecaprenol. Polymerization of the polysaccharide occurs at the level of the lipid intermediate.

Particulate enzyme preparations of *M. lysodeikticus* are involved in the biosynthesis of a cell-wall polysaccharide as shown by the recent work of Page and Anderson (1972) and Anderson *et al.* (1972). The polysaccharide contains D-glucose and *N*-acetylmannosaminuronic acid, and the *in vitro* synthesis of the polysaccharide was dependent on the presence of UDP-*N*-acetylhexosaminuronic acid fraction isolated from *M. lysodeikticus* and UDP-*N*-acetyl-glucosamine. Pre-incubation of the particulate enzyme with the appropriate amino sugar nucleotides eliminated the lag in the incorporation of [^{14}C]glucose from UDP-D-[^{14}C]glucose into the polymer. So far there is no evidence to support the role of a carrier lipid in this biosynthetic reaction (Page and Anderson, 1972).

C. Biosynthesis of Teichoic Acids

The biosynthesis of ribitol- and glycerol teichoic acids of bacterial cell walls has been investigated principally in the laboratories of Baddiley, Glaser, and Strominger, and much of the evidence has been reviewed recently by Osborn (1971) and Baddiley (1972). With the growing prominence of the membrane-bound lipoteichoic acids (Knox and Wicken, 1973), and the development of methods for distinguishing between them and the cell-wall glycerol teichoic acids, some re-evaluation of specific details of the biosynthesis of these polymers may be necessary. This may be especially important in distinguishing between incorporations into polyol-phosphate chains linked to membrane lipid (glycolipid?) and attachment to polyisoprenyl intermediates.

Investigations in Glaser's laboratory with *Bacillus* species and *Lactobacillus plantarum* (Burger and Glaser, 1964; Glaser, 1963, 1964) indicated involvement of membrane-wall fractions in the successive transfer of polyol-phosphate units from CDP-ribitol or CDP-glycerol to an endogenous acceptor according to the following reaction:

$$n\text{CDP-polyol} + \text{acceptor} \rightarrow (\text{polyol-phosphate})_n\text{-acceptor} + n\text{CMP}$$

A similar conclusion was drawn by Ishimoto and Strominger (1966) from experiments with the teichoic acid from *Staph. aureus*. Chain extension of the polymer was shown to take place by addition of glycerol phosphate units to the "glycol end" of the chain by Kennedy and Shaw (1968).

The polyglycerolphosphate in the cell walls of *B. subtilis* is linked to the peptidoglycan, and Mauck and Glaser (1972a) have made the important discovery that the polyglycerophosphate is "exported" from the cell already linked to peptidoglycan strands. The steps involved in the assembly and export of the complex have not yet been elucidated but an important advance has been made in extracting the polyglycerophosphate polymerase from *B. subtilis* membranes and the membrane-localized "acceptor" (Mauck and Glaser, 1972b). The acceptor fraction contained glycerolphosphate, glucosamine and fatty acids but its chemical structure has not yet been fully elucidated (Mauck and Glaser, 1972b). As with many membrane enzymes, the activity of the polyglycerolphosphate polymerase upon extraction with Triton X-100 was about double that obtained with membranes assayed in the absence of the surface-active agent (Mauck and Glaser, 1972b).

At the present time, there does not appear to be a direct participation of polyisoprenol lipid intermediates in the synthesis of the polyribitol phosphate and polyglycerolphosphate teichoic acids (Osborn, 1971). However, Baddiley (1972) has presented evidence for the involvement of undecaprenol phosphate compounds in the biosynthesis of the cell-wall teichoic acid from *B. licheniformis* and the glycerolphosphate-*N*-acetylglucosamine teichoic acid of the cell wall of *Staph. lactis*. Accordingly, Baddiley (1972) has concluded that biosynthesis of peptidoglycan and wall teichoic acid (glycerol–phosphate–glucose polymer) in *B. licheniformis* is interdependent with both pathways sharing undecaprenol phosphate.

In all of these studies involving teichoic acid biosynthesis, membrane or particulate enzymes appear to be involved, and the elucidation of the steps involved in wall and membrane teichoic acids will be eagerly awaited. Although partial purification of some of the enzymes has been achieved, none appears to have been resolved to the point where the lipid requirements can be specified. Another membrane-bound reaction involved in glycerol teichoic acid synthesis in *Lactobacillus casei* is a D-alanine membrane acceptor ligase recently discovered by Linzer and Neuhaus (1972).

D. Biosynthesis of the Poly (γ-D-Glutamyl) Capsule in *Bacillus licheniformis*

The existence of a unique class of γ-glutamyl polypeptide capsules of *Bacillus* species has been known for a very long time to the microbiologists, but the site and mechanism of synthesis of this capsular structure have been little understood and virtually unexplored. Early evidence suggested that the mechanism for γ-glutamyl polypeptide synthesis involved a series of transamidation reactions for D-glutamic acid synthesis, and transpeptidation reactions for polymer synthesis (Williams and Thorne, 1954). However, recent investigations by Troy (1973a,b) have clearly established a unique system in *B. licheniformis* for synthesis of the poly(γ-D-glutamyl) capsular polypeptide.

Particulate cell-envelope fractions from an encapsulated strain of *B. licheniformis* contain a polyglutamyl synthetase which catalyses polymerization of L-glutamic acid to form a high molecular-weight polymer of γ-D-glutamic acid. The reaction was specific for L-glutamic acid, required ATP and Mg^{2+}, and was stimulated by K^+ and dithiothreitol; D-glutamic acid was not incorporated into the polymer nor did it affect polymerization of the L-isomer (Troy, 1973a). The synthetase did not require an RNA template and was not inhibited by chloramphenicol, actinomycin D, puromycin or rifampicin. Thus, the mechanism of polymer formation is clearly different from that involved in protein synthesis. The polymer synthesized by the membrane-mediated enzyme was structurally identical with that found for the native capsular γ-D-glutamyl polymers synthesized *in vivo*. The latter polymers differed only in exhibiting a greater degree of polydispersity (Troy, 1973b). Membrane-bound γ-glutamyl transpeptidase occurs in the larvae of the domestic fly but its functions and properties (Bodnaryk, 1972) appear to be quite different from the bacterial polyglutamyl synthetase studied by Troy (1973a).

VI. Electron-Transport Components

Since the early work of Weibull (1953b) it has been known that the bacterial membrane is the site of the organized electron-transport chain in these organisms. Thus the cytochromes, oxidases and dehydrogenases have been located in the plasma membranes of numerous Gram-positive bacteria and in the envelopes of Gram-negative species (Gel'man *et al.*, 1967; Hendler, 1968; Salton, 1971a,b). With the recent development of methods for the separation of the outer and inner cytoplasmic (plasma) membranes of the Gram-negative cell envelope, it has been possible to confirm what was really a foregone conclusion, namely that the electron-transport components are indeed localized in the plasma membrane. Thus Osborn *et al.* (1972) found that the total cytochromes, NADH oxidase, succinate dehydrogenase and D-lactate dehydrogenase were recovered almost exclusively in the cytoplasmic membrane of *Salmonella typhimurium*. The specific activities of these components in the outer membrane fractions were 1 to 5% of the values obtained for the cytoplasmic membranes (Osborn *et al.*, 1972).

Work on the dissection of the electron-transport chain in bacterial membranes into its individual components has not been as extensive as the studies with mitochondria. The approaches to understanding the organization of the electron-transport systems in bacteria have been essentially the same as those used in mitochondrial studies, involving characterization of membrane sub-particles, attempts to solubilize components, dissociation and reconstitution studies (Gel'man *et al.*, 1967; Razin, 1972). Bacterial cytochromes have been widely investigated, with some success having been achieved in their purification. A discussion of these components of the electron-transport chain is considered to be beyond the scope of this article. The bacterial cytochromes have been reviewed in considerable detail by Bartsch (1968) and Horio and Kamen (1970).

Although much knowledge has been gained from studies of the dissociation and reconstitution of the bacterial membrane electron-transport components, little headway appears to have been made in the purification of individual components to a state of homogeneity. No attempt will be made to review the many dissociation studies nor the properties of the great variety of particulate fractions obtained from disrupted whole bacteria or directly from isolated membranes. As with mitochondrial components, the purification of enzymes such as succinate dehydrogenase has presented great technical difficulties, and it is clear that similar sustained efforts will be needed for purification of the bacterial electron-transport components. The use of chaotropic agents in the purification of mammalian succinate dehydrogenase (Davis and Hatefi, 1971) and the successful application of these methods to a bacterial system (Hatefi *et al.*, 1972) offer hope for the resolution of some of the tightly integrated complexes into their individual components. In the latter study, the succinate dehydrogenase of chromatophore membranes from *Rhodospirillum rubrum* has been purified 80-fold, obtained in a soluble form, and judged to be about 70% pure (Hatefi *et al.*, 1972).

The loosening and weakening of the cohesive forces between membrane complexes prior to extraction with various agents such as the chaotropic compounds and detergents may provide suitable particulate fractions for further purification studies. Thus, where fairly selective release or dissociation of electron-transport components of the bacterial membranes can be achieved, such steps contribute significantly to the task of purification and eventual reconstitution, restoration of specific functions, and a fuller understanding of the molecular organization of the respiratory chains of bacteria. The literature abounds in studies of dissociation of membrane electron-transport components in bacteria, and only a few illustrations of the types of useful selective separations will be given. Yu and Wolin (1972) were able selectively to extract the primary dehydrogenase ($NADH_2$-2,6-dichlorophenolindophenol oxido-reductase) from sonicated membranes from *B. megaterium* with 0·4% sodium deoxycholate. Cytochrome was not present in the selectively extracted dehydrogenase fraction. Restoration of the functional oxidase could be achieved by recombination of extracts and cytochrome fraction, but prior treatment of either fraction with phospholipase A prevented the restoration (Yu and Wolin, 1972). Rather similar studies have been performed by Eisenberg (1972) who reported solubilization of the membrane L-malate- and reduced $NADH_2$ dehydrogenase enzymes from *M. lysodeikticus* with deoxycholate. The insoluble residues contained cytochromes of the *b*, *c* and *a* types. Reconstitution of the two fractions could be achieved with Mg^{2+} ions. Of interest is the observation that black-light irradiation inhibited both the native and reconstituted oxidase activities of the membrane and that this could be partially restored by exogenous naphthaquinones (Eisenberg, 1972). Gel'man *et al.* (1970) was able to resolve two types of cytochrome *b* by extracting *M. lysodeikticus* membranes with deoxycholate, and Salton *et al.* (1968) obtained by deoxycholate extraction a lipid-depleted membranous fraction containing the cytochromes *b*, *c* and *a* present in the original membrane. The selective release of a particulate fraction with the $NADH_2$ dehydrogenase activity of the membranes of *M. lysodeikticus* by treatment with EDTA had been achieved by Nachbar and Salton (1970b). This particulate fraction was rich in lipid, and partial restoration of activity following removal of lipids by extraction with *n*-butanol was observed upon the addition of lipid dispersions (Nachbar and Salton, 1970b).

Although some selectivity has been achieved in the release and separation of bacterial membrane dehydrogenases, it is evident that they are still relatively large complexes (Ostrovsky *et al.*, 1968; Nachbar and Salton, 1970b) and heterogeneous with respect to variety of polypeptide chains as determined by electrophoresis of dissociated fractions in polyacrylamide gels. Thus Ostrovsky *et al.* (1969) found up to 16 components (phenol–acetic acid–water system) in the $NADH_2$ dehydrogenase and malate dehydrogenase fractions examined by polyacrylamide-gel electrophoresis.

Succinate dehydrogenase is localized in the bacterial membrane, and appears to be one of the more tightly integrated proteins remaining with the insoluble cytochromes after extraction with deoxycholate (Salton *et al.*, 1968). In this respect and in their general properties, the bacterial succinate dehydrogenases resemble the mammalian enzyme (Scholes and Smith, 1968; Pollock *et al.*, 1971; Kim and Bragg, 1971; Davis and Hatefi *et al.*, 1972). Varying degrees of "activation" of bacterial-membrane succinate dehydrogenases by substrate, heat and detergents have been reported, although not all preparations have responded in the same manner (Kim and Bragg, 1971; Owen and Freer, 1970; Pollock *et al.*, 1971). Prior to the very successful use of chaotropic agents in solubilizing succinate dehydrogenase (Davis and Hatefi, 1971) and its application to the chromatophore enzyme in *R. rubrum* (Hatefi *et al.*, 1972), early studies of purification of the bacterial enzymes were reported for several species including *Corynebacterium diphtheriae* (Pappenheimer and Hendee, 1949), *Propionibacterium pentosaceum* (Lara, 1959) and *Micrococcus lactilyticus* (Warringa *et al.*, 1958). Preparations of the former two species contained *b*-type cytochrome, as did some of the partially purified fractions of *M. lysodeikticus* succinate dehydrogenase (Pollock *et al.*, 1971). The *E. coli* enzyme was devoid of membrane cytochrome b_1 (Kim and Bragg, 1971). With all of these studies it is hard to judge the degree of purification and homogeneity of the preparations because of the absence of data on sodium dodecyl sulphate–polyacrylamide gel electrophoresis examinations and the paucity of information on the extent of phospholipid depletion of the enzyme fractions. Thus, the most complete data available to date are for the succinate dehydrogenase of the chromatophore membranes from *R. rubrum* which is about 70% pure (Hatefi *et al.*, 1972). It is evident from the extensive experience with the mitochondrial enzyme that purification of the succinate dehydrogenase is a difficult task although the chaotropic agents and the use of dithiothreitol to prevent loss due to oxygen lability may provide the necessary methodologies for the future resolution of these membrane enzymes to the state of homogeneity.

Some success has been achieved with the partial purification of other membrane enzymes involved in the bioenergetics of the bacterial cell, as in the nitrate reductase complex (Azoulay *et al.*, 1967, 1969; Villarreal-Moguel *et al.*, 1973) and formate dehydrogenase–cytochrome b_1 complex (Itagaki *et al.*, 1962). Cytochrome oxidase from *Pseudomonas aeruginosa* has recently been purified to homogeneity and obtained in crystalline form (Kuronen and Ellfolk, 1972). It has a molecular weight of 119,000 daltons and possesses two subunits each of molecular weight 63,000 daltons (Kuronen and Ellfolk, 1972). Another respiratory-chain component involved in D-lactate-dependent transport is the *E. coli* ML 308-225 D-lactate dehydrogenase which has recently been purified to homogeneity by Kohn and Kaback (1973). This membrane-bound flavoprotein was extracted with the chaotropic agent, sodium perchlorate, and finally purified by DEAE-cellulose chromatography in the presence of Triton X-100. The D-lactate dehydrogenase contained approximately one mol of FAD per mol of enzyme, and gave a molecular weight of about 75,000 daltons by sodium dodecyl sulphate–polyacrylamide gel electrophoresis (Kohn and Kaback, 1973). An interesting feature of this enzyme is that the solubilized enzyme can be reconstituted into membrane vesicles prepared from a D-lactate dehydrogenase-deficient mutant with restoration of ability to oxidize D-lactate and carry out D-lactate-dependent transport (Reeves *et al.*, 1973).

VII. Conclusions

Since bacterial membranes are multifunctional structures it is to be expected that they are the cellular site of many enzymes functioning in the biosynthetic and energy transduction processes of the cell. Membrane-associated enzymes involved in active transport, electron transport, phospholipid and glycolipid biosynthesis, and the biosynthesis of external macromolecular structures such as cell-wall peptidoglycan, lipopolysaccharides, wall and membrane teichoic acids and specific enzymes involved in undecaprenol metabolism have all been detected in and investigated in isolated membranes or particulate fractions. Phospholipases (A and D) also occur in certain bacterial membranes but their physiological roles are not clearly understood at the present time. Other hydrolytic enzymes including membrane-bound proteinase of *Strep. lactis* (Sorrells *et al.*, 1972) and peptidase in *Mycoplasma laidlawii* membranes have been reported; the latter hydrolyses alanine oligopeptides. Enzymes such as those destined for export often have a transient residence as membrane-bound forms (e.g. penicillinase; Sargent and Lampen, 1970), but a discussion of these enzymes has not been included in this review.

The bacterial membrane ATPases have proved to be the most readily purifiable class of membrane-associated enzymes. A number of these have been purified to homogeneity and their subunit structures established. The purified enzymes do not appear to have lipid requirements, but their activities *in situ* on the membrane could be modulated by membrane lipid and/or protein components. Mutants deficient in this enzyme may resolve the question as to what their functions are in the bacterial membrane.

Unlike the ATPases, most of the other membrane-bound enzymes appear to have more intimate associations with lipids and, in the few rare instances where these have been purified to homogeneity, specific lipid requirements are needed for the apoprotein (e.g. isoprenoid alcohol phosphokinase; Sanderman and Strominger, 1972).

Virtually all of the membrane enzymes exhibit the property of "latency", requiring some perturbation of the isolated membrane structure for "unmasking" enzymic activity. Perturbations can be as mild as divalent-cation depletion, heat stimulation, or as drastic as gross membrane disruption with surface-active agents. Many assay procedures for membrane enzymes thus require the presence of detergents and, of those used, the non-ionic surface active agents (e.g. Triton X-100, Nonidet P-40, Cutscum, BRIJ series) appear to be the most widely useful agents. The mechanisms of *in vivo* "unmasking" of membrane enzymes and the manner in which enzymic activities can be regulated are little understood at present and virtually unexplored at the experimental level. As a consequence of the need for "unmasking", initial activities in membranes will depend on the extent of prior perturbation of the structures during isolation and, thus, purification factors will differ markedly.

It can be anticipated that bacterial mutants will contribute much to our future understanding of the functions of membrane enzymes not clearly understood at the present time (e.g. ATPases, phospholipases) and the manner in which lipid composition and phase transitions can affect specific membrane enzyme activities. Such studies will usher in a new era of ability to approach the complex problems of regulation of membrane-enzyme activities by protein conformational changes and interactions, and will require a higher level of sophistication than that needed in the difficult, messy task of membrane enzyme purification. Both aspects will, however, lead ultimately to a very clear understanding of the molecular architecture and mechanisms of functioning of the most remarkable biomembrane structure.

References

Abram, D. (1965). *Journal of Bacteriology* **89**, 885.

Abrams, A. (1965). *Journal of Biological Chemistry* **240**, 3675.

Abrams, A. and Baron, C. (1967). *Biochemistry, N.Y.* **6**, 225.

Abrams, A. and Baron, C. (1970). *Biochemical and Biophysical Research Communications* **41**, 858.

Abrams, A. and Nolan, E. A. (1972). *Biochemical and Biophysical Research Communications* **48**, 982.

Abrams, A., McNamara, P. and Johnson, F. (1960). *Journal of Biological Chemistry* **235**, 3659.

Ambron, R. T. and Pieringer, R. A. (1971). *Journal of Biological Chemistry* **246**, 4216.

Anderson, J. S., Page, R. L. and Salo, W. L. (1972). *Journal of Biological Chemistry* **247**, 2480.

Astrachan, L. (1973). *Biochimica et Biophysica Acta* **296**, 79.

Azoulay, E., Puig, J. and Pichinoty, F. (1967). *Biochemical and Biophysical Research Communications* **27**, 270.

Azoulay, E., Puig, J. and Couchoud-Beaumont, P. (1969). *Biochimica et Biophysica Acta* **171**, 238.

Baddiley, J. (1972). *In* "Essays in Biochemistry", (P. N. Campbell and F. Dickens, eds.), vol. 8, pp. 35–77. Academic Press, London and New York.

Baron, C. and Abrams, A. (1971). *Journal of Biological Chemistry* **246**, 1542.

Bartsch, R. G. (1968). *Annual Review of Microbiology* **22**, 181.
Bell, R. M., Mavis, R. D., Osborn, M. J. and Vagelos, P. R. (1971). *Biochimica et Biophysica Acta* **249**, 628.
Benedetti, E. L. and Emmelot, P. (1968). *In* "The Membranes", (A. J. Dalton and F. Haguenau, eds.), vol. 4, pp. 33–120. Academic Press, New York and London.
Biryuzova, V. L., Lukoyanova, M. A., Gel'man, N. S. and Oparin, A. L. (1964). *Doklady Akademii Nauk, USSR* **156**, 198.
Bodnaryk, R. P. (1972). *Canadian Journal of Biochemistry* **50**, 524.
Brennan, P. and Ballou, C. E. (1967). *Journal of Biological Chemistry* **242**, 3046.
Burger, M. M. and Glaser, L. (1964). *Journal of Biological Chemistry* **239**, 3168.
Burham, J. C. and Hageage, C. J., Jr. (1967). *Journal of Bacteriology* **93**, 191.
Butlin, J. D., Cox, G. B. and Gibson, F. (1971). *Biochemical Journal* **124**, 75.
Carter, J. R., Jr. (1968). *Journal of Lipid Research* **9**, 748.
Chang, Y. Y. and Kennedy, E. P. (1967a). *Journal of Lipid Research* **8**, 447.
Chang, Y. Y. and Kennedy, E. P. (1967b). *Journal of Lipid Research* **8**, 456.
Cheng, K-J., Ingram, J. M. and Costerton, J. W. (1971). *Journal of Bacteriology* **107**, 325.
Cho, K. Y. and Salton, M. R. J. (1966). *Biochimica et Biophysica Acta* **116**, 73.
Cole, H. A. and Hughes, D. E. (1965). *Journal of General Microbiology* **40**, 81.
Cronan, J. E., Jr. (1968). *Journal of Bacteriology* **95**, 2054.
Cronan, J. E., Jr. and Vagelos, P. R. (1972). *Biochimica et Biophysica Acta* **265**, 25.
Cronan, J. E., Jr., Ray, T. K. and Vagelos, P. R. (1970). *Proceedings of the National Academy of Sciences of the United States of America* **65**, 737.
Danielli, J. F. and Davson, H. (1935). *Journal of Cellular and Comparative Physiology* **5**, 495.
Davies, P. L. and Bragg, P. D. (1972). *Biochimica et Biophysica Acta* **266**, 273.
Davis, K. A. and Hatefi, Y. (1971). *Biochemistry, N.Y.* **10**, 2509.
De Siervo, A. J. (1969). *Journal of Bacteriology* **100**, 1342.
De Siervo, A. J. and Salton, M. R. J. (1970). *Bacteriological Proceedings* 69.
De Siervo, A. J. and Salton, M. R. J. (1971). *Biochimica et Biophysica Acta* **239**, 280.
De Siervo, A. J. and Salton, M. R. J. (1973). *Microbios* (in press).
de-Thé, G. (1968). *In* "The Membranes", (A. J. Dalton and F. Haguenau, eds.), vol. 4, pp. 121–150. Academic Press, New York and London.
Drapeau, G. R. and MacLeod, R. A. (1963). *Journal of Bacteriology* **85**, 1413.
Eisenberg, R. C. (1972). *Journal of Bacteriology* **112**, 445.
Ellar, D. J. and Freer, J. H. (1969). *Journal of General Microbiology* **58**, vii.
Esfahani, M., Barnes, E. M. and Wakil, S. J. (1969). *Proceedings of the National Academy of Sciences of the United States of America* **64**, 1057.
Esfahani, M., Ioneda, T. and Wakil, J. (1971). *Journal of Biological Chemistry* **246**, 50.
Evans, D. J., Jr. (1969). *Journal of Bacteriology* **100**, 914.
Evans, D. J., Jr. (1970). *Journal of Bacteriology* **104**, 1203.
Fairbanks, G., Steck, T. L. and Wallach, D. F. H. (1971). *Biochemistry, N.Y.* **10**, 2606.
Farias, R. N., Londero, L. and Trucco, R. E. (1972). *Journal of Bacteriology* **109**, 471.
Finnerty, W. R. (1971). *Bacteriological Proceedings*, 151.
Fischer, W., Ishizuka, I., Landgraf, H. R. and Herrmann, J. (1973). *Biochimica et Biophysica Acta* **296**, 527.
Fitz-James, P. C. (1960). *Journal of Biophysical and Biochemical Cytology* **8**, 507.

Fitz-James, P. C. (1967). *Protides of the Biological Fluids* **15**, 289.

Frerman, F. E. and White, D. C. (1967). *Journal of Bacteriology* **94**, 1868.

Fukui, Y. and Salton, M. R. J. (1972). *Biochimica et Biophysica Acta* **288**, 65.

Fukui, Y., Nachbar, M. S. and Salton, M. R. J. (1971). *Journal of Bacteriology* **105**, 86.

Gel'man, N. S., Lukoyanova, M. A. and Ostrovskii, D. N. (1967). *In* "Respiration and Phosphorylation of Bacteria", pp. 1–238. Plenum Press, New York.

Gel'man, N. S., Tikhonova, G. V., Simakova, I. M., Lukoyanova, M. A., Taptykova, S. D. and Mikelsaar, H. M. (1970). *Biochimica et Biophysica Acta* **223**, 321.

Georgi, C. E., Militzer, W. E. and Decker, T. S. (1955). *Journal of Bacteriology* **10**, 716.

Ghuysen, J-M. (1968). *Bacteriological Reviews* **32**, 425.

Glaser, L. (1963). *Biochimca et Biophysica Acta* **71**, 237.

Glaser, L. (1964). *Journal of Biological Chemistry* **239**, 3178.

Goldfine, H. (1968). *Annual Review of Biochemistry* **37**, 303.

Goldman, R. and Strominger, J. L. (1972). *Journal of Biological Chemistry* **247**, 5116.

Gorchein, A., Neuberger, A. and Tait, G. H. (1968). *Proceedings of the Royal Society, London, Series B* **170**, 229.

Green, E. W. and Schaechter, M. (1972). *Proceedings of the National Academy of Sciences of the United States of America* **69**, 2312.

Greenawalt, J. W. and Weigand, R. A. (1972). *Federation Proceedings* **31**, 1102.

Gross, R. and Coles, N. W. (1968). *Journal of Bacteriology* **95**, 1322.

Hachimori, A., Muramatsu, N. and Nosoh, Y. (1970). *Biochimica et Biophysica Acta* **206**, 426.

Haest, C. W. M., De Gier, J., Op den Kamp, J. A. F., Bartels, P. and van Deenen, L. L. M. (1972). *Biochimica et Biophysica Acta* **255**, 720.

Hafkenscheid, J. C. M. and Bonting, S. L. (1968). *Biochimica et Biophysica Acta* **151**, 204.

Harold, F. M. (1972). *Bacteriological Reviews* **36**, 172.

Harold, F. M. and Papineau, D. (1972). *Journal of Membrane Biology* **8**, 45.

Harold, F. M., Baarda, J. R., Baron, C. and Abrams, A. (1969). *Journal of Biological Chemistry* **244**, 2261.

Hatefi, Y., Davis, K. A., Baltscheffsky, H., Baltscheffsky, M. and Johansson, B. C. (1972). *Archives of Biochemistry and Biophysics* **152**, 613.

Hayashi, M. and Uchida, R. (1965). *Biochimica et Biophysica Acta* **110**, 207.

Hechemy, K. and Goldfine, H. (1971). *Biochemical and Biophysical Research Communications* **42**, 245.

Hendler, R. W. (1968). *Ed.* "Protein Biosynthesis and Membrane Biochemistry", pp. 1–344. John Wiley and Sons, New York.

Heydanek, M. G., Jr. and Neuhaus, F. C. (1969). *Biochemistry, N.Y.* **8**, 1474.

Heydanek, M. G., Jr., Struve, W. G. and Neuhaus, F. C. (1969). *Biochemistry, N.Y.* **8**, 1214.

Higashi, Y. and Strominger, J. L. (1970). *Journal of Biological Chemistry* **245**, 3691.

Hill, E. E. and Lands, W. E. M. (1970). *In* "Lipid Metabolism", (S. J. Wakil, ed.), pp. 185–277. Academic Press, New York and London.

Hirschberg, C. B. and Kennedy, E. P. (1972). *Proceedings of the National Academy of Sciences of the United States of America* **69**, 648.

Hokin, L. E. (1969). *In* "Membrane Proteins", Proceedings of a Symposium Sponsored by the New York Heart Association, pp. 327–353. J. and A. Churchill Ltd., London.

Horio, T. and Kamen, M. D. (1970). *Annual Review of Microbiology* **24**, 399.
Hostetler, K. Y., Van den Bosch, H. and van Deenen, L. L. M. (1972). *Biochimica et Biophysica Acta* **260**, 507.
Houtsmuller, U. M. T. and van Deenen, L. L. M. (1964). *Biochimica et Biophysica Acta* **84**, 96.
Ishida, M. and Mizushima, S. (1969a). *Journal of Biochemistry, Tokyo* **66**, 33.
Ishida, M. and Mizushima, S. (1969b). *Journal of Biochemistry, Tokyo* **66**, 133.
Ishikawa, S. (1966). *Journal of Biochemistry, Tokyo* **60**, 598.
Ishikawa, S. (1970). *Journal of Biochemistry, Tokyo* **67**, 297.
Ishikawa, S. and Lehninger, A. L. (1962). *Journal of Biological Chemistry* **237**, 2401.
Ishikawa, S., Yamatashita, S., Araki, S.-I. and Shimazono, N. (1965). *Journal of Biochemistry, Tokyo* **57**, 235.
Ishimoto, N. and Strominger, J. L. (1966). *Journal of Biological Chemistry* **241**, 639.
Itagaki, E., Fujita, T. and Sado, R. (1962). *Journal of Biochemistry, Tokyo* **52**, 131.
Joo, C. N. and Kates, M. (1969). *Biochimica et Biophysica Acta* **176**, 278.
Kaback, H. R. (1972). *Biochimica et Biophysica Acta* **265**, 367.
Kagawa, Y. and Racker, E. (1966). *Journal of Biological Chemistry* **241**, 2461.
Kanfer, J. and Kennedy, E. P. (1964). *Journal of Biological Chemistry* **239**, 1720.
Kates, M. (1964). *Advances in Lipid Research* **2**, 17.
Kates, M. and Wassef, M. K. (1970). *Annual Review of Biochemistry* **39**, 323.
Kates, M., Palameta, B., Joo, C. N., Kushner, D. J. and Gibbons, N. E. (1966). *Biochemistry*, N.Y. **5**, 4092.
Kates, M., Park, C. E., Palameta, B. and Joo, C. N. (1971). *Canadian Journal of Biochemistry* **49**, 275.
Kaufman, B., Kundig, F. D., Distler, J. and Roseman, S. (1965). *Biochemical and Biophysical Research Communications* **18**, 312.
Kennedy, L. D. and Shaw, D. R. D. (1968). *Biochemical and Biophysical Research Communications* **32**, 861.
Kent, C. and Lennarz, W. J. (1972). *Proceedings of the National Academy of Sciences of the United States of America* **69**, 2793.
Kim, I. C. and Bragg, P. D. (1971). *Canadian Journal of Biochemistry* **49**, 1104.
Klemme, B., Klemme, J.-H. and San Pietro, A. (1971). *Archives of Biochemistry and Biophysics* **144**, 339.
Knowles, A. F. and Penefsky, H. S. (1972). *Journal of Biological Chemistry* **247**, 6617.
Knox, K. W. and Wicken, A. J. (1973). *Bacteriological Reviews* **37**, 214.
Kobayashi, H. and Anraku, Y. (1972). *Journal of Biochemistry, Tokyo* **71**, 387.
Kohn, L. D. and Kaback, H. R. (1973). *Journal of Biological Chemistry*, **248**, 7012.
Kuronen, T. and Ellfolk, N. (1972). *Biochimica et Biophysica Acta* **275**, 308.
LaBach, P. J. and White, D. C. (1969). *Journal of Lipid Research* **10**, 528.
Lara, F. J. S. (1959). *Biochimica et Biophysica Acta* **33**, 565.
Lark, K. G. (1969). *Annual Review of Biochemistry* **38**, 569.
Lastras, M. and Muñoz, E. (1971). *Federation of European Biochemical Societies Letters* **14**, 69.
Law, J. H. (1967). *In* "The Specificity of Cell Surfaces", (B. D. Davis and L. Warren, eds.), pp. 87–105. Prentice-Hall, Inc., Englewood Cliffs, New Jersey.
Law, J. H. and Snyder, W. R. (1972). *In* "Membrane Molecular Biology", (C. F. Fox and A. Keith, eds.), pp. 3–26. Sinauer Associates Inc., Stamford, Conn., U.S.A.
Lennarz, W. J. (1970). *In* "Lipid Metabolism", (S. J. Wakil, ed.), pp. 155–184. Academic Press, New York and London.

Lennarz, W. J. and Scher, M. G. (1972). *Biochimica et Biophysica Acta* **265**, 417.
Lennarz, W. J. and Talamo, B. (1966). *Journal of Biological Chemistry* **241**, 2207.
Lennarz, W. J., Nesbitt, J. A. and Reiss, J. (1966). *Proceedings of the National Academy of Sciences of the United States of America* **55**, 934.
Lev, M. and Milford, A. F. (1973). *Archives of Biochemistry and Biophysics*, **157**, 500.
Leyh-Bouille, M., Ghuysen, J.-M., Bonaly, R., Nieto, M., Perkins, H. R., Schleifer, K. H. and Kandler, O. (1970). *Biochemistry, N.Y.* **9**, 2961.
Leyh-Bouille, M., Coyette, J., Ghuysen, J.-M., Idczak, I., Perkins, H. R. and Nieto, M. (1971). *Biochemistry, N.Y.* **10**, 2163.
Linzer, R. and Neuhaus, F. C. (1972). *Bacteriological Proceedings*, 44.
Macfarlane, M. G. (1961). *Biochemical Journal* **80**, 45P.
Macfarlane, M. G. (1962). *Nature, London* **196**, 136.
Macfarlane, M. G. (1964). *Advances in Lipid Research* **2**, 91.
McCaman, R. E. and Finnerty, W. R. (1968). *Journal of Biological Chemistry* **243** 5074.
Maddy, A. H. (1964). *Biochimica et Biophysica Acta* **88**, 448.
Marchesi, V. T. and Palade, G. E. (1967). *Journal of Cell Biology* **35**, 385.
Mauck, J. and Glaser, L. (1972a). *Journal of Biological Chemistry* **247**, 1180.
Mauck, J. and Glaser, L. (1972b). *Proceedings of the National Academy of Sciences of the United States of America* **69**, 2386.
Metcalfe, S. M., Metcalfe, J. C. and Engelman, D. M. (1971). *Biochimica et Biophysica Acta* **241**, 422.
Mindich, L. and Dales, S. (1972). *Journal of Cell Biology* **55**, 32.
Mirsky, R. and Barlow, V. (1971). *Biochimica et Biophysica Acta* **241**, 835.
Mirsky, R. and Barlow, V. (1972). *Biochimica et Biophysica Acta* **274**, 556.
Mirsky, R. and Barlow, V. (1973). *Biochimica et Biophysica Acta* **291**, 480.
Mitchell, P. (1970). *Symposium of the Society for General Microbiology* **20**, 121.
Morton, R. K. (1950). *Nature, London* **166**, 1092.
Muñoz, E., Nachbar, M. S., Schor, M. T. and Salton, M. R. J. (1968a). *Biochemical and Biophysical Research Communications* **32**, 539.
Muñoz, E., Freer, J. H., Ellar, D. J. and Salton, M. R. J. (1968b). *Biochimica et Biophysica Acta* **150**, 531.
Muñoz, E., Salton, M. R. J., Ng, M. H. and Schor, M. T. (1969). *European Journal of Biochemistry* **7**, 490.
Nachbar, M. S. and Salton, M. R. J. (1970a). *In* "Surface Chemistry of Biological Systems", (M. Blank, ed.), pp. 175–190. Plenum Press, New York.
Nachbar, M. S. and Salton, M. R. J. (1970b). *Biochimica et Biophysica Acta* **223**, 309.
Nachbar, M. S., Winkler, W. J. and Salton, M. R. J. (1972). *Biochimica et Biophysica Acta* **274**, 83.
Nakane, P. K., Nichoalds, G. E. and Oxender, D. L. (1968). *Science, N.Y.* **161**, 182.
Neuhaus, F. C. (1971). *Accounts of Chemical Research* **4**, 294.
Neujahr, H. Y. (1970). *Biochimica et Biophysica Acta* **203**, 261.
Neville, D. M., Jr. (1971). *Journal of Biological Chemistry* **246**, 6328.
Nicolson, G. L. and Singer, S. J. (1971). *Proceedings of the National Academy of Sciences of the United States of America* **68**, 942.
Nisonson, I., Tannenbaum, M. and Neu, H. C. (1969). *Journal of Bacteriology* **100**, 1083.
Oelze, J. and Drews, G. (1972). *Biochimica et Biophysica Acta* **265**, 209.
Okuyama, H. and Nojima, S. (1969). *Biochimica et Biophysica Acta* **176**, 120.

O'Leary, W. M. (1967). "The Chemistry and Metabolism of Microbial Lipids" pp. 93–131. The World Publishing Co., Cleveland, Ohio.
Ono, Y. and White, D. C. (1970). *Journal of Bacteriology* **104**, 712.
Op den Kamp, J. A. F., Bonsen, P. P. M. and van Deenen, L. L. M. (1969). *Biochimica et Biophysica Acta* **176**, 298.
Oppenheim, J. D. and Salton, M. R. J. (1973). *Biochimica et Biophysica Acta* **298**, 297.
Osborn, M. J. (1971). *In* "The Structure and Function of Biological Membranes", (L. Rothfield, ed.), pp. 343–400. Academic Press, New York.
Osborn, M. J., Gander, J. E., Parisi, E. and Carson, J. (1972). *Journal of Biological Chemistry* **247**, 3962.
Ostrovsky, D. N., Pereversev, N. A., Zhukova, I. G., Trutko, S. M. and Gelman, N. S. (1968). *Biokhimaya, U.S.S.R.* **33**, 319.
Ostrovsky, D. N., Tsfasman, I. M. and Gel'man, N. S. (1969). *Biokhimiya, U.S.S.R.* **34**, 993.
Owen, P. and Freer, J. H. (1970). *Biochemical Journal* **120**, 237.
Owen, P. and Freer, J. H. (1972). *Biochemical Journal* **129**, 907.
Page, R. L. and Anderson, J. S. (1972). *Journal of Biological Chemistry* **247**, 2471.
Pangborn, M. C. and McKinney, J. A. (1966). *Journal of Lipid Research* **7**, 627.
Pappenheimer, A. M. and Hendee, E. D. (1949). *Journal of Biological Chemistry* **180**, 597.
Patterson, P. H. and Lennarz, W. J. (1969). *Federation Proceedings* **28**, 403.
Patterson, P. H. and Lennarz, W. J. (1971). *Journal of Biological Chemistry* **246**, 1062.
Paulus, H. and Kennedy, E. P. (1960). *Journal of Biological Chemistry* **235**, 1303.
Pecht, M., Giberman, E., Keysary, A., Yariv, J. and Katchalski, E. (1972). *Biochimica et Biophysica Acta* **290**, 267.
Penefsky, H. S. and Warner, R. C. (1965). *Journal of Biological Chemistry* **240**, 4694.
Phillips, D. R. and Morrison, M. (1970). *Biochemical and Biophysical Research Communications* **40**, 284.
Phillips, D. R. and Morrison, M. (1971). *Biochemistry, N.Y.* **10**, 1766.
Pieringer, R. A. (1968). *Journal of Biological Chemistry* **243**, 4894.
Pless, D. D., Jonah, M. M. and Neuhaus, F. C. (1972). *Federation Proceedings* **31**, 1096.
Pollocks, J. J., Linder, R. and Salton, M. R. J. (1971). *Journal of Bacteriology* **107**, 230.
Pollock, J. J., Ghuysen, J-M., Linder, R., Salton, M. R. J., Perkins, H. R., Nieto, M., Leyh-Bouille, M., Frère, J. M. and Johnson, K. (1972). *Proceedings of the National Academy of Sciences of the United States of America* **69**, 662.
Pollock, J. J., Disteche, M., Linder, R., Salton, M. R. J. and Ghuysen, J-M. (1973). *Annual Meeting of the American Society for Microbiology (Abstract)*, 153.
Popkin, T. J., Theodore, T. S. and Cole, R. M. (1971). *Journal of Bacteriology* **107** 907.
Proulx, P. R. and Fung, C. K. (1969). *Canadian Journal of Biochemistry* **47**, 1125.
Proulx, P. R. and van Deenen, L. L. M. (1966). *Biochimica et Biophysica Acta* **125**, 591.
Pugh, E. L., Wassef, M. K. and Kates, M. (1971). *Canadian Journal of Biochemistry* **49**, 953.
Racker, E. (1963). *Biochemical and Biophysical Research Communications* **10**, 435.
Racker, E. (1967). *Federation Proceedings* **26**, 1335.

Racker, E. (1970). *In* "Membranes of Mitochondria and Chloroplasts", (E. Racker, ed.), pp. 127–171. Van Nostrand Reinhold Company, New York.

Raetz, C. R. H., Hirschberg, C. B., Dowlan, W., Wickner, W. T. and Kennedy, E. P. (1972). *Journal of Biological Chemistry* **247**, 2245.

Rampini, C., Barbu, E. and Polonovski, J. (1970). *Comptes Rendue de l'Academy des Sciences, Paris, Serie D* **270**, 882.

Razin, S. (1972). *Biochimica et Biophysica Acta* **265**, 241.

Reaveley, D. A. (1968). *Biochemical and Biophysical Research Communications* **30**, 649.

Reaveley, D. A. and Rogers, H. J. (1969). *Biochemical Journal* **113**, 67.

Redwood, W. R., Muldner, H. and Thompson, T. E. (1969). *Proceedings of the National Academy of Sciences of the United States of America* **64**, 989.

Reeves, J. P., Hong, J.-S. and Kaback, H. R. (1973). *Proceedings of the National Academy of Sciences of the United States of America*, **70**, 1919.

Remsen, C. C., Valois, F. W. and Watson, S. W. (1967). *Journal of Bacteriology* **94**, 422.

Roisin, M-P. and Kepes, A. (1972). *Biochimica et Biophysica Acta* **275**, 333.

Rosenthal, S. L. and Matheson, A. (1973). *Biochimica et Biophysica Acta*, **318**, 252.

Rothfield, L. I. (1971). *In* "The Dynamic Structure of Cell Membranes", (D. F. H. Wallach and H. Fisher, eds.), pp. 165–179. Springer-Verlag, New York.

Rothfield, L. and Romeo, D. (1971). *Bacteriological Reviews* **35**, 14.

Ryter, A. and Jacob, F. (1966). *Annales de l'Institut Pasteur, Paris* **110**, 801.

Salton, M. R. J. (1967a). *Annual Review of Microbiology* **21**, 417.

Salton, M. R. J. (1967b). *Transactions of the New York Academy of Sciences* **29**, 764.

Salton, M. R. J. (1971a). *In* "Biomembranes", (L. A. Manson, ed.), vol. 1, pp. 1–65. Plenum Press, New York.

Salton, M. R. J. (1971b). *Critical Reviews in Microbiology* **1**, 161.

Salton, M. R. J. and Nachbar, M. S. (1970). *In* "Autonomy and Biogenesis of Mitochondria and Chloroplasts", (N. K. Boardman, A. W. Linnane and R. M. Smillie, eds.), pp. 42–52. North-Holland, Amsterdam.

Salton, M. R. J. and Schor, M. T. (1972). *Biochemical and Biophysical Research Communications* **49**, 350.

Salton, M. R. J., Freer, J. H. and Ellar, D. J. (1968). *Biochemical and Biophysical Research Communications* **33**, 909.

Salton, M. R. J., Schor, M. T. and Ng, M. H. (1972). *Biochimica et Biophysica Acta* **290**, 408.

Sanderman, H., Jr. and Strominer, J. L. (1972). *Journal of Biological Chemistry* **247**, 5123.

Sargent, M. G. and Lampen, J. O. (1970). *Archives of Biochemistry and Biophysics* **136**, 167.

Scandella, C. J. and Kornberg, A. (1971). *Biochemistry, N.Y.* **10**, 4447.

Scher, M. G. and Lennarz, W. J. (1969). *Journal of Biological Chemistry* **244**, 2777.

Scher, M., Lennarz, W. J. and Sweeley, C. C. (1968). *Proceedings of the National Academy of Sciences of the United States of America* **59**, 1313.

Schnaitman, C. A. (1970). *Journal of Bacteriology* **140**, 890.

Schnaitman, C. A. (1971). *Journal of Bacteriology* **108**, 553.

Schnebli, H. P. and Abrams, A. (1970). *Journal of Biological Chemistry* **245**, 1115.

Schnebli, H. P., Vatter, A. E. and Abrams, A. (1970). *Journal of Biological Chemistry* **245**, 1122.

Scholes, P. B. and Smith, L. (1968). *Biochimica et Biophysica Acta* **153**, 363.

Sedar, A. W. and Burde, M. R. (1965). *Journal of Cell Biology* **24**, 285.

Shapiro, A. L., Vinuela, E. and Maizel, J. V., Jr. (1967). *Biochemical and Biophysical Research Communications* **28**, 815.

Shaw, N. and Baddiley, J. (1968). *Nature, London* **217**, 142.

Shaw, N., Smith, P. F. and Verheij, H. M. (1970). *Biochemical Journal* **120**, 439.

Short, S. A. and White, D. C. (1972). *Journal of Bacteriology* **190**, 820.

Silbert, D. F. (1970). *Biochemistry, N.Y.* **9**, 3631.

Simoni, R. E. (1972). *In* 'Membrane Molecular Biology", (C. F. Fox and A. Keith, eds.), pp. 289–322. Sinauer Associates Inc. Publishers, Stamford, Conn., U.S.A.

Singer, S. J. (1971). *In* "Structure and Function of Biological Membranes", (L. I. Rothfield, ed.), pp. 145–222. Academic Press, New York.

Singer, S. J. and Schick, A. F. (1961). *Journal of Biophysical and Biochemical Cytology* **9**, 519.

Skou, J. C. (1965). *Physiological Reviews* **45**, 596.

Sorrells, K. M., Cowman, R. A. and Swaisgood, H. E. (1972). *Journal of Bacteriology* **112**, 474.

Stanacev, N. Z. and Stuhne-Sekalec, L. (1970). *Biochimica et Biophysica Acta* **210**, 350.

Stanacev, N. Z., Chang, Y. Y. and Kennedy, E. P. (1967). *Journal of Biological Chemistry* **242**, 3018.

Stanacev, N. Z., Davidson, J. B., Stuhne-Sekalec, L. and Domazet, Z. (1973). *Canadian Journal of Biochemistry* **51**, 286.

Steck, T. L. and Fox, C. F. (1972): *In* "Membrane Molecular Biology", (C. F. Fox and A. Keith, eds.), pp. 27–114. Sinauer Associates Inc. Publishers, Stamford, Connecticut.

Stiles, J. W. and Crane, F. L. (1966). *Biochimica et Biophysica Acta* **126**, 179.

Strominger, J. L. (1969). *In* "Inhibitors, Tools in Cell Research", (Th. Bircher and H. Sies, eds.), p. 187. Springer-Verlag, Berlin.

Thomas, T. D. and Ellar, D. J. (1973). *Biochimica et Biophysica Acta*, in press.

Troy, F. A. (1973a). *Journal of Biological Chemistry* **248**, 305.

Troy, F. A. (1973b). *Journal of Biological Chemistry* **248**, 316.

Troy, F. A., Frerman, F. E. and Heath, E. C. (1971). *Journal of Biological Chemistry* **246**, 118.

Tzagoloff, A. (1971). *In* "Current Topics in Membranes and Transport", (F. Bronner and A. Kleinzeller, eds.), vol. 2, pp. 157–205. Academic Press, New York.

Umbreit, J. N. and Strominger, J. L. (1972a). *Proceedings of the National Academy of Sciences of the United States of America* **69**, 1972.

Umbreit, J. N. and Strominger, J. L. (1972b). *Federation Proceedings* **31**, 1100.

Uribe, E. G. (1970). *Biochemistry, N.Y.* **9**, 2100.

Vambutas, V. K. and Racker, E. (1965). *Journal of Biological Chemistry* **240**, 2660.

van Deenen, L. L. M. (1965). *Progress in Chemistry of Fats and Other Lipids* **8**, pt. 1.

van den Bosch, H., Williamson, J. R. and Vagelos, P. R. (1970). *Nature, London* **228**, 338.

van Iterson, W. and Leene, W. (1964). *Journal of Cell Biology* **20**, 361.

van Iterson, W. (1965). *Bacteriological Reviews* **29**, 299.

Villarreal-Moguel, E. K., Ilbarra, V., Ruíz-Herrèra, J. and Gitler, C. (1973). *Journal of Bacteriology* **113**, 1264.

Voelz, H. (1964). *Journal of Bacteriology* **88**, 1196.

Voelz, H. and Ortigoza, R. O. (1968). *Journal of Bacteriology* **96**, 1357.

Vorbeck, M. L. and Marinetti, G. V. (1965). *Biochemistry, N.Y.* **4**, 296.

Waggoner, A. S. and Stryer, L. (1970). *Proceedings of the National Academy of Sciences of the United States of America* **67**, 579.

Wallach, D. F. H. (1969). *Journal of General Physiology* **54**, 3s.

Wallach, D. F. H. (1971). *In* "The Dynamic Structure of Cell Membranes", (D. F. H. Wallach and H. Fisher, eds.), pp. 181–199. Springer-Verlag, New York.

Warringa, M. G. P. J., Smith, O. H., Giuditta, A. and Singer, T. P. (1958). *Journal of Biological Chemistry* **230**, 97.

Watson, S. W. and Remsen, C. C. (1970). *Journal of Ultrastructure Research* **33**, 148.

Weber, K. and Osborn, M. (1969). *Journal of Biological Chemistry* **244**, 4406.

Weibull, C. (1953a). *Journal of Bacteriology* **66**, 137.

Weibull, C. (1953b). *Journal of Bacteriology* **66**, 688.

Weibull, C., Greenawalt, J. and Low, H. (1962). *Journal of Biological Chemistry* **237**, 847.

Weigand, R. A. and Greenawalt, J. W. (1972). *Federation Proceedings* **31**, 414.

Weigand, R. A., Holt, S. C., Shively, J. M., Decker, G. L. and Greenawalt, J. W. (1973). *Journal of Bacteriology* **113**, 433.

Weinbaum, G. and Markman, R. (1966). *Biochimica et Biophysica Acta* **124**, 207.

Weinbaum, G., Kadis, S. and Ajl, S. J. (1971). *In* "Microbial Toxins", vol. IV. pp. 1–473. Academic Press, New York and London.

White, D. C. and Tucker, A. N. (1969). *Journal of Lipid Research* **10**, 220.

White, D. A., Albright, F. R., Lennarz, W. J. and Schnaitman, C. A. (1971). *Biochimica et Biophysica Acta* **249**, 636.

Whiteside, T. L. and Salton, M. R. J. (1970). *Biochemistry, N.Y.* **9**, 3034.

Whiteside, T. L., De Siervo, A. J. and Salton, M. R. J. (1971). *Journal of Bacteriology* **105**, 957.

Wickner, W. T. and Kennedy, E. P. (1971). *Federation Proceedings* **30**, 1119.

Wickus, G. and Strominger, J. L. (1970). *Federation Proceedings*, 707.

Williams, W. J. and Thorne, C. B. (1954). *Journal of Biological Chemistry* **210**, 203.

Wright, A., Dankerts, M., Fennessey, P. and Robbins, P. W. (1967). *Proceedings of the National Academy of Sciences of the United States of America* **57**, 1798.

Yang, S. and Criddle, R. S. (1970). *Biochemistry, N.Y.* **9**, 3063.

Yu, L. and Wolin, M. J. (1972). *Journal of Bacteriology* **109**, 59.

AUTHOR INDEX

Numbers in italics refer to the pages on which references are listed at the end of each article.

A

B

C

D

E

F

G

H

I

J

K

L

M

N

O

P

Z

SUBJECT INDEX

A

B

C

D

E

F

G

H

I

J

K

L

M

N

O

P

Q

R

S

T

U

V

W

X

Y

Z